Scientific-Research Cooperation Between Vietnam and Poland (POL-VIET 2017)

E3S Web of Conferences Volume 35 (2018)

Krakow, Poland
20-22 November 2017

Editors:

M. Borowski
J. Jarzyna
T. Lipecki

ISBN: 978-1-5108-6005-6

Printed from e-media with permission by:

Curran Associates, Inc.
57 Morehouse Lane
Red Hook, NY 12571

Some format issues inherent in the e-media version may also appear in this print version.

Printed by Curran Associates, Inc. (2018)

For additional information, please contact EDP Sciences – Web of Conferences at the address below.

EDP Sciences – Web of Conferences
17, Avenue du Hoggar
Parc d'Activité de Courtabœuf
BP 112
F-91944 Les Ulis Cedex A
France

Phone: +33 (0) 1 69 18 75 75
Fax: +33 (0) 1 69 28 84 91

contact-edps@webofconferences.org

Additional copies of this publication are available from:

Curran Associates, Inc.
57 Morehouse Lane
Red Hook, NY 12571 USA
Phone: 845-758-0400
Fax: 845-758-2633
Email: curran@proceedings.com
Web: www.proceedings.com

TABLE OF CONTENTS

sciences

Statement of Peer review

In submitting conference proceedings to E3S Web of Conferences, I certify to the Publisher that I adhere to the **Policy on Publishing Integrity** of the journal in order to safeguard good scientific practice in publishing.

1. All articles have been subjected to peer review administered by the proceedings editors.
2. Reviews have been conducted by expert referees, who have been requested to provide unbiased and constructive comments aimed, whenever possible, at improving the work.
3. Proceedings editors have taken all reasonable steps to ensure the quality of the materials they publish and their decision to accept or reject a paper for publication has been based only on the merits of the work and the relevance to the journal.

Title, date and place of the conference

SCIENTIFIC-RESEARCH COOPERATION BETWEEN VIETNAM AND POLAND (POL-VIET), 20-22 NOVEMBER 2017

AGH UNIVERSITY O SCIENCE AND TECHNOLOGY, KRAKOW, POLAND

Proceedings editor(s):

MAREK BOROWSKI
JADWIGA JARZYNA
TOMASZ LIPECKI

Date and editor's signature

05 th MARCH 2018

Geomechanical Assessments of Simultaneous Operation in the Case of Transition from Open Pit to Underground Mine in Vietnam

Zbigniew Niedbalski[1], *Phu Minh Vuong* Nguyen[2,*], and *Eleonora* Widzyk-Capehart[2]

[1] AGH University Science and Technology, Faculty of Mining and Geoengineering, A. Mickiewicza Av. 30, 30-059 Krakow, Poland

[2] AMTC, University of Chile

Abstract. Nowadays, for a number of reasons, many open pit mines are considering a transition from Open Pit (OP) to Underground (UG) to remain competitive. In OP-UG transition, UG operation is operated simultaneously with the OP operation for a certain period of time. Guidelines for the simultaneous operation of OP and UG are very difficult to establish, as there are very few case studies available. Yet, because of the OP-UG interactions; the operation has a higher safety, technical and management requirements than the OP or UG methods when considered separately. In Vietnam, Cao Son is one of many OP mines, which decided to change the operational system from OP to UG. Simultaneous operation started in 2015 and will be conducted until 2030 when the OP mine Cao Son ends its mining activities. In this paper, selected geomechanical considerations of the simultaneous operation are presented. A number of numerical modelling calculations using finite-difference software with code FLAC were carried out for calibration process, slope stability analysis and the OP-UG interaction analysis for the Cao Son – Khe Cham II–IV mine. Based on the results obtained from numerical modelling, the geomechanical assessments of simultaneous operation Cao Son – Khe Cham II–IV are discussed in this paper. **Key words** – OP-UG transition, OP-UG interaction, numerical modelling, slope stability.

1 Introduction

Over time, open pit mines tend to go deeper and become larger. However, with the increasing depth of exploitation, there are a number of difficulties that make further exploitation unsafe and/or unprofitable. Consequently, changing the way of mining is required to continue the life of the OP mines. From the technological and mining costs perspectives, the only solution for further extraction are underground methods, which would allow the access to the deposit below the bottom of the open pit [1, 2]. Therefore, in transition from OP to UG, both operations will often be run simultaneously for a certain period of time. Underground operation causes subsidence, slope movements, discontinuous deformations and change of hydrogeological conditions. In the case of underground operation located below the open pit mine, slope strain

* Corresponding author: nguyen.pmv@amtc.cl

can be expected as a result of subsidence induced by underground exploitation. Changes in physical and mechanical properties of rock mass, groundwater levels and the state of stress inside the slope can lead to slope deformation. The negative OP-UG interaction cases [3–5] provide a clear warning as to the possible impact of underground mining on open pit mines. Therefore, it is necessary to carry out more research and analysis on slope stability of open pits affected by underground mining and, due to the different geological conditions and the technology application, any such case should be analyzed individually.

Currently, the Vietnam coal mining is a very important part of the national economy and hard coal is mined mainly by the OP method. In the coal mining of Vietnam, the decision to switch from OP mining to UG mining was taken for several reasons, including: technological possibilities, economic and environmental conditions and social aspects [6]. Cao Son is an OP mine located in coal basin Cam Pha, Quang Ninh. The OP mine Cao Son is decided to change the operation system from OP to UG (Khe Cham II–IV). Simultaneous operation started in 2015 and will be conducted until 2030. To estimate the simultaneous operation Cao Son – Khe Cham II–IV from point of view of safety, the surface and subsurface strain fields generated by simultaneous operation were investigated by using numerical modelling with FDM code. Based on the results obtained, several geomechanical assessments are presented as a case study of Cao Son – Khe Cham II–IV.

2 Case study: simultaneous operation Cao Son – Khe Cham II–IV

The OP anthracite mine Cao Son is located in the Cam Pha coal basin in the Quang Ninh province of Vietnam. Cao Son is about 150 km eastward from the capital city – Hanoi, at the altitude of 100–150 m a.s.l. Cao Son mine has decided to make a transition from OP to UG in 2015. The UG operation, Khe Cham II–IV, is being operated under Cao Son. Simultaneous operation (Cao Son and Khe Cham II–IV) will be conducted until 2030 when the OP mine Cao Son ends its mining activities (Fig. 1). The Quaternary period rock mass consists mainly of gravel, sand, and rolled fill. It is approximately 20 m thick and in some places 150 m thick. The Triassic rock mass is located under the Quaternary period rock and consists of hard rocks: conglomerate (16%), sandstone (48%), mudstone (25%), claystone (3%) and coal (8%). Thickness of the Triassic period rock mass ranges from 500 m to 700 m [6].

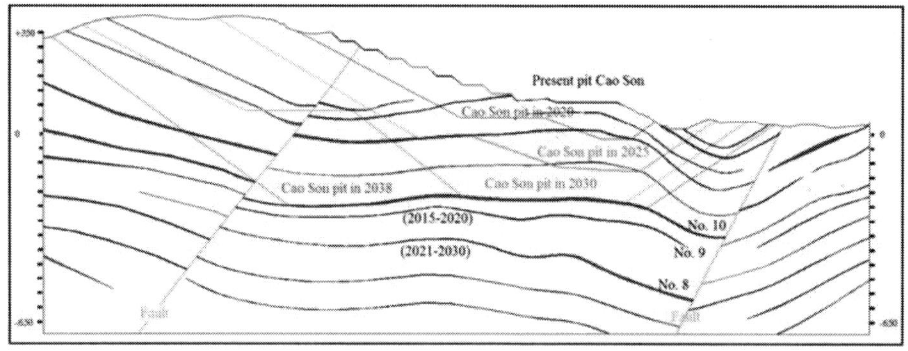

Fig. 1. Exploitation plan in region Cao Son and Khe Cham II–IV in the North-South cross section

The hydrogeological conditions of the Cao Son and Khe Cham II–IV can be characterised as follows [6]: presence of surface water from local streams (Bang Nau, Khe Cham) and from open pits due to rainfall: average rainfall is 4.7–8.0 mm/day and up to 350 mm/day in the rainy season. Surface water is the main source that supplies water to the Triassic rock mass. Groundwater in the Quaternary and dumping ground exists only in the rainy season.

Underground water in the Triassic rock mass is present in the slits of rock mass. The measured filtration ratio is 0.038 m per day. Water is mainly in conglomerate and sandstone. Water does not occur or occurs in very small quantities in the layers of mudstone and claystone and in the fault zones.

Cao Son mine operation will be completed in 2038 in the coal seam No. 10 located at depth of 300 m b.s.l. Under the bottom of the Open Pit Cao Son, UG operation is planned to extract coal seam No. 9 with a thickness of 2 m (2015–2020) and No. 8 with a thickness of 3 m (2021–2030). The average size of the UG operation Khe Cham II-IV is 4,300 × 1,600 m [6].

3 Numerical modelling

Prediction of interaction between UG and OP was carried out with the application of FLAC 2D software based on Finite Difference Method [7]. It was assumed that the rock mass was stratified with the properties according to the elastic-plastic model of Mohr-Coulomb. The initial model was solved as an elastic model to obtain the initial stress conditions. The vectors and speeds of displacements were zeroed. In the next step, the "null" value was assigned to the model of the caved zone and the fractured zone (stresses were zeroed in these areas), and re-assigning of the Mohr-Coulomb model to the rock mass with the assumed parameters was made, followed by new set of calculations. In purpose to define the mechanical properties of rock mass for numerical modelling, the calibration data process was carried out base on determining the values of mechanical parameters of rock mass caved zone and fractured zone for which the maximum values of vertical displacement obtained with assumption 'worst case' is in accordance with the Knothe theory [8]. Because of slight difference in the mechanical properties of the rock mass types in the Cao Son and Khe Cham II–IV regions, rock mass was divided into 2 main groups of rocks: a group of sandstone and conglomerate and a group of claystone, mudstone and coal. The correlation between laboratory mechanical parameters and rock mass parameters is commonly used for numerical modelling. Due to the lack of RMR studies for the Cao Son and Khe Cham II–IV regions, it was assumed that the strength parameters should be reduced to 25% and the deformation parameters to 50% [9]. The mechanical parameters used for numerical modelling is presented in Table 1.

Table 1. Mechanical parameters of rock mass used for numerical modelling

Type of rock mass	Cohesion c [MPa]	Fiction angle θ [°]	Density γ [kg/m³]	Tensile strength σ^t [MPa]	Bulk modulus K [GPa]	Shear modulus G [GPa]
Group of sandstone and conglomerate	1.27	28	2580	0.51	2.06	1.37
Group of claystone, mudstone and coal	0.52	29	2370	0.325	1.14	0.74
Coal	0.6	27	1500	0.305	1.09	0.68

To determine the geometry of the model, the average heights of caved zone and fractured zone were calculated using the following formulas [10]:

$$h_c = \frac{100g}{c_1 g + c_2} \text{ and } h_f = \frac{100g}{c_3 g + c_4} \qquad (1)$$

where: h_c – height of caved zone, m; h_f – height of fractured zone, m; g – average thickness of coal seam, m; c_1, c_2, c_3, c_4 – coefficients are shown in Table 2.

It was assumed that the height of caved zone and fractured zone are the same for all calculation variants and their values were determined according to Equation 1. For coal seam No. 9 the height of caved zone and fractured zone are 10 m and 45 m, respectively. Deformation and strength parameters of caved and fractured zones were selected according to the investigation of Nawrot [11].

Table 2. Coefficients for average heights of caved zone and fractured zone [10]

Type of rock mass	Compressive strength, [MPa]	Coefficients			
		c_1	c_2	c_3	c_4
Strong and hard	> 40	2.1	16	1.2	2
Medium strong	20–40	4.7	19	1.6	3.6
Soft and weak	< 20	6.2	32	3.1	5

For caved zone: the values of strength parameters are 0 and the values of deformation parameters are 20–40 times lower than the values of roof rock. For fractured zone: the value of strength parameters are 1.25–2 times lower, and the value of deformation parameters are 2–5 times lower than the values of roof rock. Roof rocks located above coal seam No. 9 are mainly conglomerate and sandstone [3]; thus, the mechanical parameters of caved and fractured zones were determined based on the mechanical parameters of the sandstone and conglomerate rock group. Density of fractured zone is assumed to be close to the value of rock mass density. Density of caved zone is calculated using the following formulas [12]:

$$\gamma_c = \frac{y_{rm}}{b} \text{ and } b = \frac{c_1 + c_2}{100} + 1 \tag{2}$$

where: γ_c – density of caved zone, [kg/m^3]; y_{rm} – density of rock mass, [kg/m^3]; b – bulking factor, –; g, c_1, c_2 – defined earlier in Equation 1.

Computational mechanical parameters of caved and fractured zones are shown in Table 3 after the calibration process, which consists in finding the values of the mechanical parameters of fractured zone and caved zone to obtain the subsidence value close to the coal seam thickness.

Table 3. Mechanical properties of caved zone and fractured zone used for numerical modelling

Type of rock mass	Cohesion c [MPa]	Fiction angle θ [°]	Density γ [kg/m^3]	Tensile strength σ' [MPa]	Bulk modulus K [GPa]	Shear modulus G [GPa]
Group of sandstone and conglomerate	1.27	28	2580	0.51	2.06	1.37
Caved zone	**0**	**10**	**2100**	**0**	**0.085**	**0.056**
Fractured zone	**0.908**	**19**	**2500**	**0.365**	**0.793**	**0.527**

The strength parameters of fault for numerical modelling were selected: friction angle 200, cohesion 4 kPa, tensile strength 0 kPa [6]. Due to the lack of data on deformation parameters of faults in the Cao Son and Khe Cham II–IV regions, deformation parameters of the fault were calculated according to the formula developed by Itasca [7]:

$$k_s = k_n = 10 \cdot \max\left[\frac{K + \frac{4}{3}G}{\Delta z_{\min}}\right] \tag{3}$$

 https://doi.org/10.1051/e3sconf/20183501001

where: K and G are the bulk and shear moduli, respectively, GPa; Δz_{min} is the smallest width of an adjoining zone in the normal direction, m; k_n and k_s are the normal and shear stiffness, respectively, GPa/m.

After the calculation according to Equation 3, a set of mechanical properties of faults used for numerical modelling are determined: k_n = k_s =2.6 GPa/m, friction angle 200, cohesion 4 kPa, tensile strength 0 kPa.

4 Results, analysis and discussions

Based on the exploitation plan (Chapter 2) for Cao Son and Khe Cham II–IV, the following observations can be drawn for coal seam No. 9: by the end of 2016 (Fig. 2a), only two panels with an average mine face length of 150 m would have been mined. Therefore, the most important issue is to determine the location, where to start the UG operation to ensure maximum slope stability and continuous progress of UG operation. By 2020 (Fig. 2b), coal seam No. 9 will be completely exploited; thus, the pit in 2020 may be affected by UG operation in coal seam No. 9. Numerical calculations were performed to show the impact of UG operation on coal seam No. 9 (Khe Cham II–IV) on advanced phases of OP (Cao Son) using FDM software program with the 2D FLAC code. For each variant, characteristic points were set on the surface of the slope during underground operation. From those characteristic points, values of displacements, velocity vectors were read and used with plasticity indicator to analyse the slope stability at each advance phase of OP. Calculations were made for the pit in 2016 and 2020 year.

a)

b)

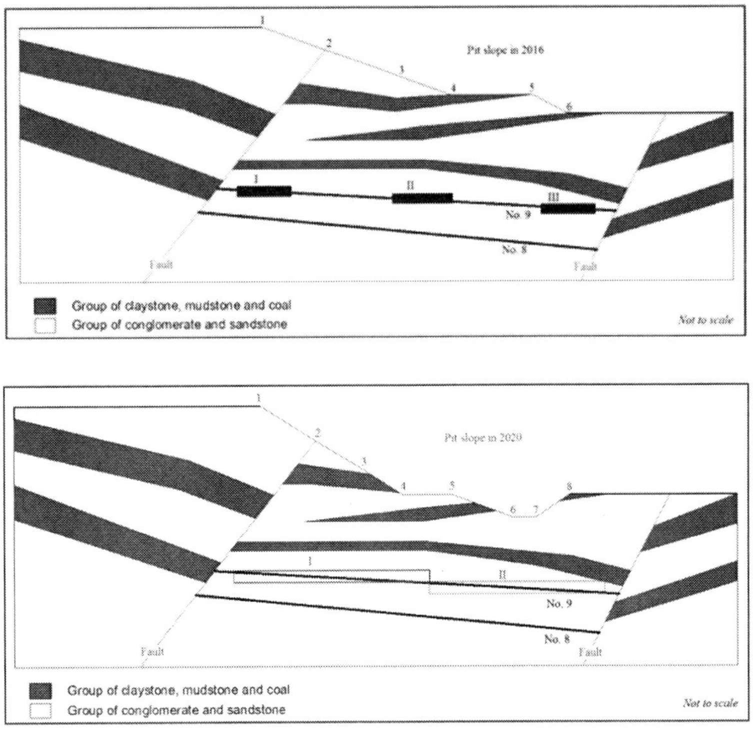

Fig. 2. 2D numerical model for the stage of UG exploitation a) to 2016, b) to 2020

The purpose of this analysis was to determine the location of where to start the UG operation and to indicate the direction of exploitation in the coal seam No. 9 (Fig. 2a).

Numerical calculations were carried out to show the influence of UG operation on the two panels (average mining face length is 150 m) on the slope in 2016 with different panel locations (left – I, central – II, right – III). Table 4 shows displacements and velocity vectors read from the characteristic points located on the slope surface with different locations of two UG long wall panels. Velocity vectors measured on the slope surface are low (much less than transitional value indicating a possibility of unstable state: $1 \cdot 10^{-6}$) with different locations of two long wall panels. Vertical displacements in case of location I are approx. 1 m (upper part of the pit) while vertical displacements in case of location II and III are lower with obtained value of 0.6 m. Horizontal displacements of all cases are approx. 20 cm (Tab. 4). These displacements do not signify instability, however, the displacements induced by Case I would cause difficulties during OP exploitation, such as, OP transportation. Generally, obtained results in Table 4 indicate that initial exploitation (two long wall panels) has a slight influence on slope stability of the 2016 pit. The location where the UG operation begins should be selected with the shortest distance to the first working (shafts) and further UG exploitation will have a minimum affect on slopes stability over time. Location II and III have been shown to be better than location I from the point of view of their influence on slope stability; however, location III is centered below the bottom of the mine in 2016. Therefore, it is recommended to start underground operation at location II.

Table 4. Displacements and velocity vectors along the slope surface in 2016 with different locations of the two UG panels

Location UG panels	**Measured parameters**	**Chosen points located on the slope surface**				
		1	**2**	**4**	**5**	**6**
I	Vertical displacements [m]	–1.077	–1.048	–0.047	–0.007	–0.006
	Horizontal Displacements [m]	0.228	–0.070	0.080	0.090	0.080
	Vertical velocity vectors [10^{-6}]	–0.172	–0.036	–0.010	–0.001	–0.006
	Horizontal velocity vectors [10^{-6}]	0.051	0.092	0.148	0.092	0.080
II	Vertical displacements [m]	–0.116	–0.246	–0.593	–0.121	–0.028
	Horizontal Displacements [m]	0.197	0.237	–0.074	–0.161	–0.043
	Vertical velocity vectors [10^{-6}]	–0.193	–0.073	–0.010	–0.020	–0.002
	Horizontal velocity vectors [10^{-6}]	0.247	0.177	0.133	0.075	0.037
III	Vertical displacements [m]	–0.008	–0.007	–0.100	–0.442	–0.555
	Horizontal Displacements [m]	0.031	0.048	0.145	0.213	0.014
	Vertical velocity vectors [10^{-6}]	–0.124	–0.022	–0.080	–0.015	–0.060
	Horizontal velocity vectors [10^{-6}]	0.147	0.191	0.193	0.119	0.087

The aim of next analysis is to determinate a sequence of UG operation that minimise the UG impact on pit slope 2020 (Fig. 2b). Numerical calculations were carried out to show the influence of UG operation in coal seam No. 9 on slope in 2020 with two different sequences. The adequate direction of UG operation in coal seam No. 9 was determined to minimise its impact on the stability of the slope 2020. Two options were taken into consideration: first right half then left half (right-left) and first left half then right half (left-right). Displacements and velocity vectors on the slope surface in 2020 with two directions of UG exploitation are presented in Table 5. Velocity vectors obtained from all calculations are low (much less than transitional value indicating a possibility of unstable state $1 \cdot 10^{-6}$). Initially, slope 2020 can be considered to be stable during UG operation in coal seam No. 9. First step is exploitation of first half. With the UG operation in left half (I), the maximum vertical displacement reached 1.7 m (point 2) while the horizontal displacement is 0.8 m (point 1). With the UG operation in right half (II), the vertical displacement was about 1.2 m (points 6 and 7), while the horizontal displacement was approximately 0.45 m (point 4)

(Tab. 5). This means that the displacements caused by the UG operation in the left half (I) are greater the displacements caused by the UG operation in the right half (II). Thus, the UG operation in the right half (II) affects the slope lower than the UG operation in the left half (I). This can be explained by the fact that the left part of slope is larger than the right part of the slope. Second step is exploitation of first half. In the left-right UG operation, the vertical displacements reached 0.8–2 m along the slope, while the maximum horizontal displacement was about 1.2 m (point 1). In the "right-left" UG operation, the vertical displacements obtained were 0.9–1.9 m along the slope, while the maximum horizontal displacement was 1 m (point 1) (Tab. 5). It shows that the "right-left" UG operation is slightly more advantageous than the "left-right" UG operation. Thus, the final impact induced by the "left-right" UG operation is less than the final impact induced by the "right-left" UG operation. Obtained results indicate that UG exploitation in coal seam No. 9 influence slightly the slope stability of the 2020 pit and the left-right UG operation is recommended.

Table 5. Displacements and velocity vectors along the slope surface in 2020 with 2 different UG exploitation directions

Sequence of operation		Measured parameters	Chosen points located on the slope surface				
			1	3	4	6	7
From left hand side to right hand side	Left half (I)	Vertical displac. [m]	–1.382	–1.584	–0.930	–0.033	–0.006
		Horizontal displac. [m]	0.802	0.292	–0.173	0.079	0.081
		Vert. velocity vectors [10^{-6}]	–0.124	–0.266	–0.047	–0.003	–0.002
		Horiz. velocity vectors [10^{-6}]	0.290	0.470	0.110	0.140	0.078
	Right half (II)	Vertical displac. [m]	–1.614	–2.058	–1.783	–1.235	–1.114
		Horizontal displac. [m]	1.150	0.820	0.366	0.253	0.050
		Vert. velocity vectors [10^{-6}]	–0.346	–0.174	–0.156	–0.083	–0.070
		Horiz. velocity vectors [10^{-6}]	0.398	0.451	0.216	0.307	0.143
From right hand side to left hand side	Right half (II)	Vertical displac. [m]	–0.061	–0.163	–0.434	–1.150	–1.126
		Horizontal displac. [m]	0.112	0.230	0.416	0.088	–0.131
		Vert. velocity vectors [10^{-6}]	–0.133	–0.070	–0.050	–0.005	–0.012
		Horiz. velocity vectors [10^{-6}]	0.183	0.200	0.111	0.067	0.040
	Left half (I)	Vertical displac. [m]	–1.559	–1.890	–1.688	–1.313	–1.171
		Horizontal displac. [m]	1.020	0.592	0.338	0.147	–0.038
		Vert. velocity vectors [10^{-6}]	–0.220	–0.407	–0.193	–0.037	–0.082
		Horiz. velocity vectors [10^{-6}]	0.538	0.518	–0.189	0.184	0.046

5 Conclusions

The obtained results showed that all pit slopes were considered stable for various arrangements of UG exploitation. However, this does not entirely eliminates the possibility of slope displacement, which could negatively affect an OP operation. The impact of UG operation on OP slope stability is a continuous process related with the advance of both OP and UG exploitations. During the simultaneous operation, such as, Cao Son and Khe Cham II–IV, changes to slope geometry can also prevent the impact of UG exploitation. Hence, an appropriate scheduling and coordination of exploitation plan is required. Simultaneous operation is not a common mining technique, especially, for coal mining. Therefore, special actions should be taken to guide these types of operation, including: carrying out slope stability analysis for any advancing phase of open pit with the advance of UG operation; applying monitoring system to ensure slope stability by detecting any in-ground and surface movements; carrying out water drainage from pit bottom to prevent water inrush into UG workings;

minimising blasting in both OP and UG Operation; minimising subsidence surface by using UG methods.

References

1. W.F. Visser, *Optimization of the OP/UG Transition. Development of a Software Tool for Optimization of the Transition Depth and the Open Pit Slope Angle – Main report.* Technische Universiteit Delft (2006)
2. P.M.V. Nguyen, Z. Niedbalski, *Numerical modelling of open pit (OP) to underground (UG) transition in coal mining.* Studia Geotechnica et Mechanica, Vol. 38, No. 3, p.p. 35-48 (2016)
3. R. Brummer, H. Li, A. Moss, *The Transition from Open Pit to Underground Mining: an Unusual Slope Failure Mechanism at Palabora.* Proceedings Int. Symposium on Stability of Rock Slopes, Cape Town (2006)
4. A.D. Campbell, E. Mu, C.R. Lilley, *Cave propagation and open pit interaction at the Ernest Henry mine.* Seventh international conference and exhibition on mass mining, Sydney. Hoek E., Bray J., 1981: Rock Slope Engineering, 3rd eds. IMM. London (2016)
5. D. Díaz, *Geomechanical status and action plans for interaction between Andina subsidence crater and Los Bronces pit*, MSc thesis, Universidad de Los Andes, Santiago (2014)
6. Institute of Mining Science and Technology, Unpublished materials, Hanoi, Vietnam (2012)
7. Itasca, User's Manual of FLAC v. 7.0. Minneapolis, MN, Itasca Consulting Group Ltd. Available at www.itascacg.com (2011)
8. S. Knothe, *Asymmetric function of distribution of mining exploatation influences in the medium with changing properties.* Archives of Mining Sciences, Vol. 50, Issue 4, pp. 401–415 (2005)
9. N. Mohammad, D.J. Reddish, L.R. Stace, *Technical note: the relation between in Situ and laboratory rock properties used in numerical modelling.* Int. J. Rock Mech. Min. Sci. Vol. 34, No. 2, pp. 289–297 (1997)
10. M. Bai, F. Kendorski, D. Van Roosendaal, *Chinese and North American high-extraction underground coal mining strata behavior and water protection experience and guidelines.* Proceedings of the 14th International Conference on Ground Control in Mining, Morgantown (1995)
11. W. Nawrot, *Właściwości mechaniczne utworów skalnych o strukturze blokowo-warstwowej*, Praca Doktorska, Kraków (1972)
12. H. Yavuz, *An estimation method for cover pressure re-establishment distance and pressure distribution in the goaf of longwall coal mines.* International Journal of Rock Mechanics and Mining Science, Vol. 41, pp. 193–205 (2004)

Integrated Approach to Drilling Project in Unconventional Reservoir Using Reservoir Simulation

Jerzy Stopa[1,*], *Rafał* Wiśniowski[1], *Paweł* Wojnarowski[1],
Damian Janiga[1], and *Krzysztof* Skrzypaszek[1]

[1] AGH University Science and Technology, Faculty of Drilling, Oil and Gas,
A. Mickiewicza Av. 30, 30-059 Krakow, Poland

Abstract. Accumulation and flow mechanisms in unconventional reservoir are different compared to conventional. This requires a special approach of field management with drilling and stimulation treatments as major factor for further production. Integrated approach of unconventional reservoir production optimization assumes coupling drilling project with full scale reservoir simulation for determine best well placement, well length, fracturing treatment design and mid-length distance between wells. Full scale reservoir simulation model emulate a part of polish shale – gas field. The aim of this paper is to establish influence of technical factor for gas production from shale gas field. Due to low reservoir permeability, stimulation treatment should be direct towards maximizing the hydraulic contact. On the basis of production scenarios, 15 stages hydraulic fracturing allows boost gas production over 1.5 times compared to 8 stages. Due to the possible interference of the wells, it is necessary to determine the distance between the horizontal parts of the wells trajectories. In order to determine the distance between the wells allowing to maximize recovery factor of resources in the stimulated zone, a numerical algorithm based on a dynamic model was developed and implemented. Numerical testing and comparative study show that the most favourable arrangement assumes a minimum allowable distance between the wells. This is related to the volume ratio of the drainage zone to the total volume of the stimulated zone. **Key words** – hydraulic fracturing, shale reservoir, numerical simulation.

1 Introduction

Hydraulic fracturing in shales formation should be aimed to maximizing hydraulic contact between wells and rock matrix by production network of micro fractures. By increasing the drainage area well performance could be increased over times, however efficient hydraulic treatment process is a complex process in which simultaneous phenomena directly affect the effect and scale, making it difficult to predict the overall project efficiency. Owing to the heterogeneity of reservoir layers, the shape of the fracture can be properly predicted only on the basis of a properly selected fracture model and quality of geological-reservoir

* Corresponding author: stopa@agh.edu.pl

properties [1]. The complexity of this process causes that the analysis of expected results cannot be done quickly. In this case complex numerical modeling of this operation is very useful as it accounts for geomechanical properties of rocks, varying distribution of stresses, gradient of geostatic and reservoir pressure as well as filtration parameters connected with exploitation model of the reservoir [2]. In the case of tight or shale gas reservoirs the 3D numerical models have to be involved where from total stress in reservoir fracture dimension can be calculated [3–5]. Fracturing treatment design comes to three technological parameters: slurry injection rates and density and viscosity of used fluids. Advanced fracturing models are based on stress balance, energy balance, fluid flow and transport of the proppant, as well as equation of fracturing fluid filtration to the rock matrix. Fracturing models based on 3D geometry enable modeling of an irregular set of fractures, the shape of which is conditioned by local properties of reservoir. Process efficiency can be evaluated by applying simple analytical solutions or reservoir simulations, performed as an individual calculation process based only on the fracturing results. This solution is limited by the lack of full integration of the project with the predicted production, which is necessary for optimization processes.

Mechanisms of accumulation and flow of hydrocarbons in unconventional reservoirs are different in comparison to conventional reservoirs. The phenomena characteristic for unconventional gas deposits can be classified as [6]:

- natural fracture,
- sorption / desorption,
- non-stationary flows in nano-Darcy permeation matrix,
- modelling complex geometries of natural and induced gaps,
- high flow velocities in fractures,
- rock compaction,
- exposure wells with complex topology.

Integration of complex fracture geometry with mechanism of accumulation and flow can be obtain by coupling fracturing simulation with numerical flow simulation.

2 Methodology

2.1 Near wellbore simulation model

2.1.1 Static model

Numerical model of near well zone of shale reservoir using hydraulic fracturing requires the development of an accurate characteristic of the interval to be tested. It is necessary to construct a model of the interesting zone integrating geomechanical and flow properties. On the basis of geological data in the form of records of geophysical logs, a general model of the canopies zone has been developed covering the interval 2600–2915 m TVD. The local geological model used to simulate the fracturing process and their efficiency covers an area of 5 km^2 (2.5 × 2 km) and has been discretized with a grid of 25 × 25 m blocks. The distribution natural permeability in numerical simulation model is shown in Figure 1.

2.1.2 Dynamic model

The static geological model was supplemented with the fluid model as well as the initial and boundary conditions. Reservoir temperature is 85°C and is full field by methane. PVT fluid properties are calculated from the Peng-Robinson equation [7]. Desorption curve was described by Langmuir [8]. The static model assumes the existence of a double porosity system, where a mass storage matrix and fracture network coexists. The dynamic model assumes a hydrodynamic balance between those systems. The matrix does not have primary permeability and is saturated with natural gas. Natural fractures have very low permeability and are saturated with water. Due to the low permeability of the fractures, it is possible to extract the reservoir water during operation.

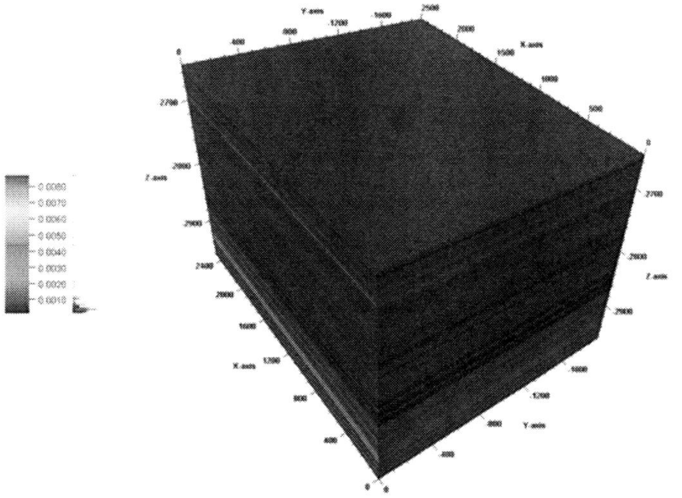

Fig. 1. Nature permeability distribution in near well simulation model

2.2 Well trajectory

The trajectory of the wellbore proposed in this study assumes the vertical section to a depth of 2623.5 m and then begins to curve with a fixed DLS of 6 °/30 m. The trajectory assumes the end of the curvature at a depth of 2910 m (TVD) with a final inclination of 90°. Final borehole length (MD) is 4450 m, of which approximately 1400 m is the horizontal section in which fracture will be performed. Well trajectory is presented in Figure 2.

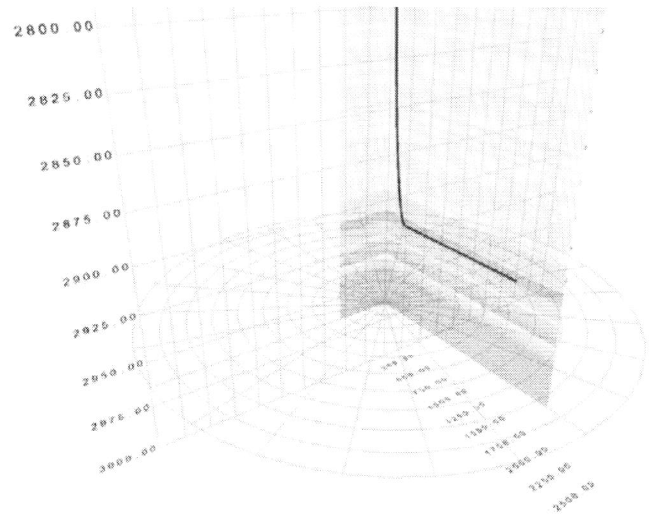

Fig. 2. Borehole trajectories in the reservoir zone model

2.3 Technological model of multi-stage fracturing

2.3.1 Selection of perforation locations for multi-stage hydraulic fracturing in horizontal wells based on the model of the reservoir zone

The purpose of modeling the hydraulic fracturing process is to achieve the best hydraulic contact of the well with the reservoir by fracturing. One of the key decisions in this process is the location of the perforation clusters, taking into account the reservoir conditions. The distance between perforation clusters should take into account the minimal impact on the stressing area of the adjacent fractures. The direct effect on the fracture have: optimal selection of the productivity and amount of proppant in each stage of fracturing, reservoir thickness and its brittleness, as well as determination of the fracture propagation limits. For the assumed well trajectory, two variants were analysed: 8 clusters of perforations (equaled 150 m) and 15 perforation clusters, (equaling 75 m). Clusters of 8 perforation is presented in Figure 3.

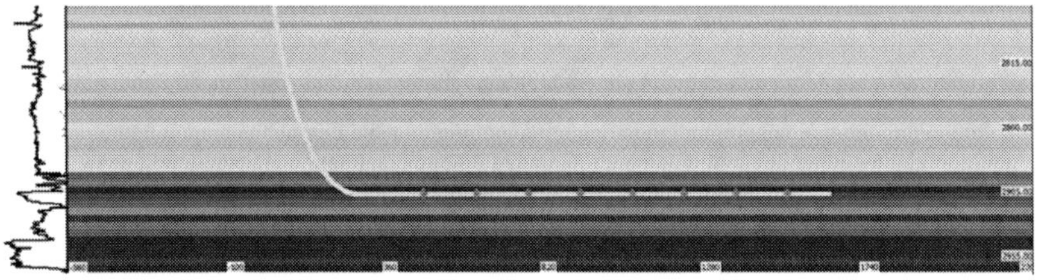

Fig. 3. Localization of the perforation in the second variant comprising 8 clusters

Performing the hydraulic fracture in the first variant minimizes the adverse effect of the individual stages on the overall effect of the treatment [9–10]. The distance between the individual clusters in the second variant allows the use of Zipper Frac [11]. For the proposed perforation locations, the influence of applied treatment liquid on the geometry of the resulting gap and the amount and type of applied proppant material was analyzed on the hydraulic conductivity of the resulting.

2.3.2 An integrated model of shale reservoir and SRV zones

Due to the superposition of phenomena occurring during the exploitation of shales stimulated by hydraulic fracturing, fast analysis of the effectiveness of analytical methods is impossible. In this case turns out to be a comprehensive numerical modeling of exploitation by integrating the results of fracture treatment with a dynamic model. The idea of integrating fracturing with a dynamic model is shown in the Figure 4. Due to the possible interference of the wells, it is necessary to determine the distance between the horizontal parts of the trajectories providing the shale reservoir. In order to determine the distance between wells allowing to maximize the rate of depletion of resources located in the stimulated zone, a numerical algorithm based on the dynamic model was developed and implemented. The SRV zone along with the models is implemented in the reservoir simulator (Fig. 5) where, at the initial stage of operation, the volume of gas occurring in the assumed zone is estimated. In the next stage of the algorithm, the exploitation P1 and P2 wells is forecasted (the wells are assumed to work with the same production parameters). At the end of the simulation stage, the resource depletion rate in the SRV is estimated, if it has improved between successive iterations, the algorithm continues to operate to reach the minimum possible distance between the wells. To determine the minimum distance between holes, it is assumed that the maximum gap xf is at least 50 meters from the second well.

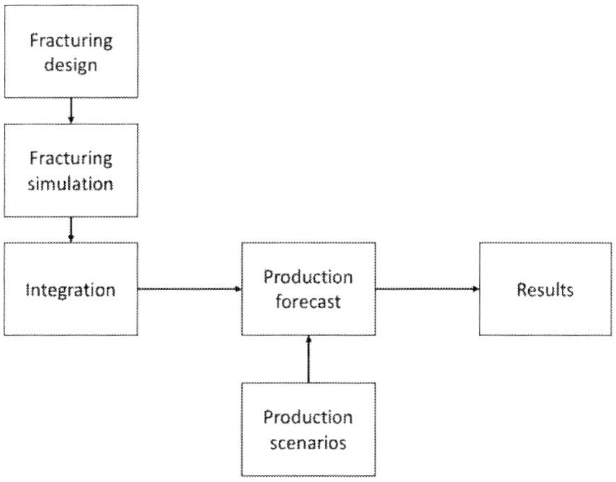

Fig. 4. Schematic diagram of the integration of fracture treatment with a reservoir simulator

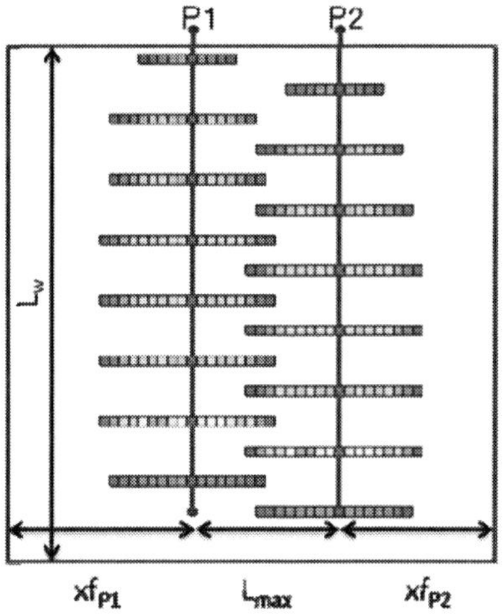

Fig. 5. SRV zone size

Numerical simulation of hydraulic fracturing was performed by utilisation of Guar 40# treatment fluid, 30/50 mesh proppant size with maximum concentration of 400 kg/m^3. The assessment of efficiency of exploitation depending on the wells location was carried out where wells P1 and P2 are control by maximum pressure depression of 90 bar with limits of water production of 40 m^3/d and 80 m^3/d respectively for case A (8 fractures) and B (15 fractures).

3 Results

For variants involving the execution of 8 states fracturing, the recovery factor in the SRV zone is from 0.904% for a distance of 600 m to 1.168% for a distance of 300m. In case B (15 fractures) SRV recovery factor is 2.235% for a distance of 600 m to 2.949% for a distance of 300 m. Results are presented in Figure 6. Extremly low matrix permeability of shales couses that depletion process takes places only in SRV zone (Fig. 7–10). Comparison of results reveal that for both cases (A and B) distance between wells is an inportant factor which have influence on SRV recovery factor. By getting well closer to each other, results can be improved from 5.50 to 29.75% (case A) or 32.50% (case B).

 https://doi.org/10.1051/e3sconf/20183501002

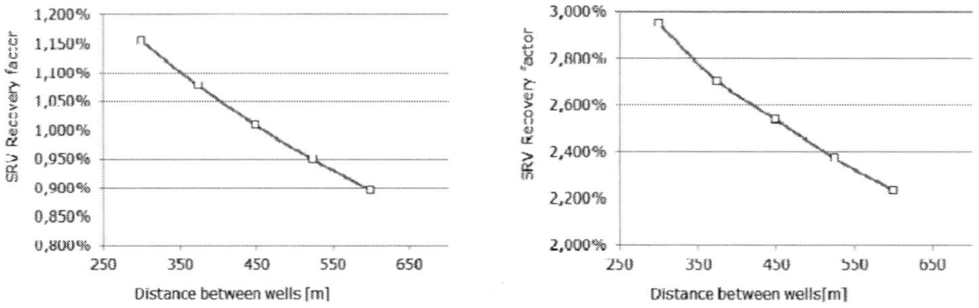

Fig. 6. Summary of the results of the recovery factor as a function of the distance between wells, for case A (left) and B (right)

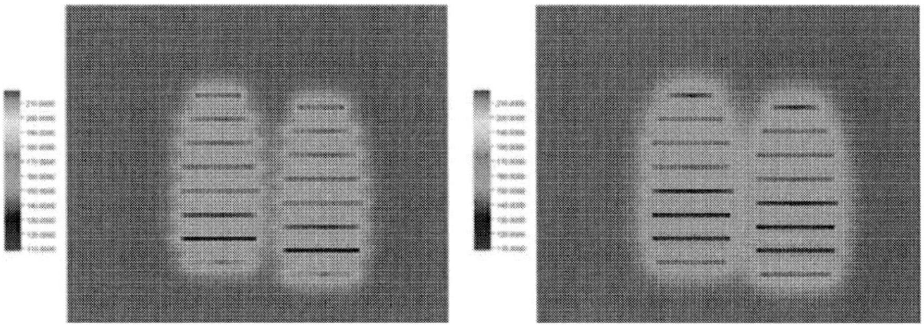

Fig. 7. Pressure distribution after 5 and 10 years of exploitation, case A (8 ftactures), 600 m distance between wells

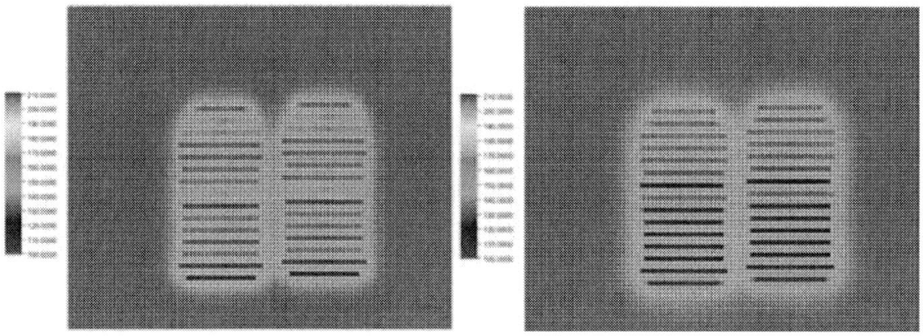

Fig. 8. Pressure distribution after 5 and 10 years of exploitation, case B (15 ftactures), 600 m distance between wells

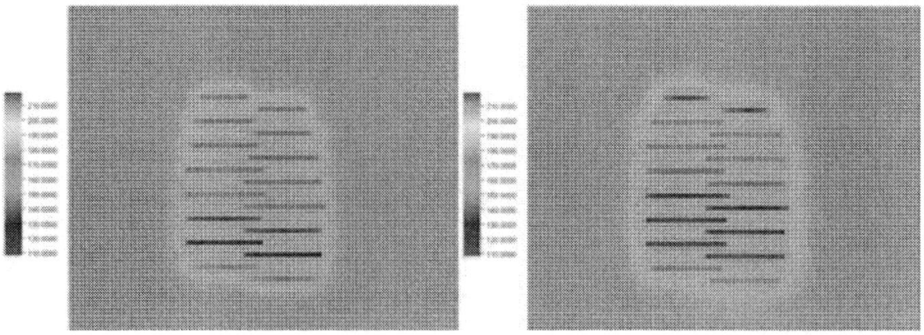

Fig. 9. Pressure distribution after 5 and 10 years of exploitation, case A (8 ftactures), 300 m distance between wells

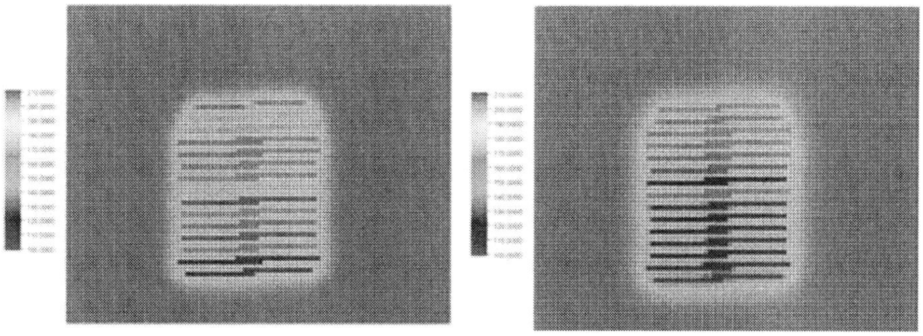

Fig. 10. Pressure distribution after 5 and 10 years of exploitation, case B (15 ftactures), 300 m distance between wells

Summary results are presented in Figure 11. Clusters of 15 perforation generates better results in comparition of 8 fractures due to better hydraulic connection (larger overal fracure area).

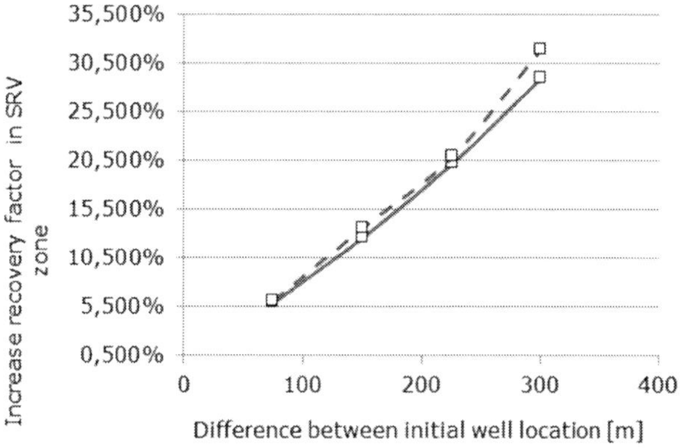

Fig. 11. Changes in the RSV recovery factor on relative to the initial position of the wells, continuous line – case A, dashed line case B

The cumulative output of natural gas in the assumed operating period was chosen as the measure of the effectiveness of the simulation. After 10 years of production for 8 fracture total gas production is over 45 mln m^3 in comparition of over 63 mln m^3 in case of 15 fractures. Results of total gas production is presented in Figure 12.

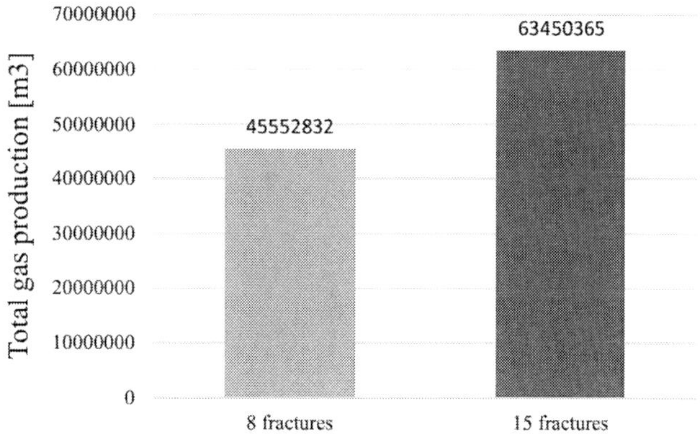

Fig. 12. Total gas production after 10 years of production

4 Conclusion

Because of the low shale reservoir permeability, the fracturing procedure should be aimed at maximizing the hydraulic contact. Based on the analyzed scenarios, in terms of operation, cluster of 15 perforation formation allows over 1.5 times the increase in cumulative production compared to the 8 cluster. Due to the possible interference of wells, it is necessary to determine the effect of distances of stimulated wells on the recovery factor in SRV zone. With regard to the recoverable resources of the stimulated zone, the most advantageous arrangement assumes the minimum permissible distance between the wells. This is related to the ratio of the volume of the drainage zone to the total volume of the stimulated zone.

References

1. J. Stopa, P. Wojnarowski, D. Janiga, AGH Drilling, Oil and Gas, 31, (2014)
2. P. Valko, M.J. Economides, Hydraulic Fracture Mechanics, (1997)
3. T.H. Leshchyshyn, K.K. Deadall, P.E. Mayer, M.W. Hagel, B.R. Mayer, SPE International Thermal Operations and Heavy Oil Symposium and Western Regional Meeting, (2004)
4. L.H. Ribeiro, M.M. Sharma, SPE Annual Technical Conference and Exhibition, (2012)
5. J. Adachi, E. Siebrits, A. Peirce, J. Desroches, Journal of Rock Mechanics and Mining Sciences, 44, 739–757, (2007)
6. M.J. Economides, K.G Nolte, Reservoir Stimulation 3rd Edition, (2000)
7. D. Peng,, D.B. Robinson, Industrial and Engineering Chemistry Fundamentals, 15, 59–64, (1976)
8. I. Langmuir, Journal of the American Chemical Society, 38, 2221–2295, (1916)
9. H. Liu, Y. Luo, X. Li, Y. Xu, K. Yang, L. Mu, W. Zhao, W. Zhou, SPE Annual Technical Conference and Exhibition, (2012)
10. M Rafiee, M.Y. Soliman, E. Pirayesh, SPE Canadian Unconventional Resources Conference, (2012)
11. M Rafiee, M.Y. Soliman, E. Pirayesh, SPE Annual Technical Conference and Exhibition, (2012)

Construction of Shale Gas Well

Aneta Sapińska-Śliwa[1,*], *Rafał* Wiśniowski[1], and *Krzysztof* Skrzypaszek[1]

[1] AGH University of Science and Technology, Krakow, Poland, Faculty of Drilling, Oil and Gas, A. Mickiewicza Av. 30, 30-059 Krakow, Poland

Abstract. The paper describes shale gas borehole axes trajectories (vertical, horizontal, multilateral). The methodology of trajectory design in a two- and three-dimensional space has been developed. The selection of the profile type of the trajectory axes of the directional borehole depends on the technical and technological possibilities of its implementation and the results of a comprehensive economic analysis of the availability and development of the field. The work assumes the possibility of a multivariate design of trajectories depending on the accepted (available or imposed) input data. **Key words** – shale gas, construction of the well, unconventional reservoirs.

1 Introduction

Due to an increased demand for energy and changing trends, resources not produced so far because of the lack of a suitable technology available started to be reached out for. Such resources are unconventional gas and oil reservois the production of which, on a larger scale, started at the end of 20th century. Presently, the largest producer of gas from unconventional reservois is the United States. In 2004, the production of shale gas in the US accounted for 5% of the amount of gas being produced and in 2016 it was as high as 60%. Gas production had been preceded by many years of research on a new technology, which would enable achievement of economically justified gas flows. The key to success became the application of horizontal drilling technology and hydraulic fracturing.

Opening a shale gas reservoir consists in drilling of a gas formation with a network of directional boreholes with end horizontal sections, with outlets located in close proximity, i.e. group wells. Grouping of well outlets at small intervals (a multiwell pad) has a number of advantages, yet, it requires very precise well design and performance. In this case, vertical sections are located very close to one another, which, together with imprecision of measurement instruments, may result in well collision. In the event that wells are operational, such collision may lead to eruption, and thus, to direct threat to life of the drilling rig staff. Well performance is preceded by a designing process, which has to take into account all factors affecting the precision of calculations and measurements. Well design may be supported by such computer programs as Compass, Petrel, WellArchitect or 3DPathDesigner, created at the Drilling, Oil and Gas Faculty. All those programs have tools calculating the course of borehole trajectory and enable estimation of the required separation between boreholes [1, 2].

* Corresponding author: ans@agh.edu.pl

2 Conventional and unconventional hydrocarbon fields, including hydraulic fracturing

Over the last dozen years, continuous growth in the number of directional boreholes in relation to the total number of drilled boreholes has been observed. This phenomenon relates to the development of technologies allowing exploitation of unconventional raw material reservoirs, which are closely related to directional boreholes. Conventional energy sources are located at a small depth, whereas permeability of rocks creating them is high. Good conditions enable effective reservoir production by means of vertical wells, which translates into low investment costs. Unconventional reservoirs (shale gas & liquid, tight oil & gas) are located much more deeply and, additionally, rock-mass is characterised by very low permeability. Opening of an unconventional reservoir by means of standard methods, characterised by low gas flow, is most frequently uneconomic. A better solution is directional wells with end straight-lined sections running along the productive level. Such location of a borehole in a reservoir layer enables increasing of the contact surface with the reservoir rock, which translates into a higher stream of volume [6, 7].

Directional boreholes are also used for opening gas storage facilities in salt caverns, coalbed methane, making wells near faults, bypassing obstacles in vertical boreholes, or liquidating eruptions.

Hydraulic fracturing is an example of a procedure increasing gas inflow to a borehole. This technology was introduced in 1948 by J.B. Clark of Standolind Oil and Gas Co. The first intensification procedures were carried out in 1960's. The technology of fracturing poorly permeable shales was developed in 1980's by George T. Mitchel. In 2000, Devon Energy Co. became the owner of the said company and combined the fracturing technology with the horizontal drilling technology. The application of this intensification procedure enables to obtain economically profitable gas flow from shale formations to the well – this method has become the reason for abrupt production growth from unconventional reservoirs.

Hydraulic fracturing consists in fluid injection under pressure, exceeding fracturing pressure of the centre, which leads to its crashing. Fluid is injected through casing perforated in indicated places of the horizontal section (fracturing stations). Fractures occurring in this way are maintained in an open state by a proppant, being quartz sand or any other material with high compressive strength. Proppant permeates inside fractures and makes their closure impossible after pressure release. Fractures propagate along the largest compressive stresses. As mentioned above, hydraulic fracturing can be performed in vertical wells; in such case, the growth of the contact surface between the well and the reservoir is not so large (Fig. 1).

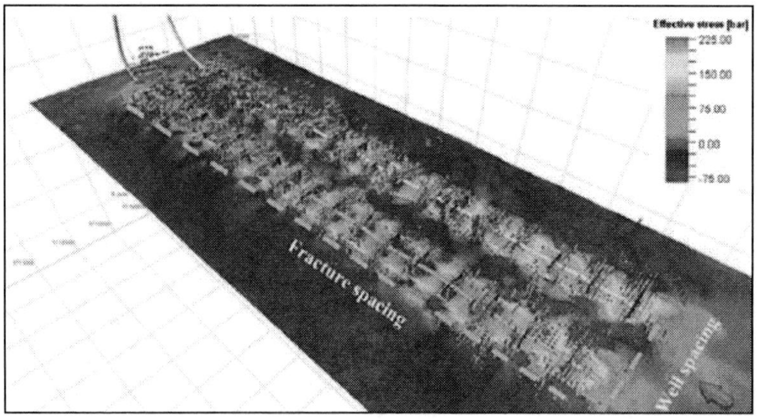

Fig. 1. Distribution of stresses and range of fracture propagation during hydraulic fracturing in horizontal sections of a drilling borehole [5]

3 Way of distributing directional boreholes in case of unconventional reservoirs

In case of opening unconventional reservoirs (e.g. shale gas), a distance between wells cannot be long. It is due to a small impact range of the well in layers of poorly permeable shales. The impact range after hydraulic fracturing is, on average, 150–250 m in a horizontal direction. For the needs of this study, the way of opening a piece of the reservoir, 3.5 km per 1.5 km in size, has been analysed. When performing a vertical well, together with an intensification procedure, one can cover the surface corresponding to the area of a circle with the radius equal to the scope of fracturing (ca. 250 m), assuming that the scope of fractures will be slightly larger and will cover the area of a square with the side equal to a double radius of fracturing, 21 vertical well will fall on the area under analysis (Fig. 2).

Fig. 2. Number of vertical boreholes covering a given area

When analysing the above arrangement, one may notice that the number of wells is large compared to the covered surface, and the investment project can be not economically viable. In order to reduce the number of well, one can make a directional well with a horizontal section in a gas layer. Assuming that the deviation from the vertical direction is about 900 m (the sum of a curvilinear section projection on the horizontal plane and the length of the horizontal section), whereas the range of fracturing will be the same as previously, 12 directional boreholes will be necessary to open a piece of the reservoir. The situation has been presented in Figure 3.

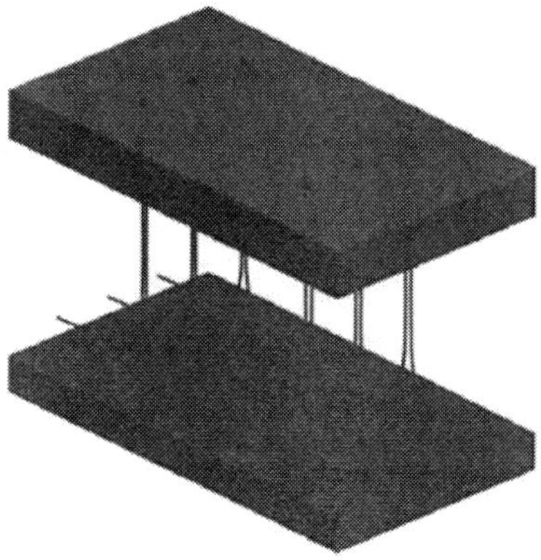

Fig. 3. Directional boreholes in a 2D space

The application of the directional drilling technology (2D) has enabled reduction of the number of boreholes (from 21 down to 12) and the working area on the surface. The available technology allows drilling of horizontal sections even up to 10 km (Extended Reach Wells), whereas in the case under analysis, 6 wells can be applied with a deviation from the vertical direction by ca. 1750 m or even longer sections, which will enable to reduce their number even more. 2D directional boreholes have enabled to reduce significantly the number of wells, however, their distribution on the surface is still not designed optimally. Drilling rigs sites are located at long distances from one another and there are plenty of them; in such situation it is necessary to transport heavy equipment. Figure 4 presents the diagram of opening a piece of the reservoir by means of 3D directional boreholes. Owing to this technology, well outlets are more grouped, which decreases the required area and the number of equipment relocations.

Fig. 4. Directional boreholes in a 3D space – Well Pad Drilling

Advantages of group well distribution are the following [8]:

- reduction of the required area on the surface for the performance of drilling works, a smaller amount of working areas, a lower negative impact upon the natural environment, a possibility to open reservoirs under inaccessible sites and protected areas,
- no need to dismantle and relocate heavy equipment many times to each well, modern drilling rigs have a stepper mechanism enabling moving among outlets of group wells, reducing road traffic and the number of necessary access roads,
- reduction of drilling time per well (by 50% or more) due to equipment concentration in one place and no need to dismantle the equipment each time,
- one water tank for hydraulic fracturing for the whole group of wells (even up to 12), a possibility to use the same water and chemical agents many times,
- after making of the first well up to date results of geological studies are obtained, which enable drilling optimisation and cementing subsequent wells, as geology will be similar, due to a short distance between wells,
- concentration of extraction instruments in one place,
- possibility to reduce costs when drilling a high number of wells.

Disadvantages of group well distribution [8]:

- higher cost of a single well, in particular, for companies not having any experience in performing wells,
- longer working time in one place and a larger area is required in one place,
- a well is more difficult to make and it requires a higher number of pipes and grout,
- special equipment and software is required for designing and drilling wells with variable azimuth, together with a module estimating the risk of collision.

The development of directional drilling technology enables to perform more and more complex boreholes. Owing to continuous development of this drilling sector, the price of directional drilling services is declining, through which it is possible to perform the above-presented diagrams of drilling unconventional reservoirs.

Summing up the advantages and disadvantages and taking into account trends on markets of the largest producers of natural gas from unconventional reservoirs, one can state that making shale gas and tight gas available is economically viable, whereas group location of well outlets enables cost reduction and improvement of environmental aspects. Group location of outlets, due to the proximity of neighbouring wells in vertical sections and the beginning of curving, requires using advanced techniques of designing drilling borehole trajectories, highly precise measurement devices, as well as tools allowing to analyse measurement data and to estimate risks of collision.

4 Computer aided design of directional wells trajectories in case of unconventional reservoirs

For the needs of supporting the designing process of directional boreholes trajectories at the Faculty of Drilling, Oil and Gas at AGH University of Science and Technology, within the frameworks of the OPTIDRILLTEC program, the authors of the article created open software utilising state-of-the-art IT technologies. The application is irrespective of the operating systems being used or the hardware platform; it was developed using php and java languages and it can work in new internet browsers. Such a solution has a number of advantages, for instance, a possibility to use the program on desktops, tablets, smartphones; scalability of the number of users; access to the collected complete knowledge base; data security on the server side (Fig. 5).

Fig. 5. Login window of the 3DPathDesigner application

The 3DPathDesigner application uses algorithms to calculate trajectories created within the frameworks of research work by the authors of this article [3, 4]. Below, Figure 6 present an interface for designing 2D-type trajectories (with a constant azimuth); it enables very comfortable preparation of even very complex trajectory shapes, taking into account most of designer's expectations.

a)

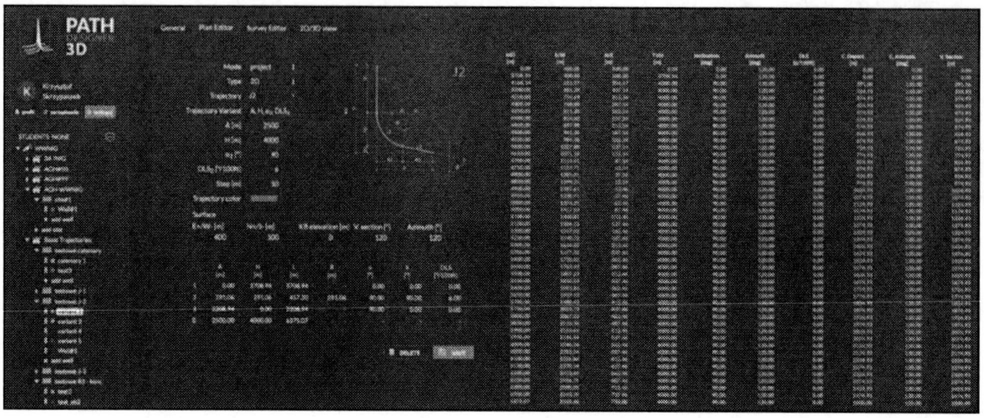

b)

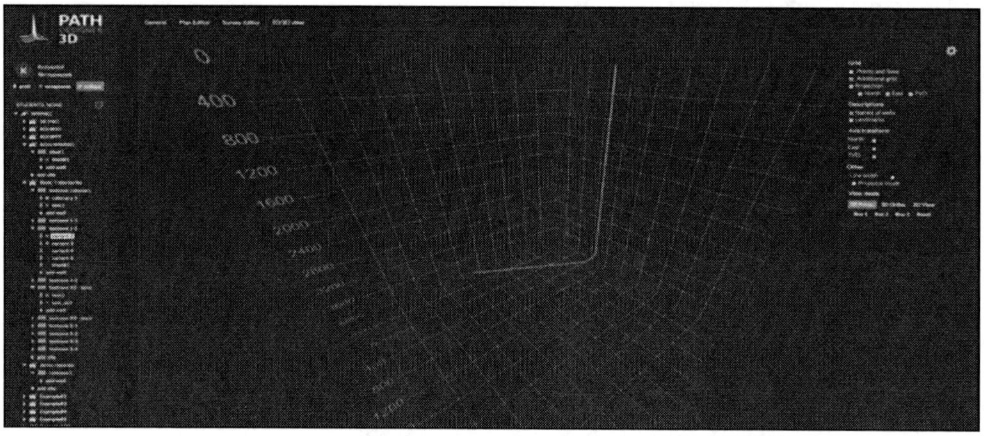

Fig. 6. Window of designing a 2D trajectory in the 3DPathDesigner application

3DPathDesigner program makes it possible to prepare complex designs of this kind. Figure 7 presents user tools within the scope of trajectory designing in a 3D space (with variable azimuth), whereas Figures 8 and 9 show application possibilities when designing ERW-type well networks and multilateral wells, respectively.

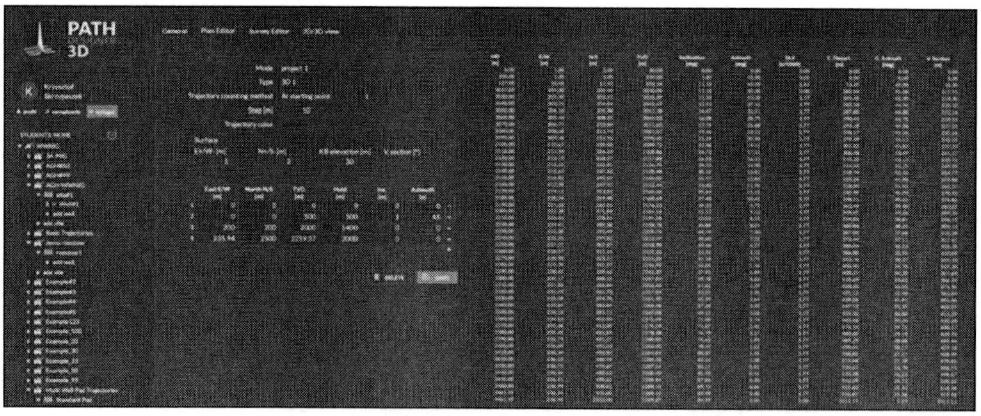

Fig. 7. Window of designing a 3D trajectory in the 3DPathDesigner application

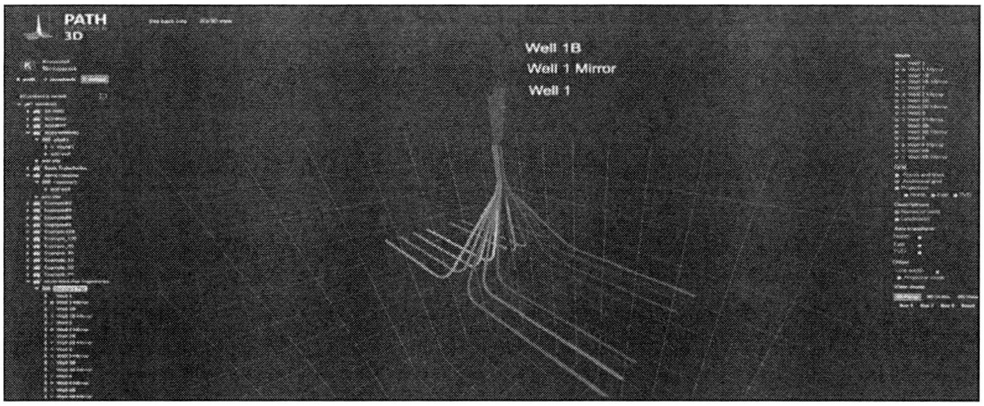

Fig. 8. Window of designing trajectories of a network of fracturing wells with a long horizontal section (ERW) in the 3DPathDesigner application

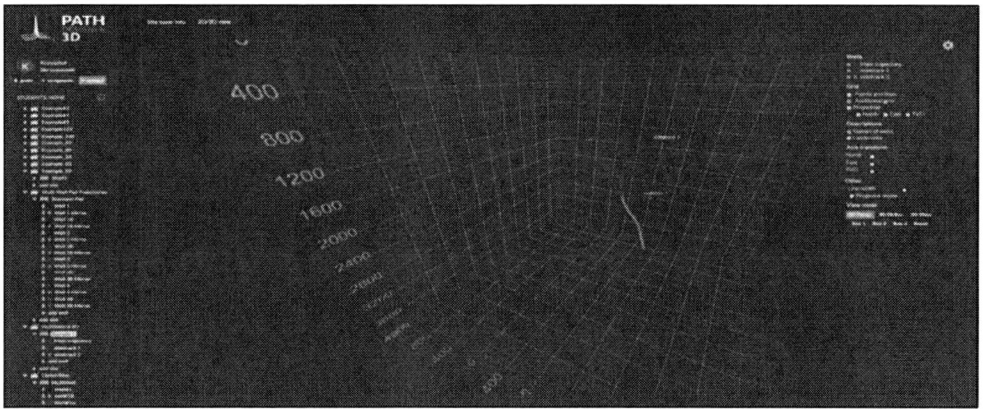

Fig. 9. Window of designing trajectories of a network of multilateral wells in the 3DPathDesigner application

Directional boreholes in a 3D space require special structures. It relates, first of all, to the intensity of such boreholes curving.

5 Conclusions

1. Grouping of borehole outlets at a short distance, i.e. on a pad, allows to reduce the surface of a necessary working area, it has a lower impact upon the natural environment than single 2D boreholes; it eliminates the necessity to relocate equipment frequently; it reduces total costs of making wells;
2. Computer programs, such as 3DPathDesigner, enable fast and simple designing of a trajectory axes of the directional borehole and the analysis of anti-collision by means of charts of separation index and the centre-to-centre distance;
3. Kind of measurement devices applied has an impact upon optimum distribution of outlets of group directional boreholes.
4. The application of grouped wells requires special drilling techniques. They enable very precise maintenance of the assumed well axis trajectory, up-to-date control and verification.
5. Wells drilled in a single pad have special structures enabling selective gas exploitation, making available a bundle of operational heads concentrated on the surface.

6. Rolling drilling devices enable relocation on the site without dismantling and installation again. It limits the time spent on performing a bundle of wells. Time reduction in the drilling industry usually leads to savings. Although a single well can be more expensive, savings due to no need to dismantle and installation of drilling equipment again lead to lower costs in total for all wells.

Acknowledgements

The paper was created within statute studies at the Faculty of Drilling, Oil and Gas at AGH University of Science and Technology in Krakow, Poland. Grant no. 11.11.190.555.

References

1. A. Jamieson (ed.), *Introduction to wellbore positioning*, University of the Highlands and Islands, (2012), (access on internet: 21.08.2017), http://www.uhi.ac.uk/en/research-enterprise/energy/wellbore-positioningdownload
2. *Why Multiple Horizontal Wells from centralized well pads should be used for the Marcellus Shale*, (access on internet: 21.08.2017): http://wvsoro.org/multiple-horizontal-wells-centralized-well-pads
3. R. Wiśniowski: *Metodyka określania przebiegu trajektorii otworu kierunkowego*, Materiały wykładowe, (no published)
4. R. Wiśniowski: *Wybrane aspekty projektowania konstrukcji otworów kierunkowych z wykorzystaniem technik numerycznych*, AGH University Press, Kraków, (2002)
5. J. Schofield, A. Rodriguez-Herrera, X. Garcia-Teijeiro, *Optimization of Well Pad & Completion Design for Hydraulic Fracture Stimulation in Unconventional Reservoirs*, SPE-174332-MS, EUROPEC, Madrid, Spain (2015)
6. *Multi-Well Pad Operations for Shale Gas Development*, (access on internet: 21.08.2017) http://gekengineering.com/Downloads/Free_Downloads/MultiWell_Pad_Operations_For_Shale_Gas_Developmen.pdf
7. S. Rychlicki (ed.), *Poradnik Górnika Naftowego, 2*, Press SITPNiG (2015)
8. M. Ostadhassan, S. Zamiran, H. Jabbari, A. Osouli, B. Bubach, B. Oster, *Stability Analysis of Multilateral High Density Pad Wells In the Three Forks Formation*, SPE-174040-MS, SPE Western Regional Meeting, Garden Grove, California, USA, (2015)

Comparison of Methane Control Methods in Polish and Vietnamese Coal Mines

Marek Borowski[1,*], and *Zbigniew* Kuczera[1]

[1] AGH University Science and Technology, Faculty of Mining and Geoengineering, A. Mickiewicza Av. 30, 30-059 Krakow, Poland

Abstract. Methane hazard often occurs in hard coal mines and causes very serious accidents and can be the reason of methane or methane and coal dust explosions. History of coal mining shows that methane released from the rock mass to the longwall area was responsible for numerous mining disasters. The main source of methane are coal deposits because it is autochthonous gas and is closely related with carbonification and forming of coal deposits. Degree of methane saturation in coal deposits depends on numerous factors; mainly on presence or lack of insulating layers in cover deposit that allow or do not on degasification and easily methane outflow into surroundings. Hence in coal mining there are coal deposits that contain only low degree of methane saturation in places where is lack of insulating layers till high in methane coal deposits occurring in insulating claystones or in shales. Conducting mining works in coal deposits of high methane hazard without using of special measures to combat (ventilation, methane drainage) could be impossible. Control of methane hazard depends also on other co-occuring natural dangers for which used preventive actions eliminate methane hazard. Safety in mines excavating coal deposits saturated with methane depends on the correct estimation of methane hazard, drawn up forecasts, conducted observations, hazard control as well as undertaken prevention measures. Methane risk prevention includes identification and control methods of methane hazards as well as means of combating the explosive accumulation of methane in longwall workings. The main preventive actions in underground coal mines are: effective ventilation that prevents forming of methane fuses or placed methane accumulation in headings ventilated by airflow created by main fans and in headings with auxiliary ventilation, methane drainage using drain holes that are drilled from underground headings or from the surface, methanometry control of methane concentration in the air; location of the sensors is defined by law, additional ventilation equipment used in places of lower intensity of ventilation and places where methane is concentrated.
Key words – methane hazard, methods of methane control, methane risk prevention, methane drainage, methanometry control.

1 Introduction

History of coal mining shows that methane released from the rock mass to the longwall area was responsible for numerous mining disasters. Especially the methane hazard causes very serious accidents and can be the reason of methane or methane and coal dust explosions.

* Corresponding author: marek.borowski@agh.edu.pl

The main source of methane are coal deposits because it is autochthonous gas and is closely related with carbonification and forming of coal deposits. Degree of methane saturation in coal deposits depends on numerous factors.

Mining activity in coal deposits of high methane content without using of special measures to combat could be impossible. Safety in mines excavating coal deposits saturated with methane depends on the correct estimation of methane content in coal seams, drawn up forecasts of methane emission, hazard control as well as undertaken prevention measures.

Methane risk prevention includes identification and control methods of methane hazards as well as means of combating the explosive accumulation of methane in longwall workings. The main preventive actions in underground coal mines are [2, 5]:

- ventilation that prevents forming methane accumulation in roadways,
- methane drainage using boreholes that are drilled from underground or surface,
- methanometry control of methane concentration in the air,
- additional ventilation equipment used in places where methane is concentrated.

The amount of released methane is closely related with the range of mining activity, actual exploitation as well as development works.

Safe work conditions can be provided only through properly adjusted actions. Selection of prevention method is closely related to longwall ventilation systems. Ventilation system should be depended on predicted methane bearing capacity of longwall workings and chosen at the design stage.

In this article the comparison has been carried out between the actions taken in Polish and Vietnamese coal mines in order to prevent of methane hazard.

2 Longwall ventilation system

The amount of released methane is closely linked to exploitation of coal seams. Ventilation is aimed at methane diluting to the level defined for ensuring safe work. When ventilation is not sufficient then methane drainage systems are used. In the case the methane concentration in the longwall face is high, the use of additional preventive measures are recommended.

In Polish coal mines the longwall retreat system is commonly used. The Vietnamese coal mines employ both longwall and shortwall mining systems. The mining system needs the selection of longwalls ventilation system. The choice of ventilation should depend on estimated natural hazards beginning with the design stage of the project. The co-occurance of methane emission, spontaneous fires and temperature hazards are decisive issues for the sector.

Presently ventilation systems of retreat mining are most frequently used (Fig. 1) [5]:

- system U,
- system Y: return side system or bleeder system.

U ventilation system is widely used in Polish coal mines, however under conditions of high methane hazard Y ventilation system is commonly employed. The Vietnamese coal mines use U ventilation system. Due to lower productivity of ventilation system in Vietnamese mines the auxiliary ventilation (i.e. booster fan) is used in longwall system. An example of its use is presented in Figure 2.

The U ventilation system is generally used in longwalls with low-predicted methane emission; with high methane emission the system can be used only including effective methane drainage. In case of spontaneous fire in goaf the system makes its containing difficult. The system is irrelevant in extremely tough conditions, when there is serious methane hazard with limited face advance rate and spontaneous combustion of coal occurs. It causes difficulties in effective combating of coexisting mining hazards. It is necessary to use the supplementary protection measures and ventilation equipments however, they do not achieve expected results.

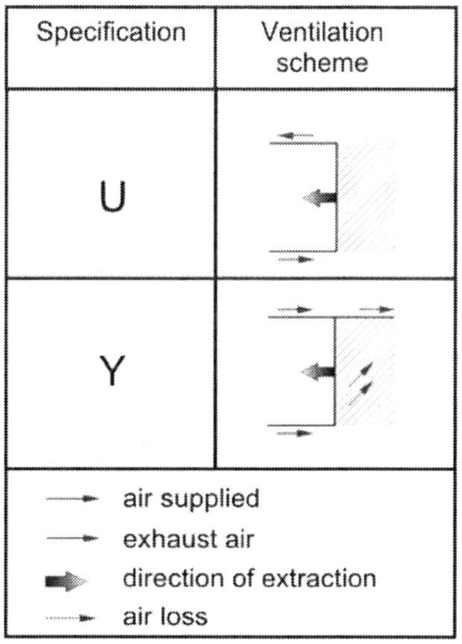

Fig. 1. Ventilation systems in hard coal mines: U ventilation system and Y ventilation system (bleeder system)

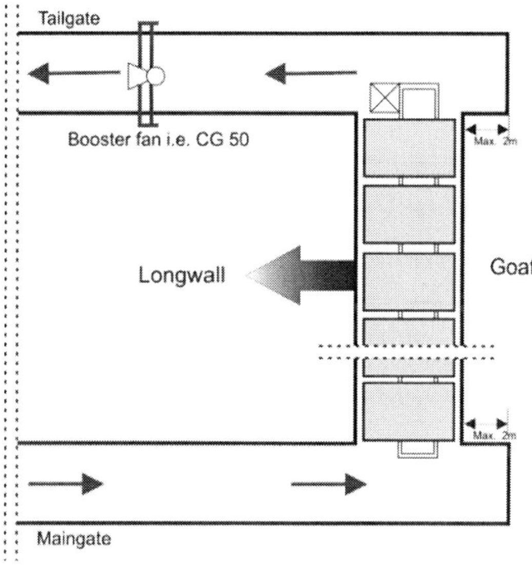

Fig. 2. Example of auxiliary ventilation use in U ventilation system for longwall system

It happens that methane emission during mining is significantly higher than estimated in the technical project. Therefore, the replacing U system with Y system is necessary but generates additional costs.

The Y ventilation system (various types) is used to fight methane hazard. The system is appropriate in coal seams highly saturated with methane; the same applies with prediction of spontaneous heating process in the goaf area or its lack. Also these longwall faces achieve the record level of advance rate and coal output.

The Y return side ventilation system is used in low thickness seams in hazardous methane conditions. It allows to "move" methane concentration towards goaf from longwall face. After the longwall face is ventilated, the gasses are mixed with fresh air and removed by returning air along the goaf. This system ensures work safety in face area in hazardous methane condition and creates good conditions for methane drainage. It is also useful in case of rock bursts because of three escape routes [5].

3 Systems of longwall methane drainage

Methane drainage of rock-mass is the most effective method of controlling methane hazard as it ensures the reduction of methane emission into the working areas. The method that has proved the most efficient is draining methane from rock-mass and goaf area and transporting it to the surface through pipelines, using the depression from drainage station pumps.

Method involves drilling boreholes, called methane boreholes, from tailgate roadway to an unstressed zone in roof or floor layers of a mined seam. The location and parameters of the boreholes depend on the exploitation system and the way of longwall ventilation.

Defining of desorption zone caused by exploitation is necessary during methane drainage of the adjacent coal seams. Drainage boreholes should be situated in stress-relieved zone while not crossing the caving zone. The method of calculating the slope angles of the drainage boreholes, presented in the figure below, produces good results (Fig. 3).

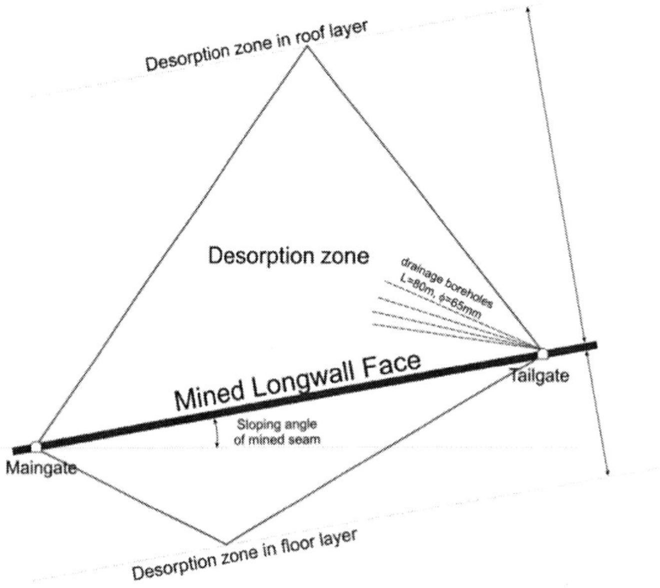

Fig. 3. Determining the methane desorpion zone during longwall exploitation [4]

The location of drilled boreholes is dependent on mining and ventilation systems of longwalls. The highest effectiveness of methane drainage has been achieved by means of Y ventilation system. In case of classic U ventilation system of longwall panel, over time, boreholes drilled from the tailgate road behind the longwall front are lost [5].

Figure 4 presents the drainage system used in Polish mines. An example of drainage system in Vietnamese conditions in Quang Hanh mine, coal seam 7, is shown in Figure 5.

Vietnamese and Polish coal mines employ similar drainage system because of similar geological conditions. Moreover parameters of drainage system and parameters of boreholes are comparable. However, in Vietnamese coal mines the intensity of methane release into the headings is lower because of lower concentration of production. Therefore, the methane hazard is usually tackled through ventilation. For this reason, the effectiveness of methane drainage is simply lower.

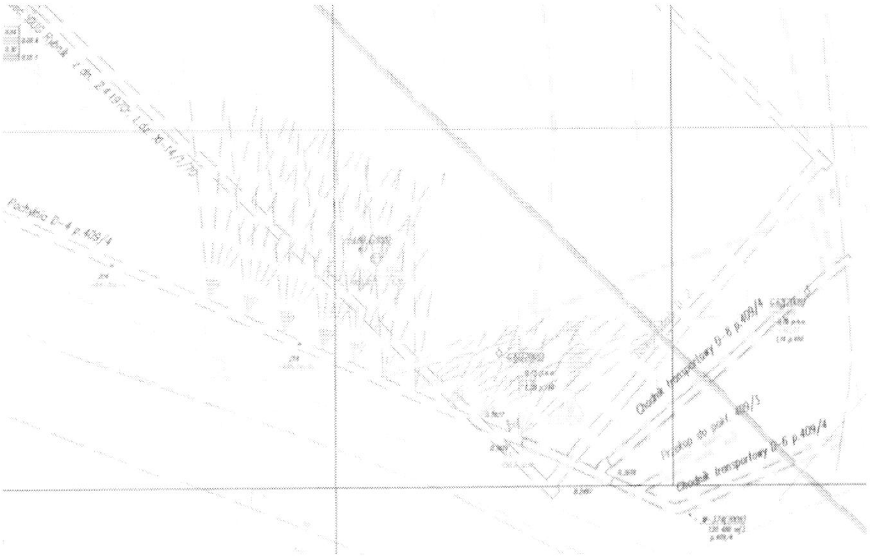

Fig. 4. The sample of boreholes location in Polish mines

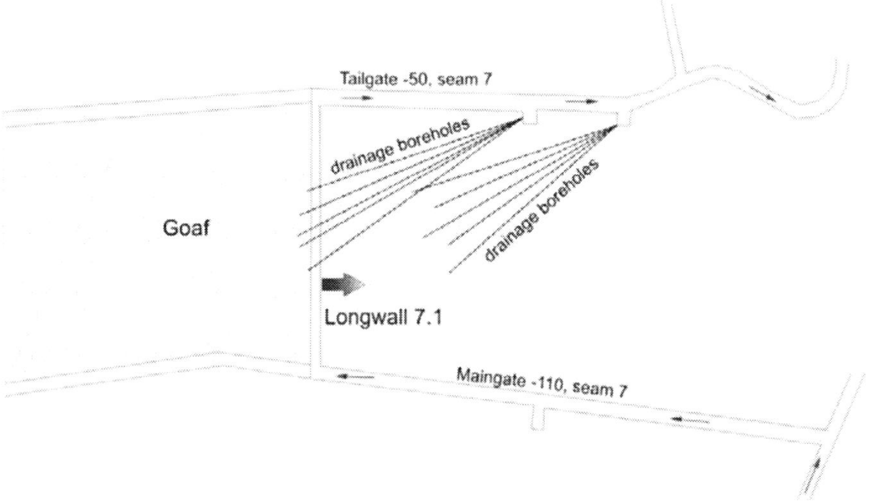

Fig. 5. Scheme of methane drainage for longwall in coal seam 7

4 Control of methane content in the mine air

In Vietnam and Poland the specific rules require to maintain the permissible concentration of methane in the mine air. They define moreover, where and how the measures should be taken. The ventilation system of the extracted longwall affects the way of deployment of the methane sensors. While their location relates to mining and geological conditions of the longwall. It also depends on the deployment of the ancillary ventilation devices that prevent accumulation of the dangerous methane [2, 3].

Full control of the methane hazard is carried out by:

- the measurement of methane content in longwall,

- the measurement of methane content in drainage pipelines,
- the measurement of air velocity in longwall,
- control of the closure of ventilation stopping that affect the ventilation of the underground longwall,
- the measurement of other gases (e.g. oxygen, carbon monoxide, carbon dioxide).

The deployment of sensors is close in Vietnamese and Polish mines. However, in Polish conditions due to high concentration of production, the area of longwall working is secured by more different sensors. The sample of the deployment of the methane sensors is presented in Figure 6.

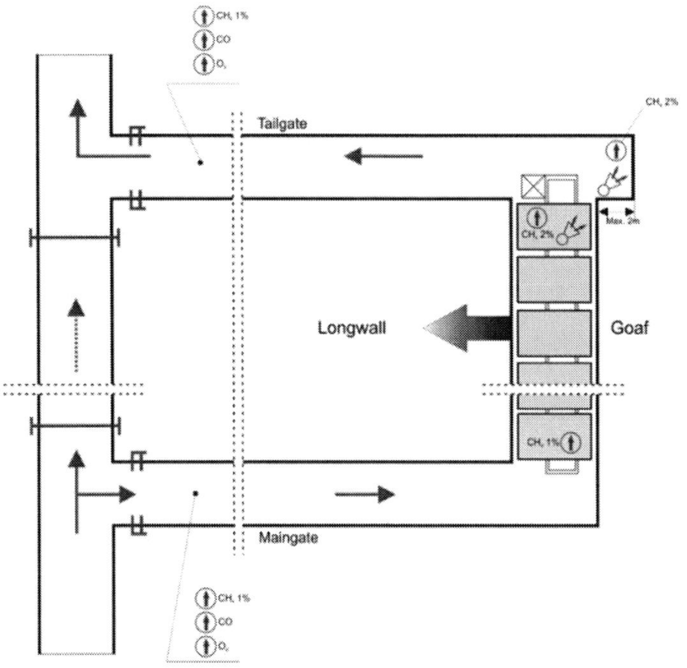

Fig. 6. The scheme of sensors deployment in the longwall

5 Auxiliary ventilation devices

The auxiliary ventilation devices are installed in order to protect against the dangerous methane concentration in the point of intersection of longwalls and longwall headings. The most frequently used are wing curtains and jet-fans [2, 3, 5].

Air injectors are used with low methane risk. Their task is to guide the part of the air from the longwall workings into the goaf. Ventilation using the wing curtains is presented in Figure 7. The effectivness of their use is limited. They are most often used with other solutions that provide protection from methane hazard and owing to that wing curtains efficiency is being increased.

In the event of a greater risk jet-fans are commonly employed. They cause higher air turbulence resulting in good mixing of methane flowing from the caving zone. The examples of jet-fans use is shown in Figure 8. The operating range of the free airflow depends primarily on its initial speed. The efficient operation of the jet-fan is affected mainly by its location and air discharge direction. The methane risk is efficiently eliminated in the case where the air discharge is pointed upwards. Both in Polish and Vietnamese conditions, auxiliary ventilation devices are used which together with other measures in the most effective way reduce the methane threat.

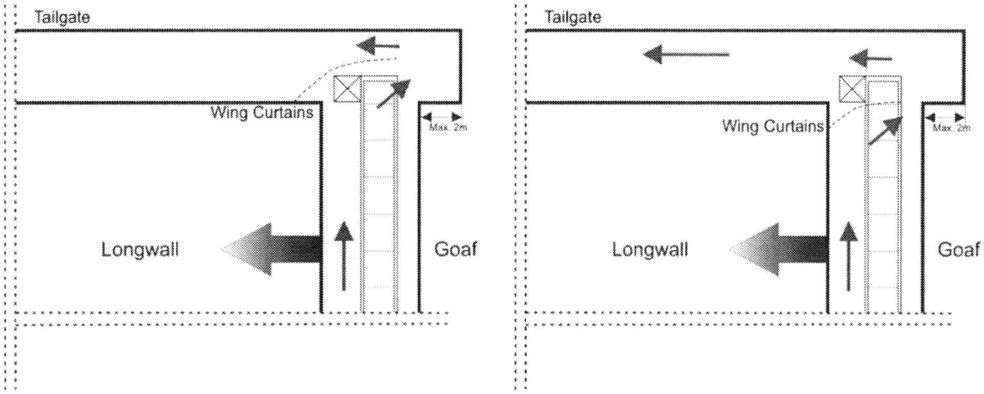

Fig. 7. The ventilation using wing curtains

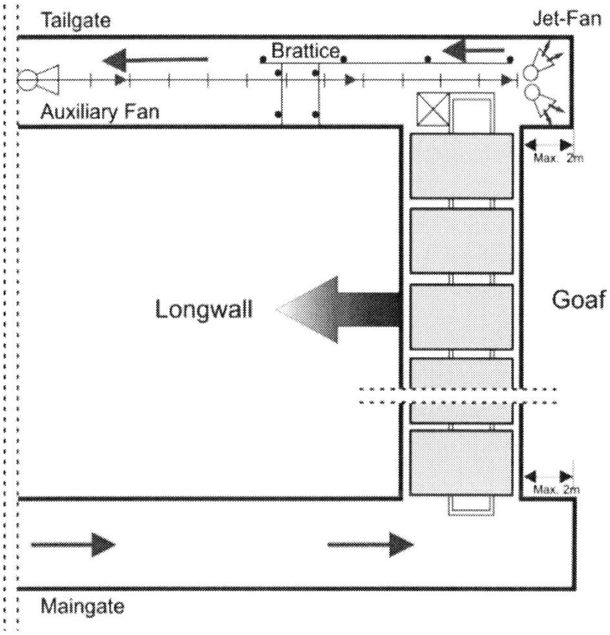

Fig. 8. The examples of jet-fans use

6 Conclusions

Prevention actions against the methane hazard comprise recognition, control and combat measures. The main efforts taken by the coal mines are:

- ventilation that prevents the methane concentration,
- methane drainage,
- methanometry that controls the methane concentration in the air,
- auxiliary ventilation devices used in places of high methane concentration.

The first way of fighting the methane hazard is effective ventilation. This involves choosing the proper way of underground working ventilation. The Polish and Vietnamese coal mines mainly employ the U ventilation system. However Polish coal mines, because of high concentration of the output, use other methods including Y ventilation system. With methane hazard increasing the methane drainage becomes necessary.

Methane drainage of rock-mass is the most effective method of preventing methane hazard as it reduces the frequency of methane inflows into working areas and prevents or reduces occurrences such as outflows, sudden outbursts of methane and rocks, etc.

The method that has proved the most efficient is draining methane from rock-mass and gobs and transporting it to the surface through separate pipelines, using the negative pressure of drainage station pumps. Although this method ensures the desired parameters of ventilation, it imposes certain requirements concerning the development of methane-bearing coal seams.

To ensure the required efficiency of methane drainage in longwalls, it is necessary to select an adequate system of ventilation and drainage at the design stage of development. Methane is captured with greatest efficiency in drainage boreholes located directly in front of the longwall front; the further a borehole is removed from the front, the smaller the amount of captured methane. When the development front moves further away and the borehole is in the unstressed zone, the amount of captured methane increases to reach its maximum about 30 m behind the longwall front. At a distance of about 200 m, the capture disappears following the destruction of the borehole. Therefore, effective methane drainage in longwalls is possible only as long as drainage boreholes are situated in a depression. This in turn is possible when a ventilation heading is situated behind the longwall front or the exploitation is conducted using a second ventilation heading. The highest methane drainage efficiency is accomplished by a ventilation system Y.

The additional measure of methane hazard control is methanometry system which might be expanded by extra sensors. This system is extended in case of methane at dangerous level or in case of coexistence of fire and rock-burst hazards.

The applied preventive measures might be supplemented by additional devices that shall help to achieve good results but only where using with other solutions.

Acknowledgments

The article was prepared within the framework of statutory activities 11.11.100.005.

References

1. M. Borowski: *Method of methane emission's prognosis in longwall excavations in coal mines*, Wydawnictwa AGH,. Kraków, (2012)
2. C.Ö. Karacan, F. A. Ruiz, M. Cotè, S. Phipps: *Coal mine methane: A review of capture and utilization practices with benefits to mining safety and to greenhouse gas reduction.* International Journal of Coal Geology 86, pp. 121–156, (2011)
3. F. Kissell (Ed.): *Handbook for Methane Control in Mining.* Information Circular No: 9486. NIOSH, Pittsburgh, USA, (2006)
4. K. Noack: *Control of gas emissions in underground coal mines.* International Journal of Coal Geology 35, pp. 57–82, (1998)
5. N. Szlązak, M. Borowski, D. Obracaj, J. Swolkień, M. Korzec: *Comparison of methane drainage methods used in Polish coal mines.* Archives of Mining Sciences, ISSN 0860-7001, vol. 59, No. 3, pp. 655–675, (2014)

Methane Content Estimation in DuongHuy Coal Mine

Van Thinh Nguyen[1,*], *Waldemar* Mijał[2], *Vu Chi* Dang[1], and *Thi Tuyet Mai* Nguyen[1]

[1] Hanoi University of Mining and Geology, Faculty of Mining, Vietnam

[2] AGH University Science and Technology, Faculty of Mining and Geoengineering, A. Mickiewicza Av. 30, 30-059 Krakow, Poland

Abstract. Methane hazard has always been considered for underground coal mining as it can lead to methane explosion. In Quang Ninh province, several coal mines such as Mạo Khe coal mine, Khe Cham coal mine, especially Duong Huy mine that have high methane content. Experimental data to examine contents of methane bearing coal seams at different depths are not similar in Duong coal mine. In order to ensure safety, this report has been undertaken to determine a pattern of changing methane contents of coal seams at different exploitation depths in Duong Huy underground coal mine.
Key words – coal mining, methane hazard, prediction.

1 Introduction to DuongHuy coal mine [1, 3]

DuongHuy coal mine located in Duong Huy commune, Cam Pha city, Quang Ninh province. Duong Huy coal mine has an area 3,6 km^2, within the limits of the coordinates (HN72 coordinate system, Vietnam):

X: From 27.600 to 30.000
Y: From 421.500 to 423.000

- The North, it borders on Mine Construction Company
- The East, it borders with KheCham coal mine
- The South, it borders on 86 coal Company
- The West, it borders with Khe Tam coal mine

The expected output of Duong Huy coal mine in 2017 is 1,400,000 tons/year.

2 Coal sampling and methane gas analysis sequence [2]

At the coal seams of Duong Huy coal mine, drill holes into the coal seams. When drilling to a depth of 2.5 m, stop drilling. Take samples of coal from the whitewalls at the bottom of the borehole, with each bore taking two samples. The first specimen was taken at a long of 2.5, with the second sample taken at a long of 4.5 m. Choice of particle size from 1 to 2 mm. Samples of coal are fed to the vibrating mill for grinding, the coal sample is crushed by the marbles in the cylinders as they move up and down. Maximum grinding time is 5 hours.

* Corresponding author: nguyenthinhtkv@gmail.com

After grinding, the gas is separated from the coal sample by the MOD-1 gas separator. All the gas in the coal sample jug is separated and transferred to the sample bag. Gas separation parameters are used to calculate exactly the volume of gas to be extracted.

The specimen after being separated by MOD-1 was introduced into an analytical gas analyzer (Fig. 1) to determine the volume of methane contained in the sample. The extracted gas composition was analyzed by VARIAN gas chromatography. Analysis results from the VARIAN machine will determine the volume of methane gas from each sample of coal (Analysis results are presented in Section 3).

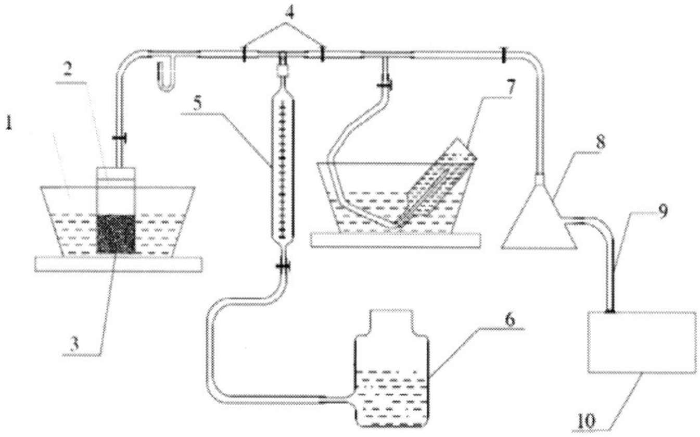

Fig. 1. Gas separation diagram by vacuum heat: 1 – Hot water bottle; 2 – Coal sample container; 3 – Electric stove; 4 – Regulating valve; 5 – Piet measured the volume; 6 – Pressure vessel; 7 – Bottle containing specimen separation gas; 8 – Intermediate gas cylinders; 9 – Tube; 10 – Vacuum machine

From the results of the analysis, establish a relationship between the trend of methane density and the depth of the coal seam by using Excel software to construct methane gas prediction graph (the least squares method). Results of methane content forecast are presented in Section 4.

3 Reasults of analysis of methane content in Duong Huy coal mine

The results of the analysis are shown in Table 1 to Table 4 with the corresponding coal seams.

Table 1. Analysis results CH_4 at No. 6, No. 7 and No. 8 coal seam

No. 6 coal seam			No. 7 coal seam			No. 8 coal seam		
STT	**Depths**	**CH_4 content, m^3/TKC**	**STT**	**Depths**	**CH_4 content, m^3/TKC**	**STT**	**Depths**	**CH_4 content, m^3/TKC**
1	+50	0.271	1	+38	0.37784	1	+100	0.1406
2	+38	1.4	2	0	0.436	2	+70	0.119
3	0	1.90045	3	–35	0.56338	3	+38	0.1392
4	–35	1.9229	4	–35	0.5782	4	0	0.28566
5	–80	2.57567	5	–100	0.61374	5	–40	0.74581
6	–100	2.82104						

Table 2. Analysis results CH_4 at No. 9, No. 10 and No. 11 coal seam

No. 9 coal seam			No. 10 coal seam			No. 11 coal seam		
STT	Depths	CH_4 content, m^3/TKC	STT	Depths	CH_4 content, m^3/TKC	STT	Depths	CH_4 content, m^3/TKC
1	+90	0.3657	1	+38	0.23	1	+38	0.009
2	+55	0.45	2	0	0.2946	2	0	0.6808
3	+38	0.57	3	–50	1.39559	3	–100	1.80868
4	0	0.5465	4	–100	2.02255			
5	–20	0.68						
6	–100	1.60856						

Table 3. Analysis results CH_4 at No. 12, No. 13 and No. 14 coal seam

No. 12 coal seam			No. 13 coal seam			No. 14 coal seam		
STT	Depths	CH_4 content, m^3/TKC	STT	Depths	CH_4 content, m^3/TKC	STT	Depths	CH_4 content, m^3/TKC
1	+70	0.179	1	+100	0.02	1	+20	1.048
2	+38	0.1884	2	+50	0.1819	2	+10	1.02347
3	0	0.13665	3	+38	0.2447	3	–20	2.02751
4	–30	0.2477	4	+25	0.23558	4	–100	3.407
5	–70	0.281	5	0	0.281			
6	–100	0.38801	6	–50	0.31125			
			7	–100	0.33033			

Table 4. Analysis results CH_4 at No. 7KN, No. 9KN and No. 12DB coal seam

No. 7KN coal seam			No. 9KN coal seam			No. 12DB coal seam		
STT	Depths	CH_4 content, m^3/TKC	STT	Depths	CH_4 content, m^3/TKC	STT	Depths	CH_4 content, m^3/TKC
1	+100	2.609	1	+80	0.08259	1	+100	0.16082
2	+70	3.747	2	+55	0.00538	2	+70	0.17933
3	+60	3.908	3	+38	0.00653	3	+65	0.37121
4	+38	4.29614				4	+38	0.88438
5	0	4.23136						
6	–15	4.32103						
7	–35	4.861						

4 Results of methane content forecast

From the methane content figures in Tables 1 to Tables 4. We use Excel software to graph the gas variability at different levels of coal seams.

The results of the analysis are shown in table 5 to table 8, and Figure 2 to Figure 13 with the corresponding coal seams.

 https://doi.org/10.1051/e3sconf/20183501005

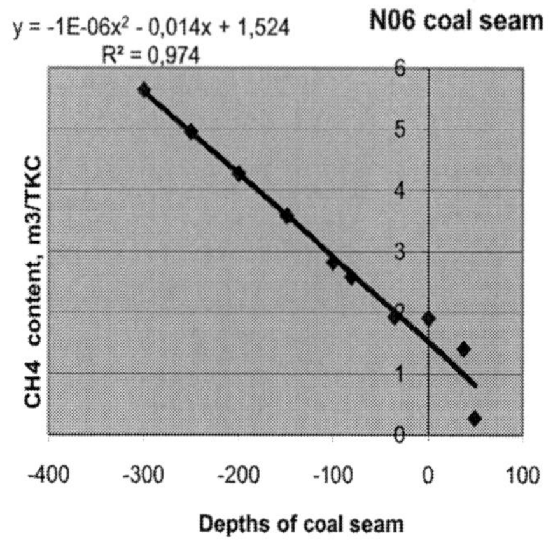

Fig. 2. Distribution chart CH_4 at No. 6 coal seam

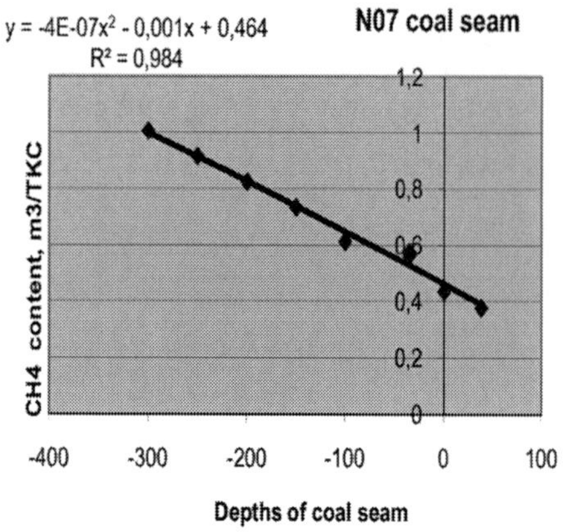

Fig. 3. Distribution chart CH_4 at No. 7 coal seam

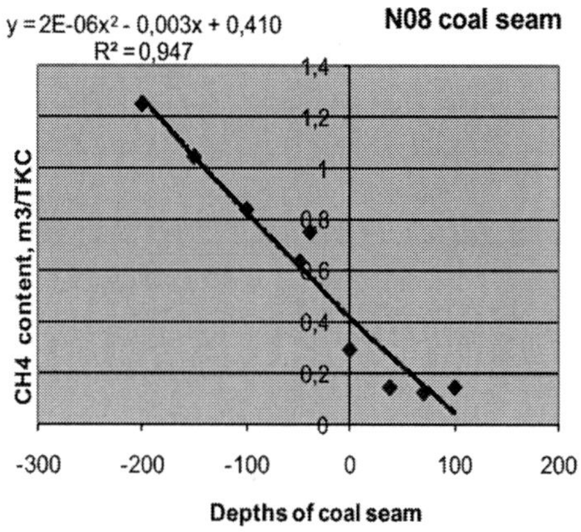

Fig. 4. Distribution chart CH_4 at No. 8 coal seam

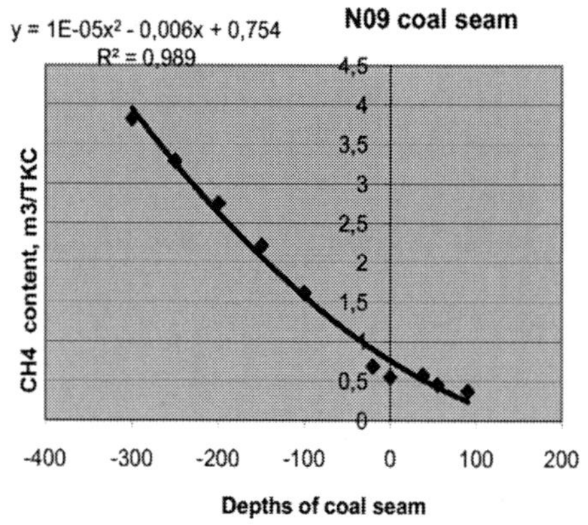

Fig. 5. Distribution chart CH_4 at No. 9 coal seam

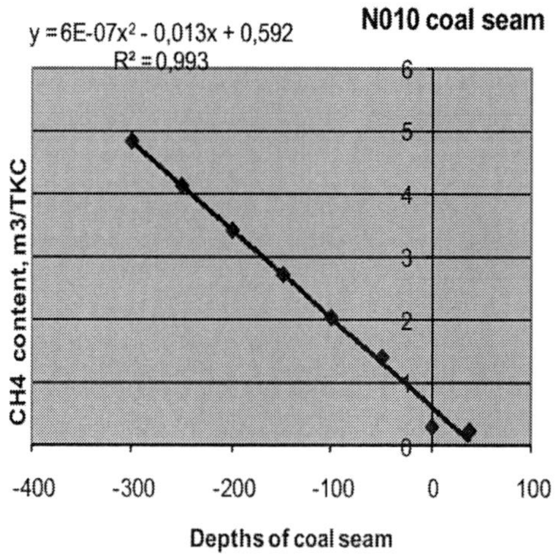

Fig. 6. Distribution chart CH_4 at No. 10 coal seam

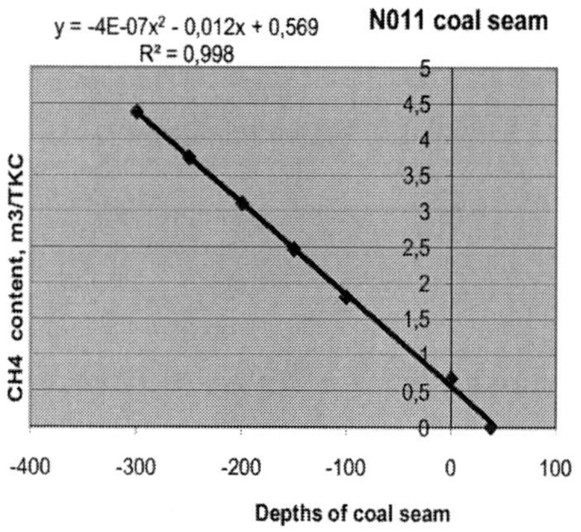

Fig. 7. Distribution chart CH_4 at No. 11 coal seam

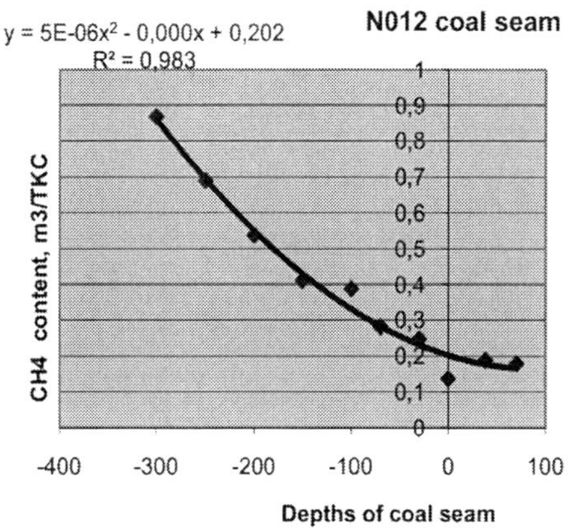

Fig .8. Distribution chart CH_4 at No. 12 coal seam

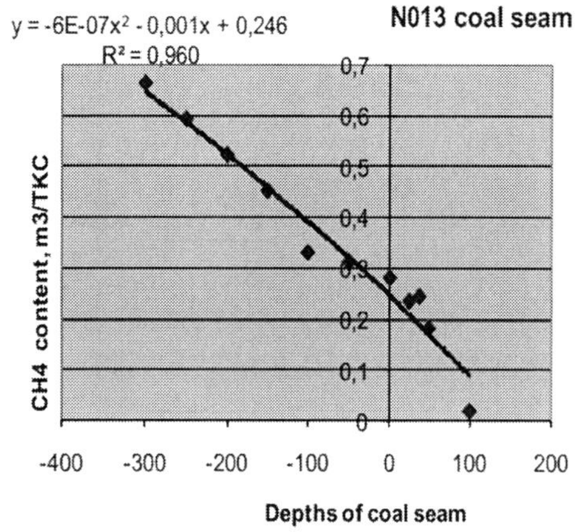

Fig. 9. Distribution chart CH_4 at No. 13 coal seam

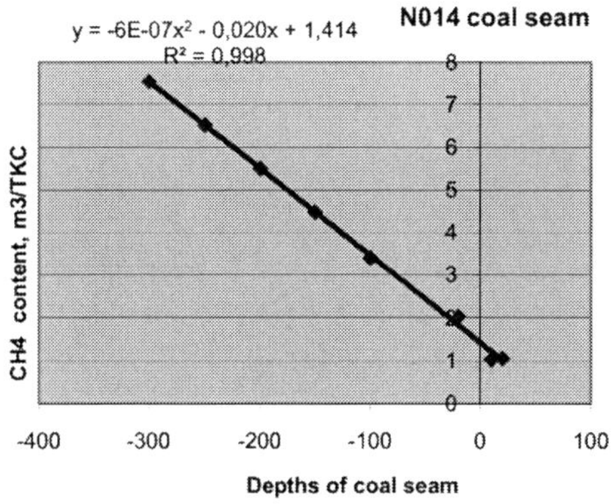

Fig. 10. Distribution chart CH_4 at No. 14 coal seam

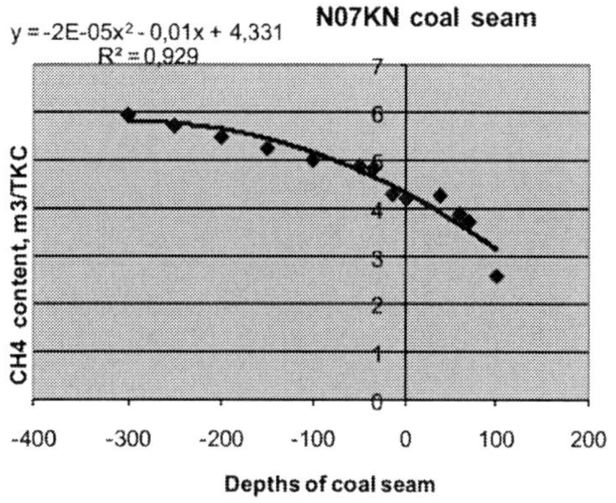

Fig. 11. Distribution chart CH_4 at No. 7KN coal seam

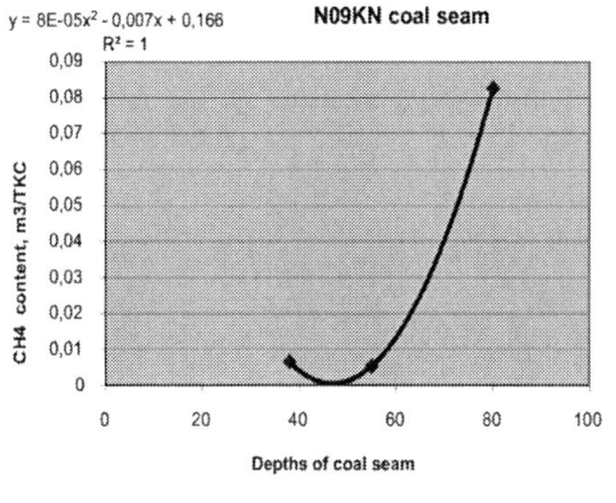

Fig. 12. Distribution chart CH_4 at No. 9KN coal seam

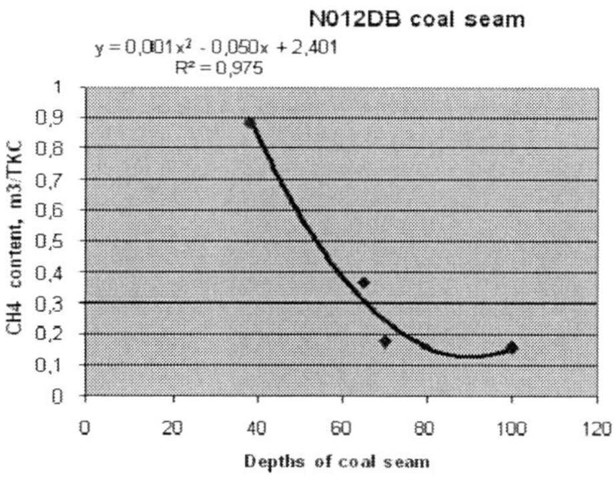

Fig. 13. Distribution chart CH_4 at No. 12DB coal seam

Table 5. Forecast results CH_4 at No. 6, No. 7 and No. 8 coal seam

No. 6 coal seam			**No. 7 coal seam**			**No. 8 coal seam**		
STT	**Depths**	**CH_4 content, m^3/TKC**	**STT**	**Depths**	**CH_4 content, m^3/TKC**	**STT**	**Depths**	**CH_4 content, m^3/TKC**
1	–150	3.5807	1	–150	0.7353	1	–50	0.6282
2	–200	4.2657	2	–200	0.8253	2	–100	0.8332
3	–250	4.9507	3	–250	0.9153	3	–150	1.0382
4	–300	5.6357	4	–300	1.0053	4	–200	1.2432
						5	–250	1.4482
						6	–300	1.6532

Table 6. Forecast results CH_4 at No. 9, No. 10 and No. 11 coal seam

No. 9 coal seam			**No. 10 coal seam**			**No. 11 coal seam**		
STT	**Depths**	**CH_4 content, m^3/TKC**	**STT**	**Depths**	**CH_4 content, m^3/TKC**	**STT**	**Depths**	**CH_4 content, m^3/TKC**
1	–150	2.2119	1	–150	2.7046	1	–150	2.4757
2	–200	2.7469	2	–200	3.4096	2	–200	3.1107
3	–250	3.2819	3	–250	4.1146	3	–250	3.7457
4	–300	3.8169	4	–300	4.8196	4	–300	4.3807

Table 7. Forecast results CH_4 at No. 12, No. 13 and No. 14 coal seam

No. 12 coal seam			**No. 13 coal seam**			**No. 14 coal seam**		
STT	**Depths**	**CH_4 content, m^3/TKC**	**STT**	**Depths**	**CH_4 content, m^3/TKC**	**STT**	**Depths**	**CH_4 content, m^3/TKC**
1	–150	0.4099	1	–150	0.4517	1	–150	4.4772
2	–200	0.5374	2	–200	0.5217	2	–200	5.4972
3	–250	0.6899	3	–250	0.5917	3	–250	6.5172
4	–300	0.8674	4	–300	0.6617	4	–300	7.5372

Table 8. Forecast results CH_4 at No. 7KN, No. 9KN and No. 12DB coal seam

No. 7KN coal seam			**No. 9KN coal seam**			**No. 12DB coal seam**		
STT	**Depths**	**CH_4 content, m^3/TKC**	**STT**	**Depths**	**CH_4 content, m^3/TKC**	**STT**	**Depths**	**CH_4 content, m^3/TKC**
1	–50	4.9	1	0	0,166	1	0	2,401
2	–100	5.031	2	–20	0,33552	2	–20	2,581
3	–150	5.2735	3	–50	0,7005	3	–50	3,151
4	–200	5.511	4	–80	1,19832	4	–80	4,081
5	–250	5.7435	5	–100	1,604	5	–100	4,901
6	–300	5.971	6	–150	2,8765	6	–150	7,651

5 Conclusions

From the results of the analysis and forecast results from Table 1 to Tables 8 and Figures 2 to Figures 13 show the degree of methane storage in the coal seams of the mines with great variation. And the greater the depth, the higher the gas density. Specific for each seam as follows:

- For No. 6 coal seam: From depths level +38 to depths level –150 the density of methane contained in the coal seam is determined and predicted from 0.271 m^3/TKC to 3.5807 m^3/TKC (This coal seam is ranked I according to QCVN / 03-BCT / 2011, Vietnam). From depths level –200 to depths level –300, The density of methane contained in the coal seam is forecasted from 4.2657 m^3/TKC to 5.6357 m^3/TKC (This coal seam is ranked II according to QCVN / 03-BCT / 2011, Vietnam [4]).
- For No. 7 coal seam: From depths level +38 to depths level –300 the density of methane contained in the coal seam is determined and forecasted from 0.37784 m^3/TKC to 1.0053 m^3/TKC (This coal seam is ranked I according to QCVN / 03-BCT / 2011, Vietnam [4]).
- For No. 8 coal seam: From depths level +100 to depths level –300 the density of methane contained in the coal seam is determined and forecasted from 0.1406 m^3/TKC to 1.6532 m^3/TKC (This coal seam is ranked I according to QCVN / 03-BCT / 2011, Vietnam [4]).
- For No. 9 coal seam: From depths level +90 to depths level –300 the density of methane contained in the coal seam is determined and forecasted from 0.3657 m^3/TKC to 3.8169 m^3/TKC (This coal seam is ranked I according to QCVN / 03-BCT / 2011, Vietnam [4]).
- For No. 10 coal seam: From depths level +38 to depths level –200 the density of methane contained in the coal seam is determined and forecasted from 0.23 m^3/TKC to 3.4096 m^3/TKC (This coal seam is ranked I according to QCVN / 03-BCT / 2011, Vietnam). From depths level –250 to depths –300, the density of methane contained in the coal seam is forecasted from 4.1146 m^3/TKC to 4.8196 m^3/TKC (This coal seam is ranked II according to QCVN / 03-BCT / 2011, Vietnam [4]).
- For No. 11 coal seam: From depths level +38 to depths level –250 the density of methane contained in the coal seam is determined and forecasted from 0.009 m^3/TKC to 3.7457 m^3/TKC (This coal seam is ranked I according to QCVN / 03-BCT / 2011, Vietnam). At depths below level –300, the methane content forecast is greater than 4.3807 m^3/TKC (This coal seam is ranked II according to QCVN / 03-BCT / 2011, Vietnam [4]).
- For No. 12 coal seam: From depths level +70 to depths level –300 the density of methane contained in the coal seam is determined and forecasted from 0.179 m^3/TKC to 0.8674 m^3/TKC (This coal seam is ranked I according to QCVN / 03-BCT / 2011, Vietnam [4]).
- For No. 13 coal seam: From depths level +100 to depths level –300 the density of methane contained in the coal seam is determined and forecasted from 0.02 m^3/TKC to 0.6617 m^3/TKC (This coal seam is ranked I according to QCVN / 03-BCT / 2011, Vietnam [4]).
- For No. 14 coal seam: From depths level +20 to depths level –100 the density of methane contained in the coal seam is determined and forecasted from 1.048 m^3/TKC to 3.407 m^3/TKC (This coal seam is ranked I according to QCVN / 03-BCT / 2011, Vietnam). From depths level –150 to depths –300, the density of methane contained in the coal seam is forecasted from 4.4772 m^3/TKC to 7.5372 m^3/TKC (This coal seam is ranked II according to QCVN / 03-BCT / 2011, Vietnam [4]).
- For No. 7KN coal seam: From depths level +100 to depths level +60 the density of methane contained in the coal seam is determined and forecasted from 2.609 m^3/TKC to

3.908 m^3/TKC (This coal seam is ranked I according to QCVN / 03-BCT / 2011, Vietnam). From depths level +38 to depths –300, the density of methane contained in the coal seam is forecasted from 4.29614 m^3/TKC to 5.971 m^3/TKC (This coal seam is ranked II according to QCVN / 03-BCT / 2011, Vietnam [4]).

- For No. 9KN coal seam: From depths level +80 to depths level –150 the density of methane contained in the coal seam is determined and forecasted from 0.08259 m^3/TKC to 2,8765 m^3/TKC (This coal seam is ranked I according to QCVN / 03-BCT / 2011, Vietnam [4]).
- For No. 12DB coal seam: From depths level +100 to depths level –50 the density of methane contained in the coal seam is determined and forecasted from 0.16082 m^3/TKC to 3,151 m^3/TKC (This coal seam is ranked I according to QCVN / 03-BCT / 2011, Vietnam). From depths level –80 to depths –150, the density of methane contained in the coal seam is forecasted from 4,081 m^3/TKC to 7,651 m^3/TKC (This coal seam is ranked II according to QCVN / 03-BCT / 2011, Vietnam [4]).

Sample analysis results show that all coal seams of Duong Huy Coal Mine currently ranked I according to QCVN / 03-BCT / 2011, Vietnam

However, the results at the deeper levels of the coal seams show that in the seams No. 6, No. 10, No. 11, No. 14, No. 7KN, No. 12DB will rank II. Particularly coal seams No. 14 and No. 12DB have methane content reaching the threshold of rank II close to rank III (methane content of $\geq$ 8 m^3/TKC)

References

1. *Mining curent status of Duong Huy coal mine in 2017*. Technical Department, Duong Huy coal Company, Quangninh-Vietnam (2017)
2. *Report on sampling results, annual methane analysis for 2016*. Mine safety center – Vinacomin, Quangninh, Vietnam (2016)
3. *Report on geological exploration results of Duong Huy coal mine, 2013*. Mining geology company –TKV, Quangninh,Vietnam (2013)
4. *Vietnam National Technical Regulation on safety in underground coal mining* – QCVN01:2011/BCT. Ministry of Industry and Trade, Hanoi, Vietnam (2011)

Evaluation of Rock Bolt Support for Polish Hard Rock Mines

Krzysztof Skrzypkowski[1,*]

[1] AGH University Science and Technology, Faculty of Mining and Geoengineering, A. Mickiewicza Av. 30, 30-059 Krakow, Poland

Abstract. The article presents different types of rock bolt support used in Polish ore mining. Individual point resin and expansion rock bolt support were characterized. The roof classes for zinc and lead and copper ore mines were presented. Furthermore, in the article laboratory tests of point resin rock bolt support in a geometric scale of 1:1 with minimal fixing length of 0.6 m were made. Static testing of point resin rock bolt support were carried out on a laboratory test facility of Department of Underground Mining which simulate mine conditions for Polish ore and hard coal mining. Laboratory tests of point resin bolts were carried out, especially for the ZGH Bolesław, zinc and lead "Olkusz – Pomorzany" mine. The primary aim of the research was to check whether at the anchoring point length of 0.6 m by means of one and a half resin cartridge, the type bolt "Olkusz – 20A" is able to overcome the load.The second purpose of the study was to obtain load – displacement characteristic with determination of the elastic and plastic range of the bolt. For the best simulation of mine conditions the station steel cylinders with an external diameter of 0.1 m and a length of 0.6 m with a core of rock from the roof of the underground excavations were used. **Key words** – point resin rock bolt support, roof class, ore mining.

1 Introduction

In Poland the underground exploitation of ore deposits is carried out in two regions. Wherein the primary source of lead are zinc-lead ores deposits of Mississippi Valley type in the Silesian-Cracow region and copper ore deposits in the Fore-Sudetic Monocline. The exclusive primary sources of zinc in Poland are zinc and lead ore deposits of Missisippi Valley type in the Triassic dolomites of the Silesian-Cracow region. In the last year there were the following deposits operated in the Olkusz region, i.e. Olkusz, Pomorzany and Klucze I [4, 5]. It is worth mentioning that the mentioned deposits are explored in the only one underground zinc and lead mine „Olkusz-Pomorzany" belonging to ZGH Bolesław S.A. in Bukowno. Copper ores are excavated in the Fore-Sudetic Monocline deposits, which contains many accompanying minerals. Currently underground exploitation is carried out in three mines "Rudna", "Polkowice – Sieroszowice" and „Lubin" of Legnica–Głogów Copper District (LGOM), that belong to KGHM Polish Copper SA. The basic type of rock mass reinforcement method for both preparatory and operational excavations in underground metal ore mines, both in Poland

* Corresponding author: skrzypko@agh.edu.pl

 https://doi.org/10.1051/e3sconf/20183501006

and in different countries across the world, is the expansion shell or adhesive-bonded rock bolt [3, 6, 8]. The annual consumption of rock bolt supports in Polish ore mining is more than 3 million units (Tab. 1).

Table 1. Division of rock bolt support in Polish ore mining with particular emphasis on the annual consumption

Criterion of division		**ZGH Bolesław S.A. zinc and lead „Olkusz – Pomorzany" mine**	**KGHM Polish Copper S.A. Copper "Rudna", "Polkowice – Sieroszowice" and „Lubin" mines**
Group	**Subgroup**		
1. Mechanism of action rock bolts	1. Grouted – installd in the rock mass by means of various types of cement binders or resin cartridges (the annual consumption in thousands)	130	about 1500
	2. Mechanical – installed in the rock mass by means of shell head spread out inside a borehole (the annual consumption in thousands)	They are not used	about 1500
	3. Friction – acting on the principle of friction occurring between bolt and borehole	They are not used	(sporadically only for tests)
2. Type of material	1. Steel – in the form of rods, pipes, sheets, strings and cables	Only rods	Rods and cables (sporadically pipes only for tests)
	2. Wooden rods	They are not used	They are not used
	3. Composite materials in the form of rods, pipes and strings	They are not used	(sporadically only for tests)
3. Method of mounting	1. Continuous - the bolt is fixed in the hole along its entire length by applying adhesive resin catridges or a suitable cement binder and on the friction principle	Full column resin bolts	Full column resin bolts and Full column slow (120 s) and fast (30 s) resin combination bolts. Additionally suitable cement binder for cable bolts
	2. Point – this type is characteristic for wedge, jaw or short encapsulated bolts, it allows the bolt be pre-tensioning	They are not used	Expansion (mechanical) bolts
4. Length	1. Short, [m]	1.6	1.2; 1.6; 1.8; 2.2; 2.6
	2. Long, [m]	2.7 (about 300 units) in form of steel rods	3.5; 5.2; 7 (about 270000 units) in form of steel connected rods and cables
5. Installation	1. Self-propelled drilling-bolting machine (the annual consumption in thousands)	6÷7	3000
	2. Manual bolting (the annual consumption in thousands)	123	They are not used

It is worth mentioning that, exploitation of zinc and lead in "Olkusz – Pomorzany" mine is carried out at depth of about 100 m below ground surface whereas exploitation of copper in Legnica–Głogów Copper District (LGOM) mines is carried out at depth over 1000 m up to 1200 m below ground surface. Therefore in the "Olkusz – Pomorzany" mine there are only rigid bolts used with satisfactory strength and only failing, when taking over the load near the limit of strength of bolt rod, with relatively small deformation capacity, whilst in the copper mines there are used rigid and yielding bolts capable of carrying significant deformation, with relatively low carrying capacity and also more and more there are used energy-absorbing of dynamic load bolts, characterized by high load carrying capacity and the ability to carry relatively high displacement and deformation.

2 Choice of rock bolt support

The variety of conditions in which the rock bolt support is used makes it suitable for securing the stability of the excavation or as reinforcement of existing rock bolt support [3]. Choice of rock bolt support in „Olkusz – Pomorzany" mine is based on the determination of roof's class which is determined on the basis of the results of geomechanical tests and on the basis evaluation of weakness factor of rock mass „c" carried out by an expert [4]. There are five classes of the roof, for which a specific parameters are determined for application of rock bolt support; way of fixing (point or continuous), the mechanism of action (resin, expansion or friction), the length of bolts, spacing in rows, with the additional security and, above all, cross-sectional shape of excavation (Tab. 2).

Table 2. Roof class for zinc and lead ores

Roof class	Description	Compressive strength Rc [MPa]	Rock mass weakening coefficient „c"	Shape of the excavation	Support type
I	very hard rocks	at least 70	more than 0.9	rectangular or trapezoidal	rock bolts or without support
II	strong stratified rocks	50–70	0.7–0.9	rectangular or trapezoidal	expansion or point resin rock bolts
III	fragile stratified rocks	30–50	0.5–0.7	flat or oval roof	expansion or point resin rock bolts
IV	weak rocks	20–30	0.4–0.55	oval roof	point resin or full column resin rock bolts and steel lagging
V	very weak rocks	less than 20	less than 0.4	oval roof	arch yielding support
The length of roof rock bolt is at least 1.6 m. The basic net bolting is a diagram 1 × 1 m to 1.2 × 1.2 m. Side walls of excavations more than 4.5 m are subjected to bolting, wherein the first row is made 4.5 m from the floor; the length of side walls rock bolt is at least 1 m.					

Selection of excavation support in copper ore mining is performed by the Head of the Mining Works Department on the basis of knowledge of geological and mining conditions (Tab. 3). In practice, this insight is implemented by determining the roof class, in accordance with the instruction [2, 10] used by the KGHM, developed by an expert. The classification of the roof rocks is based on the hierarchy of parameters' validity, which basically determines the stability of the roof of mining excavations. These parameters include: roof delamination (divisibility in the vertical direction), density of mineralized fractures in the roof of excavations, degree of fault, average throw of fault, tensile strength. Individual parameters are obtained as a result of analysis of drill cores, analysis of drilling hole walls by means of endoscopes, analysis of exposures, and research on geomechanical properties of rocks. In addition, correction coefficients have been introduced to determine the rock of the roof, taking into account the impact of the open width of the working space and the method of roof management. Furthermore, when selecting the rock bolt support, the compressive strength, intended use projected dimensions of the excavation and the experience in the application of rock bolt support are taken into account. In the case of long-term excavations, grouted-monted bolts are preferable, more resistant to corrosion and do not require obtaining and then maintaining a high pre-tension.

Table 3. Roof class for copper ores

Roof class	Description	Net bolting, [m]	Maximal expansion of excavations in the developed phase of exploitation					Maximal expansion of excavations in the developed phase of liquidation, [m]	Net bolting, [m]	Maximal width of rooms, [m]
			Heigth of room, [m]							
			up to 2.0	**up to 5.0**	**up to 7.0**	**up to 7.0**	**from 7.0**			
			Minimal length of rock bolts, [m]							
			1.2	**1.6**	**1,8**	**2.2**	**2.6**			
I	weak rocks	1 × 1		7.0	8.0	9.0	10.0	15.0	1 × 1	6.0
II	average-hard rocks I	1.5 × 1.5		9.0	10.0	11.0	12.0	17.0	1 × 1	7.0
III	average-hard rocks II	1.5 × 1.5	8.0	10.0	12.0	13.0	14.0	18.0	1.5 × 1.5	7.0
IV	hard rocks	2 × 2	9.0	12.0	14.0	16.0	18.0	20.0	2 × 2	7.0
V	very hard rocks	2 × 2	10.0	13.0	15.0	17.0	20.0	25.0	2 × 2	8.0
	Exploitation rooms								Drift excavations	

In mining fields, where the period from the completion of the excavation to its liquidation is relatively short, expansion bolts are more often used, which due to the less time consuming construction, allow for greater bolting efficiency [1]. Examples of bolting diagrams for zinc and lead and copper ore mining are shown in Figures 1 and 2.

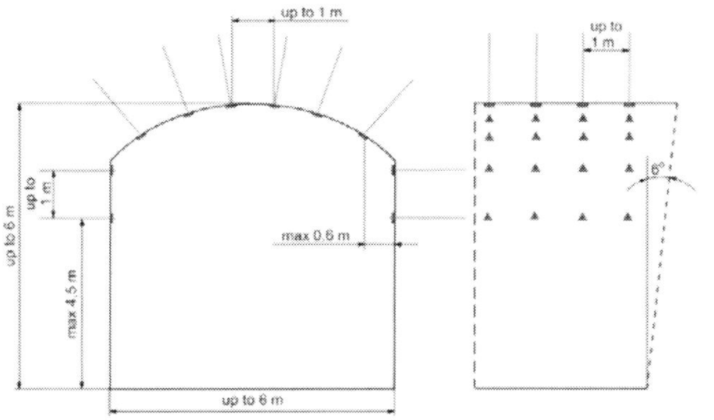

Fig. 1. Room support with oval roof by means of rock bolt support in zinc and lead ore mine

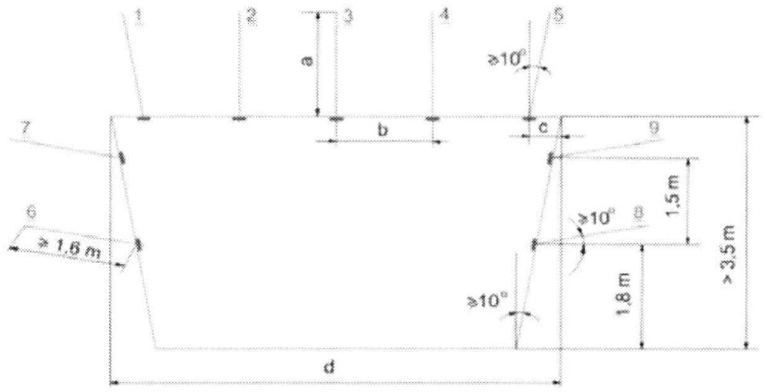

Fig. 2. Room support by means of rock bolt support for drift excavation with height more than 3.5 m in copper ore mine: a – length of bolt, b – net bolting, c – maximal distance from side wall (c ≤ d), d – width of drift, 1–9 – bolt numbers

3 Laboratory test facility

Static tests of point resin rock bolts were carried out at the laboratory test facility in simulated mining conditions, in particular for the zinc and lead "Olkusz – Pomorzany" mine. Studies were carried out using a consistent methodology. In the static load mode, the rock bolts were stretched and broken by the use of maximum force resulting from the power of the pump. Displacement and force sensors were connected to a QuantumX MX840 universal measuring amplifier via 15-pin plugs. During the tensile process the results of force and displacement measurements were recorded continuously by a program specialized in the field of measurement technology – known as „CATMAN-EASY" [9]. Resin rock bolts type "Olkusz 20A" used in the laboratory tests constitute the basis of mining excavation supports in underground zinc and lead "Olkusz – Pomorzany" mine. Rock bolt type "Olkusz 20 A" was modified for the needs of the conducted experience and consists of a ribbed bar with a 1.6 m length which has a diameter of 20 mm, at one end has a thread with M20 (Fig. 3a) and circular shaped bearing plate with a diameter of 150 mm and a thickness of 6mm (Fig. 3c). The height of the bearing plate was 24.8 mm, made of steel grade St3SAL or 18G2A. The other end of the rod is angled 30–35° in order to improve mixing of the resin. Additionally bolt was equipped with connecting sleeve (Fig. 3d) and threated extension (Fig. 3b). This configuration made it possible to install the bolt without a nut. Support's rods are made of stainless steel "EPSTAL", B500SP steel grade, which is characterized by the following parameters: yield strength (Re) of 500 MPa to 625 MPa; ratio of tensile strength to yield strength (Rm/Re) from 1.15 to 1.35; percentage elongation (A5) a minimum of 16% and the percentage elongation at maximum force (Agt) equal to at least 8%.

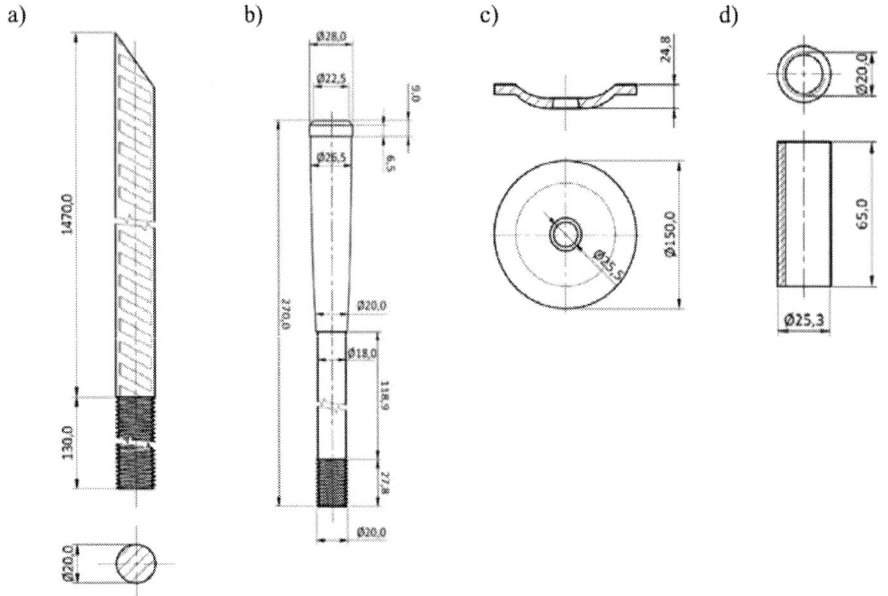

Fig. 3. Elements of the resin rock bolt support type "Olkusz 20A": a) bolt rod, b) threated extension, c) shaped bearing plate, profiled with a ring, d) connecting sleeve

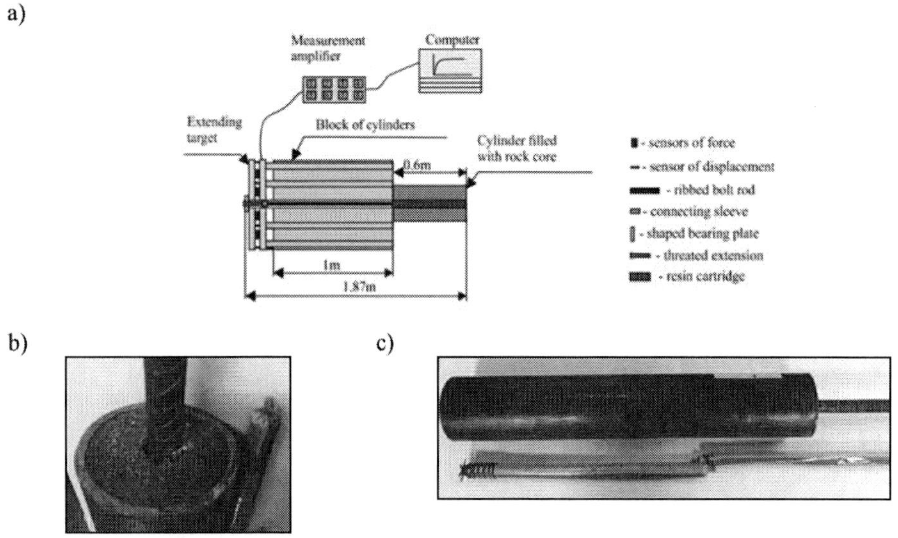

Fig. 4. Arrangement of sensors and elements of rock bolt support on the laboratory test facility; a) rock bolt support with a length 1.87 m, b) rock core with a diameter 0.028 m, c) resin cartridge with a length 0.4 m (photo: K. Skrzypkowski)

In order for laboratory studies to best satisfy mining conditions, rock cores obtained from underground excavations were used (Fig. 4). Rock cores (dolomites) had holes with a diameter 0.028 m (Fig. 4a) and characterized compressive strength up to 100 MPa. Bolt was installed into core by means of one and a half resin cartridge with a length of 0.6 m. Resin cartridge type "Lokset" has a length 0.4 m and diameter 0.24 m (Fig. 4b). It consists of binding compound, a hardener and a plastic coating. In addition, there is an overlay at one end of the cartridge, the task of which is to keep the cartridge in the hole so that it does not fall out.

3.1 Load-displacement characteristics of point resin rock bolt

The aim of this study was to obtain load-displacement characteristics of point resin rock bolt on the basis of which it will be possible to determine individual phases of load and deformation of the bolt. During tests the results of force and displacement characteristics were recorded continuously and visualization and evaluation of the measurement were tracked on-going. In addition, after the tests were completed, reports documenting the results of the measurements were created and then stored as an ASCII file extension. Examples of the characteristic parameters obtained in the tests are shown in Figure 5.

a)

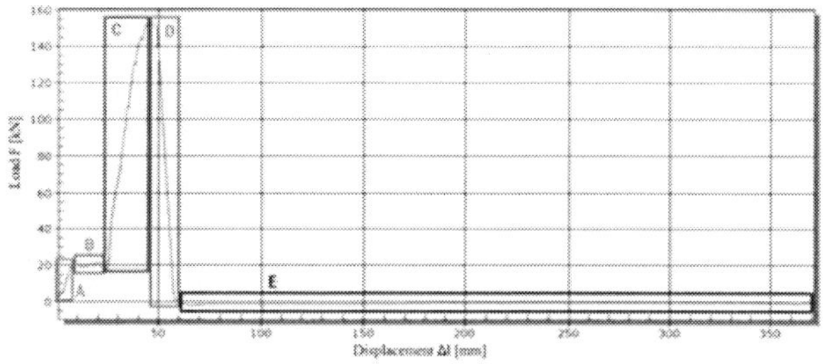

b)

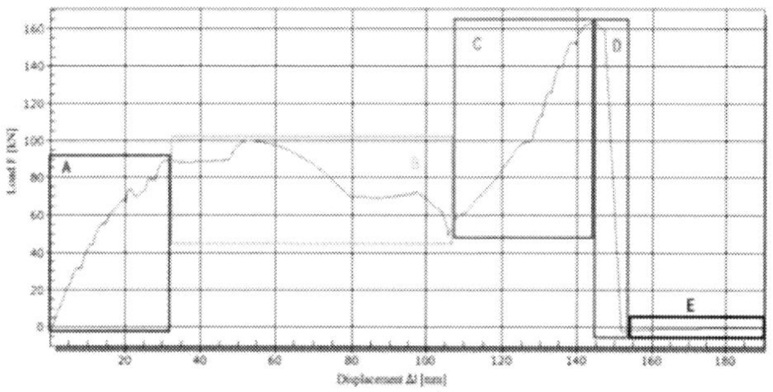

c)

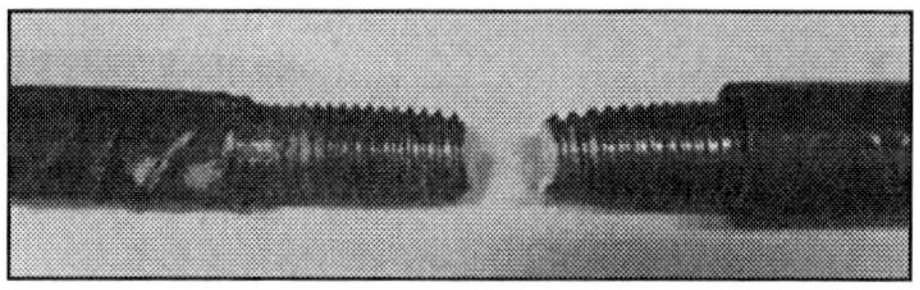

Fig. 5. Load displacement characteristic of point resin rock bolt support: a) bolt no 1, b) bolt no 2, c) break on the thread of rod bolt (photo by K. Skrzypkowski)

The obtained characteristics were divided into five phases, visible in Figure 5a (A, B, C, D, E). The marked area A together with the area C constitutes the whole and shows a similar growth character of the pull out force, relative to the displacement. The growth trend is disturbed by slipping the bearing plate – phase B, which was observed during the test. For a maximum tensile force equal 153 kN, the displacement was 45 mm. A sudden drop of load (phase D) was caused by pulled out the bolt. The next phase E is the total extension of the bolt from the hole. The load-displacement characteristic of the Olkusz-20A bolt number 2 (Fig. 5b) has been divided into five characteristic parts. The tensile force diagram in phase A is presented in accordance with the reference to tensile characteristics of the bolt. In the initial phase B, smoothing occurs, which is caused by the notch of the bearing plate which was observed in time. Then a slight increase is noticeable, followed by a drop in load relative to the increasing of displacement. This is due to the extension of the bolt from the hole, which was also noticed during the test. In phase C, the breaking force increases, up to the critical value, which was 162 kN and the bolt breaks at the thread location (Fig. 5c). Phase D presents the breaking of the bolt, while phase E shows the extension of the broken part.

4 Conclusions

At present rock bolt support in underground zinc and lead „Olkusz – Pomorzany" mine is installed in the bolt holes made by means of self-propelled drilling – bolting machines type SWK-2 Hz and Robolt 320-22C. Drilling bolt holes is one of the most dangerous operations, as is done with an unprotected roof of the excavation. The excavations, in which the roof is very fragile also is used manual bolting. W 2017 it was used about 123000 resin rock bolts support „Olkusz 16A" type at manual bolting and about 7000 resin rock bolts support „Olkusz 20A" type at self-propelled drilling-bolting mining machines. For bolts „Olkusz 16A" type, holes are drilled with a diameter of 35 mm. For fixing rod is used at least two loads of resin „Lokset" type with a diameter of 30 mm and length 600 mm for which the gel time is equal 2 minutes. Wherein for bolts „Olkusz 20A" type in the roof are drilled holes with a diameter of 33 mm. For fixing the rod typically are used four resin cartridges „Lokset" type, which have a diameter of 24 mm and length 400 mm for which is gel time is equal 30 seconds.

According to Polish Standard Norm [7] capacity of resin bolts in underground exploitation of ore deposits of lead and zinc should be at least 90 kN. Control of the bolting consists of a pull out tests with load of force up to 40 kN. In laboratory tests, the point resin rock bolts were fixed at the length of 0.6 m. Based on the obtained characteristics, it is possible to state that the support has taken over the minimum required load. This means that it is possible to propose a modification of the bolting technology, which would mean point mounting instead of the entire length in the bolting holes. Some restriction resulting from geological conditions. It should be remembered that in „Olkusz – Pomorzany" mine, roofs of excavations (room) are represented very often by oxidized and strongly fractured ore bearing dolomites.

Acknowledgements

The study was carried out within the framework of statutory work no. 11.11.100.005.

References

1. C. Matusz, K. Szczerbiński, Cuprum. **66**, 1, (2013)
2. J. Butra, *Eksploatacja złoża rud miedzi w warunkach zagrożenia tąpaniami i zawałami.* Wydawnictwo KGHM CUPRUM sp. z o.o. Wrocław (2010)
3. K. Skrzypkowski, E3S Web of Conferences. **29**, (2018)
4. K. Skrzypkowski, *Interdisciplinary Topics in Mining and Geology.* M. Hardygóra, J. Bac-Bromowicz, J. Drzymała (eds.). Wrocław (2016)
5. K. Skrzypkowski, W. Korzeniowski W, *Selected issues related to mining and clean coal technology.* M. Borowski, J. Swolkień (eds.). Kraków (2016)
6. K. Skrzypkowski, W. Korzeniowski, K. Zagórski, P. Dudek, SGeM. **39**, 3 (2017)
7. Polish Standard Norm – PN-G-15091 – Mining bolts. Requirements. (1998)
8. W. Korzeniowski, K. Skrzypkowski, K. Zagórski, SGeM. **39**, 1 (2017)
9. W. Korzeniowski, K. Skrzypkowski, Ł. Herezy, Arch. Min. Sci. **60**, 1 (2015)
10. W. Pytel, *Geomechaniczne problemy doboru obudowy kotwowej dla wyrobisk górniczych.* Wydawnictwo KGHM CUPRUM sp. z o.o. Wrocław (2012)

Mine Water Treatment in Hongai Coal Mines

Phuong Thao Dang[1,*], and *Vu Chi* Dang[1]

[1] Hanoi University of Mining and Geology, Hanoi, Vietnam
18 Vien Street, Duc Thang Ward, Bac Tu Liem District, Hanoi, Vietnam

Abstract. Acid mine drainage (AMD) is recognized as one of the most serious environmental problem associated with mining industry. Acid water, also known as acid mine drainage forms when iron sulfide minerals found in the rock of coal seams are exposed to oxidizing conditions in coal mining. Until 2009, mine drainage in Hongai coal mines was not treated, leading to harmful effects on humans, animals and aquatic ecosystem. This report has examined acid mine drainage problem and techniques for acid mine drainage treatment in Hongai coal mines. In addition, selection and criteria for the design of the treatment systems have been presented. **Key words** – acid mine drainage, coal mining, mine drainage treatment, Hongai coal mine.

1 Introduction

Coal mining can have negative impacts on the environment in surrounding area and one of these problems is acid mine drainage (AMD). Acid mine drainage (AMD) is the result of sulfide minerals, typically pyrite in coal seams, where the pyrite is exposed to oxidising conditions in coal mining to produce sulphuric acid (H_2SO_4) and metal precipitates such as Fe oxyhydroxides (Equation 1) [1].

$$FeS_2 + \frac{7}{2}H_2O + \frac{15}{4}O_2 \rightarrow Fe(OH)_3 + 2H_2SO_4 \tag{1}$$

Based on monitoring data on mine drainage in Hongai area, it has been shown that the mine drainage is high concentrations of Iron, Manganese and total suspended solids (TSS), low pH values. So that, mine drainage must be treated in order to protect stream and to help prevent any problem that can cause pollution in the area.

Until 2009, mine drainage in Hongai area was allowed to discharge into rivers then the sea. However, after that, Vietnam National Coal and Mineral Industries Group (VINACOMIN) take responsibility for dealing with environmental problems.

Drainage treatment systems have been installed to better manage the waste water from coal mines in Hongai area.

2 Study area

2.1 Case Study

Hongai coal mines are located Quang Ninh coal basin that is situated close to the coast. This coal basin contains hard coals deposited in Upper Triassic. Hongai stage has a thickness

* Corresponding author: dangphuongthao@humg.edu.vn

of 1500–1800 m containing conglomerates, sandstones, shales, coal shales and anthracite seams.

Effluents from Hongai coal mines are released into Dien Vong river and Mip river and ultimately discharge into the Halong bay, which is classified as world natural heritage by the UNESCO. Climate of this area can be defined as a tropical type, with two seasons: hot and moist summer, and dry and cold winter leading to a variable runoff. The average temperature is from 15–25°C, and annual mean precipitation of 1832 mm/year.

Hongai underground coal mines have reached the exploiting yield of 6 million tons of coal with more than 10.000 workers.

3 Materials and methods

3.1 Sampling and analytical procedure

During the dry season (November, 2016), samples of the mine drainage (before treatment) were collected at underground coal mines in Hongai area such as Bac Bang Danh, Cai Da, Dong Binh Minh, Giap Khau and Ha Lam.

The chemical characteristics of the mine drainage discharge from underground coal mines in Hongai area are shown in Table 1.

Table 1. The chemical characteristics of the mine drainage discharge (before treatment) from underground coal mines in Hongai

Parameter	Mine					Vietnamese Industrial Effluent Limits
	Bac Bang Danh	Cai Da	Giap Khau	Dong Binh Minh	Ha Lam	
pH	4.2	3.4	4.63	3.71	4.62	5.5-9
TSS	633	309	539	372	872	100
Fe	10.7	15.8	5.6	12.4	7.36	5
Mn	3.4	4.2	1.9	6.1	2.01	1

The unit for each of the parameters is mg/L, except for the pH values

It can be seen from the Table 1 the values of pH in the mine drainage at various mines in Hongai area are much below than the lower permissible limits of industrial effluent discharge, while TSS, Fe and Mn at there are higher than the permissible industrial limits.

3.2 Acid Mine Water treatment process in Hongai

Mine drainage treatment technologies are accomplished by either active or passive treatment systems, both potentially depending on site-specific parameters [2]. Active treatment is the high capital cost and high maintenance costs for continued operation, compared with passive treatment. However, in Hongai area, because of the high volume fluxes and the very limited area for drainage treatment systems, only active treatment is appropriate.

Since 2009, the mine drainage treatment systems have been installed in Hongai area to manage the drainage that comes into the water flow. The system applied is described in Figure 1. This system was proposed by Vietnam National Coal and Mineral Industries Group (VINACOMIN).

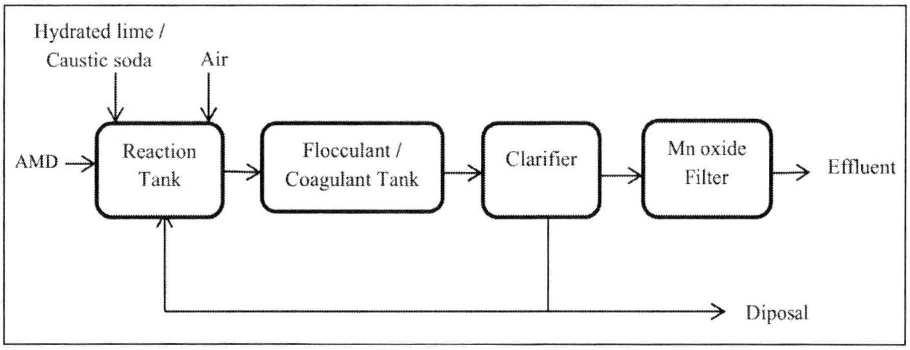

Fig. 1. Scheme of the mine drainage treatment process in Hongai

Mine water treatment is the process that adds chemicals to the water to raise pH and removes metals.

Typical system of the mine drainage treatment has some steps which are: The drainage is neutralised in a reaction tank with controlled addition of alkali to attain a pH 8 (option). Aeration is added to the reaction tank to oxidize ferrous iron to ferric. The drainage is then contacted to a settling agent and fed to a clarifier for solid/liquid separation. A proportion of sludge collected from the bottom of the clarifier is pumped back into the reaction tank. In order to limit sludge disposal costs, the process decreases sludge volumes by producing higher-density sludge. Manganese is not easily treated by controlling pH with alkali and separating precipitates; only about 50% of the Mn load could be removed in a first step at even a pH of 9.0 by sorption and oxidation in the reactor [3]. Thus the clarified effluent can always be released to manganese oxide filter in order to reduce Mn, and then the effluent can be discharged directly to stream.

This treatment process, which is the standard in the mine drainage treatment, is applied at many underground mines in Hongai area. The process is commonly applied due to its feasibility to treat huge volume of the mine drainage, cost-effect and due to its simplicity.

Based on monitoring data collected at each mine, it was decided to select appropriate treatment process for each mine. To select types of alkali to treat certain mine drainage, it is necessary to consider about acidity levels, flow, the types and concentrations of metals in the mine drainage and cost of alkali [4]. There are two types of alkali in use such as: hydrated lime ($Ca(OH)_2$), caustic soda (NaOH) in Hongai coal mines.

To determine the actual alkali requirement, experiments need to be performed. In general, the actual alkali requirement is often higher than calculated alkali, because of other chemicals present. In practice, the correction factor is added to fulfill that requirement.

In Bac Bang Danh mine, Giap Khau mine, Ha Lam mine, hydrated lime is used to neutralise mine drainage. There has to be a machine on site to mechanically add water to the powder that the hydrated lime forms and then delivers it to the mine drainage. Hydrated lime is cheap and easy to handle safely, however, hydrated lime has limited effectiveness if it is required to attain a very high pH.

Amount of lime requirement to reach pH 8 can be calculated as (Equation 2) [5]:

$$m_{\mathrm{CaO}} = \frac{1.2}{\epsilon} \cdot 500 \cdot \left(10^{-a} + 10^{-6}\right) \cdot 56 + 0.8 \cdot m_{\mathrm{CO_2}} + 1.8 \cdot m_{\mathrm{Fe}} \tag{2}$$

where:

m_{CaO} is the amount of CaO, mg/l;

a is the value pH in the mine drainage before treatment;

ϵ is the amount of effective CaO contained in lime involved in the neutralization, %;

m_{CO_2} is the amount of CO_2 in the mine drainage, mg/l;

m_{Fe} is the amount of Fe in the mine drainage, mg/l;

1.2 value is the correction factor added to adjust the theory value to the actual requirement for lime.

In Cai Da mine, Dong Binh Minh mine, the mine drainage treatment plants, which have low capacity, can be only to treat 50 to 300 m³/h. The mine drainage is very acidic, pH < 4. Caustic soda which is used to neutralise mine drainage can raise the pH very quickly. However, it is harmful in handling and transporting, also it is more expensive than hydrated lime.

Amount of caustic soda requirement to reach pH 8 can be estimated using the equation below (Equation 3) [5]:

$$m_{NaOH} = 1.2 \cdot 1000 \cdot \left(10^{-a} + 10^{-6}\right) \cdot 40 + 1.43 \cdot m_{Fe} \qquad (3)$$

Where:

m_{NaOH} is the amount of caustic soda, mg/l;

a is the pH in the mine drainage before treatment;

m_{Fe} is the amount of Fe in the mine drainage, mg/l;

1.2 value is the correction factor added to adjust the theory value to the actual requirement for caustic soda.

The settling agents, which are coagulant PAC (Poly aluminium chloride) and flocculant PAM (Polyacrylamide), are used to increase particle settling efficiency. Concentrations of PAM and PAC added in the mine drainage are about 0.2–1 mg/l and 25—45 mg/l, respectively.

The reaction tank built during period of 2009–2010 design provides for a hydraulic residence time between 6 and 10 min. After this period, the reaction tank is required to design large to in order to complete dissolution time of the solids about 10–20 min.

Due to seasonal change the quality and quantity of the mine drainage are not stable [5]. During rainy season, precipitation events in Hongai area have led both to increases and to decreases in pH, and in the concentration of metals [6]. Therefore, amount of alkali requirement for neutralization needs to adjust adequately.

In addition, in Hongai area, the amount of the mine drainage goes to treatment system is higher in the rainy season than to the dry season. The capacities of the mine drainage treatment plants are sometime overload about 117–142%. Therefore, since 2013, in the treatment plant, a stored tank has to be built in order to adjust amount of the mine drainage flow into the reaction tank adequately with extreme events caused by seasonal change.

Acid mine drainage is being treated in coal mines in Hongai area. There are treatment systems today that are installed to manage the drainage that comes into the water flow and these systems are all variably successful.

Based on monitoring data, it has been shown that the treated effluents appear clear and get runoff concentrations of 0.03–0.49 mg/L Fe, 0.07–0.47 mg/L Mn, 13–27 mg/L total suspended solids (TSS) and pH range about 7–9 [7]. Mine effluents are treated to meet Industrial Effluent Limits by Vietnamese standard.

Mine water that has been treated can be used in mining operations such as spraying dust, cleaning, etc. in mine.

4 Conclusion

Coal mining has a long history in Hongai area, however, mining activities are associated with acid drainage problem that can cause a harmful affect on the natural ecosystem of rivers.

Therefore, the treatment of the mine drainage in coal mine is required by both VINACOMIN and Viet nam government as a critical requirement of long-term operation.

Nowadays, there are systems that are installed in Hongai area with total designed capacity of 8,560 m^3/h in order to ensure that the mine drainage from mining area has to be treated appropriately before discharging into the environment.

In general the mine drainage treatment process investigated is suitable for Hongai underground coal mines. Treatment plans generally remove pollutants from mine water quite effectively, however, the operating costs, maintenance costs and the capital need to be considered.

Practical aspect needs to consider such as calculation of the needed alkalinity to raise the pH high enough, thus reduced operating costs.

In addition, seasonal change leading to a highly variable runoff that is expected to increase the risk of pollution in the drainage treatment. So that, taking seasonal change into account so as to avoid failures of the treatment plant during extreme events.

Acknowledgements

Authors would like to thank to VINACOMIN Hongai coal company Team for the support with site access and field investigation.

References

1. A. Azapagic, *Developing a Framework for Sustainable Development Indicators for the Mining and Minerals Industry*, Journal of Cleaner Production, 12 (6), pp. 639–662, (2004)
2. D. Trumm, *Selection of Active and Passive Treatment System for AMD–Flow Chart for New Zealand Conditions*. Journal of Geology and Geophysics, 53, pp. 195–210, (2010)
3. F. Bilek, K. Brömme, H. Stolpe, *Active Treatment of Fe-, Mn- and Coal dust Contaminated Mine Water as Part of the RAME-Project in Vietnam*, [in:] 11th IMWA Congress „Mine water – Managing the challenges", September 4th-11th 2011, Aachen, Germany, pp. 281–286, (2011)
4. International Network for Acid Prevention (INAP). Global Acid. Rock Drainage Guide, http://www.gardguide.com/index.php?title=Chapter_7
5. T. Mien, *Mine waste water management and treatment in coal mines in Vietnam*. Geosystem Engineering, 15 (1), pp. 66–70, (2012)
6. Công ty CP Tin học, Công nghệ, Môi trường VINACOMIN. Xây dựng định mức kinh tế, kỹ thuật vận hành các trạm xử lý nước thải mỏ than. Báo cáo tổng kết kết quả nghiên cứu của đề tài, tr. 66, (2016)
7. P.T. Dang, *Master Thesis in Vietnamese*, (2017)

Cost of Equity Estimation in Fuel and Energy Sector Companies Based on CAPM

Diana Kozieł[1], *Stanisław* Pawłowski[1], and *Arkadiusz* Kustra[1,*]

[1] AGH University Science and Technology, Faculty of Mining and Geoengineering, A. Mickiewicza Av. 30, 30-059 Krakow, Poland

Abstract. The article presents cost of equity estimation of capital groups from the fuel and energy sector, listed at the Warsaw Stock Exchange, based on the Capital Asset Pricing Model (CAPM). The objective of the article was to perform a valuation of equity with the application of CAPM, based on actual financial data and stock exchange data and to carry out a sensitivity analysis of such cost, depending on the financing structure of the entity. The objective of the article formulated in this manner has determined its' structure. It focuses on presentation of substantive analyses related to the core of equity and methods of estimating its' costs, with special attention given to the CAPM. In the practical section, estimation of cost was performed according to the CAPM methodology, based on the example of leading fuel and energy companies, such as Tauron GE and PGE. Simultaneously, sensitivity analysis of such cost was performed depending on the structure of financing the company's operation. **Key words** – cost of equity, fuel and energy sector.

1 Introduction

An accurate estimation of the cost of equity in capital groups of the fuel and energy sector is an important issue, due to the fact that in reference to extraction or production, equity is often the only available source of capital. Such situation is particularly common in mining companies in the initial stages of life cycle of an extraction project, thence at the stage of geological studies and deposit prospection, characterised by greatest risk and therefore difficulties in procuring external sources of financing.

Accuracy in equity assessment influences the weighted-average cost of capital (WACC), which translates to greater precision of estimates of such aspects as the value of the enterprise or economic efficiency of an investment.

An important aspect of estimating the cost of equity is its' changeability depending on the share in the financing structure.

In the context of significance of precise estimation of the cost of equity, the objective of the article is to present the methods of estimating such cost with special attention given to the CAPM and to estimate the cost of equity of two companies from the fuel and energy sector listed on the WSE, along with presentation of variability of the cost of equity as a function of share of this type of capital in the financing structure.

* Corresponding author: kustra@agh.edu.pl

2 Cost of Equity and Methods of Its' Estimation

The cost of equity may be identified with the rate of return expected by the owners of equity from funds invested in the company. When estimating the cost of equity, it is necessary to rely on indirect methods, based on the evaluation of investors' expectations.

An additional difficulty in estimating this value is the fact that the cost of equity is not neutral in the light of changing debt, which makes it grow along with the growth of the level of use of borrowed capital and, therefore, the financial risk. [6]

The equity in a company includes:

- entrusted capital (funds): these are the original financial or material contributions of owners, made at the moment of establishing an economic entity and they can be supplemented (increased) later;
- self-financing capital (funds): they originate from the accomplished profit retained in the company or from other sources characteristic for a given type of activity.

Therefore, it is possible to distinguish two sources of origin of equity:

- capital procured via issue of new shares;
- capital procured via retaining profit generated in the company.

The issue of shares is related to incurring costs connected with fees and commissions, preparation of an issue prospectus and, frequently, with sale of shares below their nominal value.

In the case of profit generated by a company, costs are related to the taxation. In multiple countries there is double taxation consisting in the fact that a company has to pay income tax on the generated gross profit, whereas shareholders once again pay a tax on the received dividend, which derives from the already taxed profit. Retaining a generated profit in the company therefore results in reduction of the cost of equity on account of no costs of a new issue and, on the other, no double taxation. Due to this, the cost of equity deriving from re-investment of generated profits is lower than the cost of equity procured from a new issue of shares.

Literature on the subject presents three main methods of determining the cost of equity.

- Dividend models (e.g. Gordon Growth Model),
- Compounding interest rate models (e.g. the Bond Plus Model),
- CAPM.

3 Estimation of equity cost according to CAPM methodology

The structure of the CAPM brings to mind the method of risk free rate increased by a premium for market risk; however, the important difference is the presence of beta (β) factor in the method. It is assumed in the CAPM that the market risk is the only source of risk influencing long-term average rates of return from equity. Beta is the measure of such risk. Its' value shows the degree in which changes of share rates of analysed company are related to changes in the rates of market portfolio (in the Polish conditions, the WIG index).

The value of beta depends on three factors: the area of business in which the company is conducting its' operation, the operational leverage and the financial leverage. In practice, β may assume the following values [9]:

- **β = 1**, which means that the pace of changes of the rate of return from investment in assets of a given company is exactly the same as the pace of changes of the rate of return from the market portfolio;
- **0 < β < 1**, which means that the rate of return of the company's assets is changing slower than the rate of return from the market portfolio, which means smaller risk of investing in the equity of a given company than in market instruments;

- **β > 1**, which means that the pace of changes of the rate of return from investment in assets of a given company is greater than the pace of changes of the rate of return from the market portfolio;
- **β < 0**, which means that the rate of return from investment in assets of a given company is reversely proportional to the rate of return from the market portfolio – if the market rate grows, the value of shares of the analysed company drops and vice versa.

Graphic interpretation of beta is presented in Figure 1.

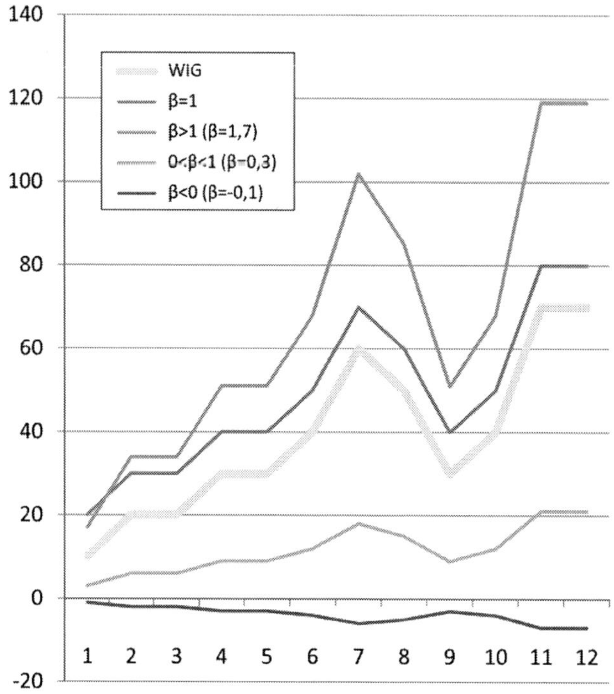

Fig. 1. Graphic interpretation of β factor

Calculation of β is made on the basis of the formula:

$$\beta_E = \frac{\operatorname{cov}(r_E, r_{WIG})}{\operatorname{var} r_{WIG}} \tag{1}$$

where:

r_E – rate of return from shares of the analysed company;

r_{WIG} – rate of return from the market portfolio.

Cost of equity in the CAPM is determined on the basis of the following formula:

$$r_E = r_{AWR} + \beta_E \cdot (r_{WIG} - r_{AWR}) \tag{2}$$

Estimation of equity of companies outside of the public trading may seem to be a problem. In such a situation, estimation is performed on the basis of the listing of companies comparable to the analysed one, operating in the same industry and with known capital structure. Beta factor is calculated for them on the basis of stock exchange data and its' value is averaged; subsequently, making use of Hamada's equation which combines the value of beta factor of equity with the capital structure, the so-called unlevered beta is calculated, deprived of the effect of financial leverage.

The mathematical formula is as follows:

$$\beta_U = \frac{\beta_{Ea}}{1+\frac{D}{E}\cdot(1-T)} \tag{3}$$

where:

β_U – unlevered beta,

β_{Ea} – beta calculated for comparable companies.

Another step is calculation of beta of equity of an analysed company, taking the structure of its' assets into account, i.e. the so-called relevered beta, which is presented in the formula below:

$$\beta_E = \beta_U \cdot \left(1+\frac{D}{E}\cdot(1-T)\right) \tag{4}$$

Beta factor received in the manner presented above is to be subsequently used in the formula for the value of cost of equity in the CAPM.

4 Estimation of Cost of Equity of Fuel and Energy Concern via CAPM Exemplified by Tauron Group and PGE

Equity is an important source to finance the operation of fuel and energy concerns due to the fact that extraction activity is characterised by the presence of expenses with a significant degree of risk (research work, prospecting the deposit), for which it is often the only available source of financing. The most precise estimation of costs of equity is therefore an important issue from the point of view of fuel and energy concerns.

Calculation of cost of equity was made on the basis of the CAPM for two fuel and energy concerns listed on the Warsaw Stock Exchange, PGE and Tauron Polska Energia. Application of the CAPM required adoption of the following principles:

- the risk-free rate r_{AWR} was adopted on the level of 3.140%, as the rate announced by the Energy Regulatory Office for companies from the gas sector and operators of electric energy systems for the second quarter of 2017,
- the time range of stock exchange data adopted for determination of beta factor encompassed dates from 01.01.2016 to 31.05.2017,
- the rate of return from the market portfolio r_{WIG} was calculated on the basis of changes in WIG in the second quarter of 2017 and amounted to 5.37%,
- beta factor of equity β_E was determined as the relation of co-variance to variance between changes in the rate of return from the WIG market portfolio r_{WIG} and the rate of return from shares of the examined concern was determined r_E, for which the cost of capital is calculated.

Table 1. Input data of CAPM

The rate of return of risk-free assets r_{AWR}	3.14%
The rate of return of the stock exchange portfolio r_{WIG}	5.37%

Based on the collected data, the estimation of the cost of equity in two largest fuel and energy concerns listed at the Warsaw Stock Exchange was made. The results are presented in Table 2.

Table 2. β factor and cost of equity of analysed companies

Company name	Factor β_E	r_{AWR}	r_{WIG}	K_E
Tauron	1.069	3.14%	5.37%	5.52%
PGE	0.984	3.14%	5.37%	5.33%

The obtained value of beta equal to 1.069 in the case of Tauron shows that the shares of this company constitute assets with risk higher than the market risk (> 1),whereas in the case of Polska Grupa Energetyczna (PGE), this interpretation is reverse, which means that the shares of this group are characterised by smaller risk from the market portfolio identified with the WIG index (< 1).

Based on the calculated values of cost of equity and balance sheet data of analysed companies, it was possible to estimate the so-called unlevered beta, devoid of the effect of financial leverage for the known capital structure, resulting from the balance sheets of analysed companies. The results are presented in Table 3.

Table 3. Unlevered beta factor of equity of analysed companies

Company name	Factor β_E	D/E financing structure factor	Unlevered beta factor β_U
Tauron	1.069	0.91953179	0.613
PGE	0.984	0.540503	0.684

On this basis, it was also possible to examine the changes in the level of equity of each company depending on the financing structure. To this aim, the so-called relevered beta is calculated for the examined capital structure. Dependence between the level of cost of equity and the capital structure is presented in Figure 2.

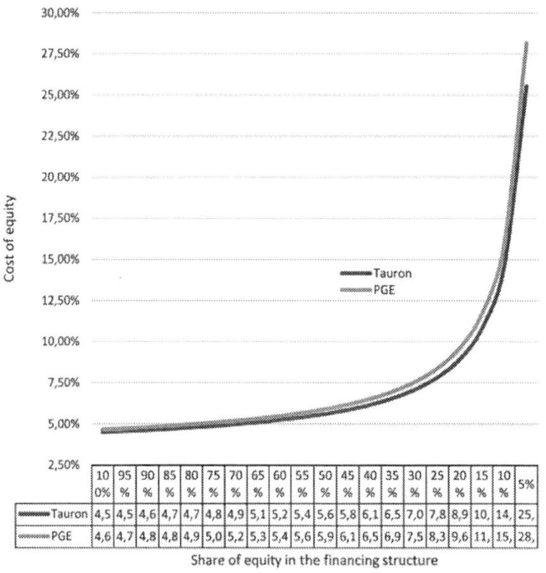

Fig. 2. Sensitivity analysis of cost of equity as a function of financing structure

As shown by the calculations, the value of unlevered beta for PGE is higher from the value of this factor for Tauron, which results from significant differences in the structure of financing for analysed companies. This also translates to greater growth dynamics of the cost of equity, along with an increase of the share of debt in the structure of financing, which is shown by the diagram above.

On account of similar values of cost of equity in both analysed companies with respect to the financing structure currently applied by them, differences resulting from the sensitivity analysis of the cost of equity from the financing structure are slight (of a few percent), similarly to the pace of changes of this value.

5 Conclusion

An accurate estimation of the cost of equity in capital groups of fuel and energy sector is an important issue on account of the significance of this value in weighted average cost of capital (WACC).

There are multiple methods of estimating cost of equity, among which the CAPM is used most often in practice; it relies on designation of beta factor of a listed company, showing the dependence between the fluctuation of share rates of a given company and the market index (in Poland, usually WIG or WIG20 index).

It is possible and justified to estimate the cost of equity of holdings from the fuel and energy industry listed on the Warsaw Stock Exchange, such as Tauron and PGE, with the application of the CAPM. Determination of beta factor and the cost of equity for the above-listed companies simultaneously allowed for performance of sensitivity analysis of their cost of equity depending on changes in the structure of financing.

The obtained results show that with respect to the current structure of financing, higher cost of equity characterises Tauron, even though differences in the level of this cost between the companies are slight. On the other hand, sensitivity analysis of the cost of equity depending on the financing structure showed a greater pace of changes and a higher level of border cost of equity for PGE. The performed analyses allowed for implementing the basic objective of the article.

Analyses presented in the article may be expanded in the future to calculation of the WACC for the described capital groups based on the hitherto equity structure and an attempt may be made at designating an optimal financing structure for determined cost levels of equity and borrowed capital.

References

1. A. Duliniec, *Struktura i koszt kapitału w przedsiębiorstwie*, PWN, Warszawa, (2001)
2. M. Czekała, A. Grześkowiak, *Szacowanie kosztu kapitału w warunkach dużej zmienności stóp zwrotu*, [in:] Zarządzanie finansami w transformacji przedsiębiorstw, collective work edited by T. Jajuga and T. Słoński, Wyd. Akademii Ekonomicznej im. Oskara Langego we Wrocławiu, Wrocław, p. 83. (1999)
3. M. Kufel, *Metody wyceny przedsiębiorstw*, Wyd. Park, Bielsko-Biała, p. 37–38, (1992)
4. J.W. Petty, A.J. Keown, D.F. Scott, J.D. Martin, *Basic Financial Management*, Englewood Cliffs, p. 267. (1993)
5. R. Ranosz, A. Kustra, *Koszt kapitału własnego spółek górniczych w Polsce*, [in:] Zeszyty Naukowe Uniwersytetu Szczecińskiego Finanse, Rynki Finansowe, Ubezpieczenia, No. 60, No. 761, (2013)

 https://doi.org/10.1051/e3sconf/20183501008

6. K.S. Rao Ramesh, *Financial Management, Concepts and Applications*, South-Western College Publishing, Cincinnati, (1995)
7. M. Sierpińska, T. Jachna, *Ocena przedsiębiorstwa według standardów światowych*, PWN, Warsaw, p. 246, (2003)
8. P. Szczepankowski, *Wycena i zarządzanie wartością przedsiębiorstwa*, Wydawnictwa Naukowe PWN, Warsaw, (2007)
9. J.F. Weston, T.E. Copeland, *Managerial Financial Management*, p. 438. (1988)

Total Factor Productivity Growth, Technical Progress & Efficiency Change in Vietnam Coal Industry – Nonparametric Approach

Vu Hung Phuong[1,*]

[1] Vietnam National Coal-Mineral Group, Vietnam

Abstract. This research applies Data Envelopment Analysis (DEA) approach to analyze Total Factor Productivity (TFP) and efficiency changes in Vietnam coal mining industry from 2007 to 2013. The TFP of Vietnam coal mining companies decreased due to slow technological progress and unimproved efficiency. The decadence of technical efficiency in many enterprises proved that the coal mining industry has a large potential to increase productivity through technical efficiency improvement. Enhancing human resource training, technology and research & development investment could help the industry to improve efficiency and productivity in Vietnam coal mining industry.
Key words – Technical & Scale Efficiency; Total Factor Productivity-TFP; Nonparametric Techniques.

1 Introduction

Total merchandise coal production in the 20 years was 525 million tons, of which the highest yearly production is 44.7 million tons in 2011, equaling 7.5 times of that in 1995 when Vietnam Coal Corporation was established. Since 2012, due to the effects of financial crisis and economic downturn in the country and all over the world, the business efficiency of the Vinacomin has much decreased compared to 2011, but production, employment and income for workers have remained stable. Overall labor productivity in 2011 (when Vietnam coal mining industry reached the highest production so far) was 4 times that in 2011 (merchandise coal production increased from 6 million tons to 45 million tons, an increase of 7.5 times, while the number of employees only increased by 1.8 times). However, after 2011, labor productivity not increased because of several reasons: proportion of coal exploited by mine tunnel technology has been increasingly raised (labor expense in tunnel technology is much higher than that in open-cast mine), from 2011 to 2015 proportion of tunnel coal increased from 45 to 56%; production output has been constrained by market, while some steps in mining process must be maintained to preserve mine tunnels (particularly ventilation, drainage, etc.); gravelly soil volume that needs to be removed in open-cast mine has increased because mines have been increasingly deeper, investment in mechanization has been limited, investment projects has been late in progress, etc.

* Corresponding author: hungphuongvu@yahoo.com

Since Vietnam economy reform started in 1986, the influence of these reforms on productivity and efficiency of coal mining industry has been cared about and investigated by many researchers, economic and enterprise managers. Prior studies suggested that institution changes are an essential cause of the increase in productivity of coal mining industry in the reform period. In these studies, effects of technological changes were either ignored or considered as insignificant.

Most of the studies on coal mining industry productivity followed the traditional parameter approach to calculate total factor productivity by estimating an aggregate production function, a cost function, or a profit function. However, the traditional parameter methods are distrusted by many economists because of their aggregate hypotheses.

Large coal mining companies listed in Enterprise survey from 2000–2014 were sampled in the research. These companies are members of Vietnam National Coal-Mineral under State owned monopoly. However, the research only analyzes TFP, technology and efficiency changes in coal mining in Vietnam from Asian financial crisis in 2007.

This research adopts a data envelopment analysis (DEA) approach to investigate total factor productivity (TFP) growth in coal mining industry of Vietnam from 2007 to 2014. Using Malmquist productivity index), the TFP increase in the coal industry is decomposed into technical and efficiency changes [2]. The technical change enables them to identify contributions of improved technical productivity and technological progress to productivity growth of coal mining industry.

In Vietnam, the nonparametric method was developed to decompose TFP change in provincial agriculture of Vietnam into changes in technological progress and efficiency change [5].

Besides nonparametric method mentioned above, there exists stochastic frontier production function method to decompose total factor productivity change into changes in technological progress and efficiency.

2 Methodology

2.1 Measuring total factor productivity: Malmquist indices

In order to define the Malmquist index of productivity change we consider, for each time period t = 1, 2, ...,T, a production technology H^tS^t for changing inputs, $x^t \in R_+^N$, into outputs, $y^t \in R_+^M$, is defined as:

$$H^t = \{(x^t, y^t) : x^t \text{ can produce } y^t\}, \qquad (1)$$

where H^t is assumed to satisfy certain axioms to define significant output distance functions (Făre, 1988).

According to Făre et al. (1994), an output distance function at t can be defined as

$$D_0^t(x^t, y^t) = \inf[\theta : (x^t, y^t/\theta) \in H^t] = \{\sup[\theta : (x^t\theta y^t) \in H^t]\}^{-1} \qquad (2)$$

The distance function is defined as the inverse of maximum proportional increase of output vector y^t given inputs X^t. It is also equivalent to inverse of output efficiency measure, measure of TFP "catch-up" of an observation (an enterprise in this study) with the best practice frontier in coal mining, in this study, the best practice frontier is the highest productivity observed in all enterprises with the same technology [4]. $D_0^t(x^t, y^t) = 1$ if and only if (x^t, y^t) is on the boundary or the frontier of technology and the production is technologically efficient. If $D_0^t(x^t, y^t) < 1$, the production is inside the technological frontier at t, and (x^t, y^t) is technologically inefficient. The distance function measures the degree of technical inefficiency. The output distance in the period $t + 1$, $D^{t+1}(x^t, y^t)$, can be defined as (2) with t replaced by $t + 1$.

Definition of output distance function with two different time periods is

$$D_0^t(x^{t+1}, y^{t+1}) = \inf\{\theta : (x^{t+1} / \theta) \in H^t\} \tag{3}$$

This is a mixed index measuring maximum proportional changes in outputs y^{t+1} given inputs x^{t+1}, under technology at t. Similarly, we can define a mixed distance function, $D^{t+1}(x^t, y^t)$, measuring maximum proportional changes in output y^t, given inputs x^t, under technology at $t + 1$.

According to Caves et al. (1982a), Malmquist productivity index is defined as

$$M_0^t = \frac{D_0^t(x^{t+1}, y^{t+1})}{D_0^t(x^t, y^t)} \tag{4}$$

This ratio index measures productivity changes originated from changes in technical efficiency at t and $t + 1$ under technology at t. Changes in technical efficiency from t to $t + 1$ also can be measured under technology at $t + 1$. This Malmquist index is defined as

$$M_1^{t+1} = \frac{D_1^{t+1}(x^{t+1}, y^{t+1})}{D_1^{t+1}(x^t, y^t)} \tag{5}$$

Output oriented Malmquist productivity change index can be specified as the geometric average of (4) and (5) and decomposed into two components:

$$\begin{aligned} M_0\left(x^{t+1}, y^{t+1}, x^t, y^t\right) &= \left\{\left[\frac{D_0^t(x^{t+1}, y^{t+1})}{D_0^t(x^t, y^t)}\right]\left[\frac{D_1^{t+1}(x^{t+1}, y^{t+1})}{D_1^{t+1}(x^t, y^t)}\right]\right\}^{1/2} = \\ &= \frac{D_1^{t+1}(x^{t+1}, y^{t+1})}{D_0^t(x^t, y^t)}\left\{\left[\frac{D_0^t(x^{t+1}, y^{t+1})}{D_1^{t+1}(x^{t+1}, y^{t+1})}\right]\left[\frac{D_0^t(x^t, y^t)}{D_1^{t+1}(x^t, y^t)}\right]\right\}^{1/2} = \\ &= E\left(x^{t+1}, y^{t+1}, x^t, y^t\right) T\left(x^{t+1}, y^{t+1}, x^t, y^t\right) \end{aligned} \tag{6}$$

where $E(*)$ is relative efficiency change index under constant return-to-scale technology, measuring catching up with the best practice frontier for each observation between periods t and $t + 1$, and $T(*)$ expresses technical change index, measuring technological frontier shift (or innovation) between the two periods evaluated at x^t and x^{t+1}.

Decomposition of Malmquist productivity index enables us to identify the contribution of efficiency catching up and technology innovation to total factor productivity growth. The Malmquist index greater than 1 indicates an increase in productivity. The Malmquist index less than 1 indicates a decrease in productivity. In addition, increases in any of the two components of Malmquist productivity index are connected with values greater than 1, and any its decreases are connected with values less than 1.

2.2 Intermediate Problems – Data Envelopment Analysis (DEA)

Some conventional methods were listed for calculating Malmquist productivity index [3]. Most of them, however, require specification of a functional form of technology. A data envelopment analysis approach (DEA) was proposed to build a best practice frontier without specification of production technology. Unlike conventional analysis techniques that look for an average path through middle points of data series, DEA seeks directly a best practice frontier with these data. Using a nonparametric linear programming technique, DEA takes account of all inputs and outputs as well as differences in technology, ability, competitiveness and population, and compares individual performance to the best practice (efficient) frontier [2].

In this research, a DEA approach developed by Făre et al. is used to construct a best practice frontier for each technology categories at each period in the coal industry. Comparison of exploited volumes of each enterprise with the best practice frontier provides a measure of catch-up in its efficiency with that frontier and a measure of the shift of that frontier (or technology innovation). Then, Malmquist index, which measures changes in total factor productivity, can be computed as a product of these components.

Assume that there are $k = 1, \ldots, K$ provinces producing $m = 1, \ldots, M$ outputs $y^t_{k,m}$ using $n = 1, \ldots, N$ inputs $x^t_{k,n}$ at each period $t = 1, \ldots, T$. In DEA approach, reference technology with constant return-to-scale at each period t from data can be defined as

$$
\begin{aligned}
&G^t = (x^t, y^t): y^t_m \le \sum_{k=1}^{K} Z^t_k y^t_{k,m} && m = 1, \ldots, M, \\
&\sum_{k=1}^{K} Z^t_k x^t_{k,n} \le x^t_n && n = 1, \ldots, N, \\
&z^t_k \ge 0 && k = 1, \ldots, K
\end{aligned} \tag{7}
$$

where z indicates weights on each cross-sectional observation. Constant return-to-scale assumption can be relaxed by adding the following constraint [1]:

$$
\sum_{k=1}^{K} Z^t_k = 1, \qquad (VRS) \tag{8}
$$

We use a higher decomposition of Malmquist index to analyze productivity growth of the coal industry in Vietnam. We decompose efficiency change component, which is calculated under constant return-to-scale technology, into a pure efficiency change component (which is calculated under variable return-to-scale technology-VRS) and a component of scale change capturing different internal changes between variable return-to-scale technology and constant return-to-scale technology.

To construct Malmquist productivity index of enterprise k' between t and t+1, we use DEA approach to compute four distance functions $D^t_0(x^t, y^t)$, $D^{t+1}_0(x^t, y^t)$, $D^t_0(x^{t+1}, y^{t+1})$ and $D^{t+1}_1(x^{t+1}, y^{t+1})$. These distance functions are inverses of Farrell technical efficiency measures. Non-parametric programming models to calculate output-oriented measure of technological efficiency for each province $k' = 1, \ldots, K$, can be defined as

$$
\left[D^t_0\left(x^t_k, y^t_k\right)\right]^{-1} = \max \lambda^k \tag{9}
$$

$$\lambda^{k'} y^t_{k',m} \le \sum_{k=1}^{K} z^t_k y^t_{k,m}, \qquad m = 1, \ldots, M,$$
$$\sum_{k=1}^{K} z^t_k x^t_{k,n} \le x^t_{k',n} \qquad n = 1, \ldots, N, \tag{10}$$
$$z^t_k \ge 0 \qquad k = 1, 2, \ldots, K$$

Calculation of $\left[D_0^{t+1}(x^{t+1}, y^{t+1})\right]$ is similar to (10), in which t is replaced by $t + 1$.

Construction of Malmquist Index also requires calculating two mixed distance functions, calculated by comparing observations in one period with the best practice frontier of another period. The inverse of the mixed distance function for observation k' can be obtained from

$$\left[D_0^t\left(x_k^{t+1}, y_k^{t+1}\right)\right]^{-1} = \max \lambda^k \tag{11}$$

subject to

$$\lambda^{k'} y^{t+1}_{k',m} \le \sum_{k=1}^{K} z^t_k y^t_{k,m}, \qquad m = 1, \ldots, M,$$
$$\sum_{k=1}^{K} z^t_k x^t_{k,n} \le x^{t+1}_{k',n} \qquad n = 1, \ldots, N, \tag{12}$$
$$z^t_k \ge 0 \qquad k = 1, 2, \ldots, K$$

To gauge changes in scale efficiency, inversed output distance functions with variable return-to-scale technology are calculated by adding (8) into constraints (10) and (12). Technological change (TECHCH) is calculated under constant return-to-scale technology. Scale efficiency change (SCH) at each period is constructed as the ratio of the distance function with constant return-to-scale to the distance function under variable return-to-scale technology, while pure efficiency change (PEFFCH) is defined as the ratio of distance functions at each period under variable return-to-scale technology. With these two distance functions under variable return-to-scale technology, decomposition of (6) becomes

$$\begin{aligned} M_0\left(x^{t+1}, y^{t+1}, x^t, y^t\right) = T\left(x^{t+1}, y^{t+1}, x^t, y^t\right) = \\ = TECHCH \cdot EFFCH = \\ = TECHCH \cdot PEFFCH \cdot SCH \end{aligned} \tag{13}$$

where EFFCH denotes efficiency change calculated under constant return-to-scale technology.

3 Estimated results and decomposition of total factor productivity (TFP)

As indicated by [3], the distance function is equivalent to the inverse of Farrell measure of output efficiency. We use this index, defined as the inverse of (2), to measure the technological efficiency of enterprises in the coal industry in the period 2007–2013. Technical efficiency indices under the return-toscale variables of 29 enterprises from 2007 to 2013 are reported in Table 1.

Table 1. Technical efficiency indices under variable return-to-scale of coal mining enterprises, 2007–2013

	Number of Observations	Mean	Standard Deviation	Min	Max
crste2007	29	0.3308	0.2601	0.01	1
vrste2007	29	0.5861	0.2711	0.122	1
scale2007	29	0.6069	0.3353	0.015	1
crste2009	29	0.5511	0.3327	0.063	1
vrste2009	29	0.7449	0.2898	0.22	1
scale2009	29	0.7512	0.3100	0.07	1
crste2011	29	0.2921	0.2663	0.002	1
vrste2011	29	0.6414	0.2883	0.037	1
scale2011	29	0.4584	0.3326	0.004	1
crste 2013	29	0.5046	0.2915	0.028	1
vrste2013	29	0.6754	0.2699	0.127	1
scale2013	29	0.7328	0.2702	0.055	1

where:
– crste: Technical efficiency under constant return-to-scale hypothesis,
– vrste: Pure technical efficiency under variable return-to-scale hypothesis,
– scale: Scale efficiency.

Source: Author's calculations from Enterprise Survey data 2007–2014. General Statistic Office

Table 1 shows averages of efficiency under constant return-to-scale, variable return-to-scale and scale efficiency of sample enterprises in the coal industry and reports only estimates of these efficiency measures for 4 years: 2007, 2009, 2011 and 2013. The results display strong fluctuations of efficiency measures. For instance, averages of efficiency under constant return-to-scale in 2007, 2009, 2011 and 2013 are 0.3308, 0.5511, 0.2921, 0.5046, respectively. In general, all efficiency indices show that performance of enterprises in the industry is still inefficient. The highest scale efficiency (in 2009) reached only 75.12%.

Table 2. Annual average changes of Malmquist indices, 2007–2013

Year	Effch	Techch	Pech	Sech	Tfpch
2007–2008	1.032	0.84	1	1.032	0.867
2008–2009	0.944	0.869	1	0.944	0.82
2009–2010	1.183	0.722	1	1.183	0.854
2010–2011	1.121	0.906	1	1.121	1.015
2011–2012	0.723	1.241	1	0.723	0.898
2012–2013	1.847	0.525	1	1.847	0.97
Average	1.095	0.823	1	1.095	0.901

where:
– Effch: Efficiency change
– Techch: Technological change
– Pech: Pure efficiency change
– Sech: Scale efficiency change
– Tfpch: Total factor productivity change

Source: Author's calculations from Enterprise Survey data 2007–2014. General Statistic Office

Table 2 shows results of decomposition of total factor productivity in the coal industry in the sample. The average efficiency change of this period increased (1.095), but there were 2 stages with low efficiency change, namely in 2008–2009 (reached only 94.4%) and especially in 2011–2012 with even average efficiency change decreased (reached only 72.3%). Technological changes decrease in general, reached only 82.3% in 2007–2013. The total factor productivity change in this period was low, reached only 90.1%. The main cause of TFP decreased was a very low level of application of technical innovation and new technology.

The year 2011 was the year with the highest TFP in the studied period, it reached 1.015. This estimated result was in accord with the report on performance of the industry in 20 years. According to the report, total merchandise coal production in 2011 reached the highest level, which equals 7.5 times of the level of the year of establishment of Vietnam Coal Corporation in 1995.

The reason for labor productivity growth of the coal industry in this stage was liberalization of production capacity by reasonable rearrangement of organization model in coal firms, application of new mining technologies, adoption of contractual mechanisms which created motivation and spring forces for labor productivity. Another equally important reason is the high growth of coal market in the period 2007–2011 which stimulated production.

The question why total factor productivity in the coal industry decreased (except for 2011) can be explained as follows: (1) Exploitation activities of coal mining industry are different from production in factories, mining has been shifting deeper and further day by day into the earth's womb, gravelly soil volume that needs to be removed in both open-cast mine and tunneling, transport distance increased (for 1 ton of exploited coal, gravelly soil volume that needs to be removed increased 3.1 times from 3.4 m^3/ton in 1995 to 10.7 m^3/ton in 2014, transport distance increased 3.72 times from 1.03 km in 1995 to 3.83 km in 2014, these two factors caused exploitation cost increase 5 times), high mine pressure, big volume of air and water,... made ventilation, drainage and transportation costs highly increase... (2) Competition has been increasingly severe; (3) Demand for coal is one of the important factors; (4) Exploitation technology can be not really advanced yet; (5) There are still many inadequacies in management...

4 Summary and conclusion

This study applies DEA approach to measure total factor productivity (TFP), technical change and technological efficiency in Vietnam coal mining industry from 2007 to 2013. Malmquist productivity index is used to measure productivity growth in this study. With data envelopment analysis approach (DEA), productivity growth can be decomposed into two components: Technical change and efficiency change. This decomposition enables us to identify contributions of technical progress and improvement in technical efficiency to productivity growth in coal production.

In the study, data envelopment analysis method (DEA) is used to calculate component distance functions of Malmquist index and construct the best practice frontiers. Technical change index and efficiency change index is obtained by comparing the efficiency of each enterprise to the best practice frontier with the same production technology. Then Malmquist productivity index is calculated as the product of these indices.

Calculated results show that change in TFP of the coal industry decreased, the main reason is that investment in technological progress of the industry is still low, or asynchronous, or inefficient, except for 2011 which is the year with high productivity growth due to increased demand stimulating production.

Policy recommendations: To raise efficiency, TFP growth, it is necessary:

1) To enhance qualification of management staff;
2) To raise technical level of workers and technical staff, enable them to stabilize their life and keep their mind on their work;
3) To strengthen technological innovation to improve productivity;
4) To implement well marketing activities, expand into new markets and maintain traditional markets;

5) To enhance competitiveness by cost reduction via good management and technical application, technological improvement.

References

1. S.N. Afriat, *Efficiency Estimation of Production Functions*, International Economic Review **13**, 568–98 (1972)
2. A. Charnes, W.W. Cooper and E. Rhodes, *Measuring the efficiency of decision making units*, European Journal of Operational Research **2**, 429–444 (1978)
3. R. Färe, S. Grosskopf and C.A.K. Lovell, Production frontiers (Cambridge University Press, Cambridge) (1994)
4. M.J. Farrell, *The measurement of productive efficiency*, Journal of the Royal Statistical Society, Series A (General) **120**, 253–289 (1957)
5. Nguyen Khac Minh & Pham Van Khanh, *A chance-constrained data envelopment analysis approach to problem provincial productivity growth in Vietnamese agriculture from 1995 to 2007*, Open Journal of Statistics, **1**, 217–235 (2011)
6. *Vietnam Coal-Mineral Group, Solutions to Increase Production and Efficiency in Period 2015–2020*, Internal report (2016)

Multidimensional Analysis of Copper Ore Flotation in Terms of Applied Hydrophobizing Agents

Paulina Pięta[1,*], *Tomasz* Niedoba[1], and *Agnieszka* Surowiak[1]

[1] AGH - University of Science and Technology, Faculty of Mining and Geoengineering, Department of Environmental Engineering and Mineral Processing, Cracow, Poland

Abstract. Flotation is a method of enrichment used to distribute particles, which differ in their surface properties. Hydrophobic solids intrinsically create contact at the solid-liquid-gas interface. However, not all minerals, including copper minerals, can be characterized by this crucial ability. In that case it is necessary to use the collector reagents which guarantees a high efficiency of the enrichment process. The main aim of the paper was to examine the impact of selected collector types and dosages on the results of Polish sandstone copper ore flotation and to find optimal parameter values for products that meet quality and quantity requirements. The laboratory tests were carried out with an application of two types of collectors (Hostaflot, sodium ethyl xanthate aqueous solution) in dosages 100 and 150 g/Mg. Data analysis was based on the use of the taxonomy methods in order to select optimal conditions of collector dosage and type. Based on the indexes, it was found that the best enrichment effects were obtained with a sodium ethyl xanthate aqueous solution 150 g/Mg. **Key words** – flotation, copper ore, taxonomy methods, multidimensional comparative analysis.

1 Introduction

Flotation is a separation method used to separate particles which can be characterized by differences in surface properties. In a single flotation act of hydrophobic grains, a naturally occurring aggregate with an air bubble is formed and subsequently it floats to the froth. Hydrophobic solids intrinsically create contact at the solid-liquid-gas interface. However, not all minerals, including copper minerals, exhibit this key hydrophobic ability. The endurance of the occurred mineral-air bubble aggregate in the flotation pulp can be modified by the addition of suitable reagents. For hydrophilic particles, the formation of a three-phase system depends primarily on the addition of a hydrophobic reagent. The dosed reagent type may affect both the change in hydrophobicity and the final effect of flotation in the form of content and recovery of the metal in the separation products [3, 4].

A group of reagents that reduce the surface tension of a solid is broad. In general, ion collectors and nonionic collectors can be distinguished.

* Corresponding author: ppieta@agh.edu.pl

The use of these reagents is conditioned by the properties of the minerals and guarantees a high efficiency of the enrichment process. Hence it is important to analyse their impact on the quality of the flotation products thoroughly. The application of chemical reagents is not the only significant factor affecting the final flotation effect. As indicated works [14, 15] the particle size distribution of feed material also plays a key role in the process. Properly selected communition and sizing processes provide a high level of copper minerals liberation, which guarantees better enrichment effects.

2 Scope and Methodology

The copper ores in LGOM are characterized by variable lithological composition, mineralization and floatability. The deposit is formed by an accumulation of copper sulphates occurring in sandstone, copper-bearing shales and carbonate rock. They are different in terms of physical and chemical properties and additionally they are varied locally within one type. Polish copper ores are dominated by copper sulphide minerals (chalcocite, bornite, chalcopyrite, cobalt). A major problem during the flotation process is the presence of a specific substance which is characterized by natural hydrophobicity, which gets across the concentrates and thus it deteriorates its quality. Even despite introduced collectors and frothers, it is necessary to use a multi-stage flotation process to achieve the desired results [4, 5, 12].

The circuit comprises of rougher and main flotation operations, with a section of the ore carbonate and sandstone-shale. The products obtained in the first stages are directed downstream to the cleaning flotation process to achieve concentrate having the highest possible concentrate grade. The sodium ethyl xanthate aqueous solution or its mixture with isobutyl xanthate (70 g/Mg) are used as collectors. Apart from the flotation reagents, the effects of copper ore enrichment depend on many factors which leads to the process being complicated and demanding. [4, 5].

2.1 Multidimensional comparative analysis (MCA)

The multidimensional structure of processes and studied objects, the complexity of the relationship between different properties characterize the whole reality surrounding us. Taking into account this condition is very important with regard to an optimization of processes and complete characterization of mineral resources [4, 11]. Different methods of multidimensional comparative analysis serve as common tools used in economics, social studies, biological, anthropogenic and geographic researches. In the works [1, 6, 7] taxonomic methods were successfully used to consider the issues of mineral processing. The undisputed advantage of these methods is their versatility, the simplicity of calculation and the ease of interpretation. They allow to replace the description of a multicharacteristic object in question with a synthetic variable which replaces all dependent variables with one value.

MCA is a comparison of objects defined with many diverse features. The methods to perform such an analysis include taxonomic methods, based on comparisons using the distance matrix. The $\boldsymbol{X}$ matrix (1) of diagnostic features is the basis for calculations and further analysis. Based on this, a synthetic matrix $\boldsymbol{Z}$ (2) of standardized values is calculated, ordering all n-situations in a linear manner, considering both positive (stimulant) and negative (destymulant) effects of the features on the phenomenon.

$$\boldsymbol{X} = \begin{bmatrix} x_{11} & x_{12} & \cdots & x_{1l} \\ x_{21} & x_{21} & \cdots & x_{2l} \\ \cdots & \cdots & \cdots & \cdots \\ \cdots & \cdots & \cdots & \cdots \\ x_{n1} & x_{n2} & \cdots & x_{nl} \end{bmatrix} \quad (1)$$

$$\boldsymbol{Z} = \begin{bmatrix} z_{11} & z_{12} & \cdots & z_{1l} \\ z_{21} & z_{21} & \cdots & z_{2l} \\ \cdots & \cdots & \cdots & \cdots \\ \cdots & \cdots & \cdots & \cdots \\ z_{n1} & z_{n2} & \cdots & z_{nl} \end{bmatrix} \quad (2)$$

where:
x_{ij} – the value of a property i-number for test j-number;
n – number of examined cases, i.e. flotation tests, $j = 1, \ldots, n$;
l – number of random variable, i.e. copper content, copper recovery, $i = 1, \ldots, l$;
z_{ij} – unitarized value of x_{ij} property, which were calculated according to formula:

$$z_i = \frac{x_i}{x_{\max} - x_{\min}},$$

$x_{\max}$, $x_{\min}$ – the maximum and minimum value of the property.

Three methods of MCA were introduced to investigate the effects of sandstone copper ore flotation process. The taxonomic development measure with the Euclidean distance d_j (3), the relative development index w_j (4), and the zero-unification method z_j (5) were used. A detailed description of the taxonomic measure can be found in [6–8, 10] below.

$$dj = \sqrt{\sum_{i=1}^{n}\left(1 - s_{ji}\right)^2}, \quad (3)$$

$$w_j = \frac{\sum_{j=1}^{k} u_{ij}}{\sum_{j=1}^{k} \max_j \left\{u_{ij}\right\}}, \quad (4)$$

$$z_j = \frac{1}{n}\sum_{i=1}^{n} H_{ij}, \quad (5)$$

where:

$$s_{ji} = \frac{z_{ji}}{z_{j\max}},$$

$$u_{ij} = x_{ij}^* + \left|\max_i \left\{x_{ij}^*\right\}\right|,$$

$$x_{ij}^* = \frac{x_{ij} - \overline{x}_i}{S_i},$$

$$H_{ij} = \frac{x_{ij} - x_{i\,\min}}{x_{i\,\max} - x_{i\,\min}} \text{ for stymulants,}$$

$$H_{ij} = \frac{x_{i\,\max} - x_{ij}}{x_{i\,\max} - x_{i\,\min}} \text{ for destymulants,}$$

S_i – standard deviation of features *i*-number,
$\bar{x}_i$ – mean value of the feature *i*-number.

3 Experimental

The paper presents the results of research on the influence of selected types and dosages of collectors on the effects of flotation of sandstone copper ore. Laboratory tests were performed using two anionic collectors: sodium ethyl xanthate aqueous solution and Hostaflot. They were added in two dosages: 100 and 150 g/Mg. Nasfroth in dosage 50 g/Mg was used as a frother during the test.

Selected taxonomic methods were used to analyse the experimental data. A multidimensional comparative analysis has allowed us to assess the effectiveness of the sandstone copper ore flotation process and to classify the case studies.

The testing material was a sample of Polish copper ore which was taken from ZG Polkowice KGHM Polska Miedź S.A. For flotation tests a sandstone type was used, which was previously separated from the raw material and then it was crushed and classified to obtain a 40–71 μm size fraction. The copper content in sandstone type was equal to 1.1%. The tests were carried out in a Denver flotation machine with a constant concentration flotation slurry of 200 g/dm^3, aerating and at a rotor speed. The research was carried out while maintaining uniform material preparation procedures. During a single experimental study one concentrate and tailings were obtained. The flotation products were dried and weighed and subjected to X-ray fluorescence to determine the copper content. For each experiment, the following process evaluation indicators were identified: copper content in concentrate (6), copper content in tailings (7), copper recovery in concentrate (8), copper recovery in tailings (9). The averaged values of selected flotation tests for the sandstone copper ore are presented in Table 1.

$$\beta_i = \frac{\sum_{i=1}^{n} \gamma_i \cdot \lambda_i}{\sum_{i=1}^{n} \gamma_i}, \tag{6}$$

$$\vartheta_i = \frac{\sum_{i=1}^{n} \gamma_i \cdot \lambda_i}{1 - \sum_{i=1}^{n} \gamma_i}, \tag{7}$$

$$\varepsilon_i = \frac{\beta_i}{\alpha} \sum_{i=1}^{n} \gamma_i, \tag{8}$$

$$\eta_i = \frac{\vartheta_i}{\alpha} \sum_{i=1}^{n} \gamma_i, \tag{9}$$

where:
α – a copper content in the feed [%],

$\gamma_i = \frac{m_i}{\sum_{i=1}^{n} m_i} \times 100$ – a yield of a product [%],

λ_i – content of copper in a product,

m_i – an mass of a product.

Table 1. Results of flotation tests of sandstone lithological type for process duration equal to 30 min

Size fraction, [μm]	Collector type	Collector dosage, [g/Mg]	β_i	υ_i	ε_i	η_i
125–200	X	100	0,0215	0,0004	0,8000	0,2000
125–200	H	100	0,0270	0,0010	0,5813	0,4187
125–200	X	150	0,0264	0,0003	0,9177	0,0823
125–200	H	150	0,0799	0,0003	0,9018	0,0982
100–125	X	100	0,0678	0,0005	0,8704	0,1296
100–125	H	100	0,0326	0,0014	0,5933	0,4067
100–125	X	150	0,0712	0,0002	0,9429	0,0571
100–125	H	150	0,0688	0,0003	0,9396	0,0604
71–100	X	100	0,0455	0,0003	0,9539	0,0461
71–100	H	100	0,0314	0,0016	0,6983	0,3017
71–100	X	150	0,0486	0,0002	0,9623	0,0377
71–100	H	150	0,0522	0,0004	0,9398	0,0602
40–71	X	100	0,1467	0,0005	0,9669	0,0331
40–71	H	100	0,5134	0,0003	0,9808	0,0192
40–71	X	150	0,0298	0,0001	0,9943	0,0057
40–71	H	150	0,0344	0,0002	0,9410	0,0590
20–40	X	100	0,0428	0,0004	0,9505	0,0495
20–40	H	100	0,1585	0,0007	0,9708	0,0292
20–40	X	150	0,0347	0,0003	0,9711	0,0289
20–40	H	150	0,0300	0,0004	0,9011	0,0989
0–20	X	100	0,0843	0,0009	0,9768	0,0232
0–20	H	100	0,0947	0,0012	0,9625	0,0375
0–20	X	150	0,0426	0,0007	0,9534	0,0466
0–20	H	150	0,0479	0,0007	0,9762	0,0238

symbols: H-collector of the izobutyl ksanthate family Hostaflot, X-collector of the ethyl ksanthate family

4 Analysis results

The assessment of the efficiency of copper ore enrichment was based on the following indicators: content and recovery of metal in concentrate and tailings. The results of laboratory experiments have confirmed that the highest quality of the concentrates was observed when a feed was a 40–71 μm size fraction. The optimal particle size distribution for this lithological type was based on the d_j values, which take into account β, υ, ε, η for the assessment process (Tab. 3). The further part of the analysis focused on that material. In the case of a single flotation test duration assortment, such unambiguous results were not obtained. Hence the results of the flotation after 1, 12, 30 [min] were presented and compared.

The effectiveness of the flotation process was assessed on the basis of two groups of indicators characterizing positive (β, ε) and negative (υ, η) enrichment results. To consider, the direction of influence of the variables in the MCA, it was necessary to replace the value of the copper content and the recovery of this metal in the tailings by their opposite. Thus, a matrix of variable values for which a homogeneous final evaluation of the results could be made.

Another important issue is normalization, which unifies the values of features from the point of view of a certain criterion [2]. The main goal of this kind of transformation is to fulfill the postulate of additivity, i.e. to bring features with different denomination to their comparability [2, 8–10, 13]. In the calculation process, two forms of standardization were chosen (a quotient conversion s_{ji} and standardization x_{ij}^*).

Taking into account the number of variables, the taxonomy measures were described by the following formulas (10–12):

- Euclidean distance,

$$d_j = \sqrt{\left(1-\frac{z_{\beta i}}{z_{\beta \max}}\right)^2 + \left(1-\frac{z_{\vartheta i}}{z_{\vartheta \max}}\right)^2 + \left(1-\frac{z_{\varepsilon i}}{z_{\vartheta \max}}\right)^2 + \left(1-\frac{z_{\eta i}}{z_{\eta \max}}\right)^2}, \qquad (10)$$

- relative development index

$$w_j = \frac{u_{\beta j} + u_{\vartheta j} + u_{\varepsilon j} + u_{\eta j}}{\max\limits_i \{u_{\beta j}\} + \max\limits_i \{u_{\vartheta j}\} + \max\limits_i \{u_{\varepsilon j}\} + \max\limits_i \{u_{\eta j}\}}, \qquad (11)$$

where:

$$u_{\beta j} = x_{\beta j}^* + \left|\max_i \{x_{\beta j}^*\}\right|, \quad u_{\vartheta j} = x_{\vartheta j}^* + \left|\max_i \{x_{\vartheta j}^*\}\right|,$$

$$u_{\varepsilon j} = x_{\varepsilon j}^* + \left|\max_i \{x_{\varepsilon j}^*\}\right|, \quad u_{\eta j} = x_{\eta j}^* + \left|\max_i \{x_{\eta j}^*\}\right|,$$

- zero-unification method

$$z_j = \frac{1}{n} \cdot \left(\frac{x_{\beta j} - x_{\beta \min}}{x_{\beta \max} - x_{\beta \min}} + \frac{x_{\varepsilon j} - x_{\varepsilon \min}}{x_{\varepsilon \max} - x_{\varepsilon \min}} + \frac{x_{\vartheta \max} - x_{\vartheta j}}{x_{\vartheta \max} - x_{\vartheta \min}} + \frac{x_{\eta \max} - x_{\eta j}}{x_{\eta \max} - x_{\eta \min}} \right) \qquad (12)$$

Based on the results of the taxonomic measure determined for a single flotation test it was possible to compare the level of the studied phenomenon. Euclidean distances are meters that use a pattern which is the maximum value of a feature. This is due to the fact that a higher value of the indicator is equivalent to a worse effect of the sandstone copper ore flotation. The reverse situation exists for the (relative development index, zero unification method), which are not included in the calculation process of the pattern. The values of indexes w_j and z_j are from 0–1. As the result is closer to 1, this phenomenon is better in terms of the general criterion. Table 2 summarizes the results of the overall evaluation of sandstone copper ore flotation under defined conditions for all three types of taxonomic methods.

The results of the MCA indicated that the best results of flotation were achieved with 150 g/Mg ethyl xanthate and a extended process time up to 30 minutes. This conclusion was confirmed by each taxonomic method. As the process duration was shortened, the quality of the flotation concentrates deteriorated and the optimal process conditions changed. It can be seen that the best quality of the separation products during the 12-minute working time provided a Hostaflot reagent dosed at 100 g/Mg. The evaluation for the shortest time of the separation process did not produce consistent results, as in the other cases. Two out of three indices have shown that the process is best carried out with the use of ethyl xanthate in 100 g/Mg.

Table 2. The evaluation of the effectiveness of sandstone copper ore flotation process including three taxonomic measures

Collector type**	Collector dosage, [g/Mg]	Flotation time, [min]	d_j	z_j	w_j
X	100	1	**1,4868***	**0,5440**	**0,2846***
H	100		1,5172	0,4566	0,2529
X	150		1,6778	**0,5799***	0,1814
H	150		1,7239	0,1295	0,0119
X	100	12	1,5863	0,6398	0,2632
H	100		**1,4698***	**0,7924***	**0,3376***
X	150		1,7062	0,7344	0,2391
H	150		1,7024	0,5923	0,1906
X	100	30	1,5845	0,6807	0,2721
H	100		1,4141	0,7579	0,4216
X	150		**0,9419***	**0,9218***	**0,8147***
H	150		1,6950	0,6434	0,2123

highlighted values indicate the best course of the flotation process
** symbols: H-collector of the izobutyl ksanthate family Hostaflot, X-collector of the ethyl ksanthate family

Table 3. The assessment of flotation results for size fractions

Size fraction, [μm]	d_j
40–71	1,329741
100–125	1,640433
71–100	1,661561
20–40	1,678242
0–20	1,679803
125–200	1,689951

Based on the results of the MCA conducted by means of different methods, it can be stated that both the Euclidean distance and the relative development index allow to draw the same conclusions. Certain differences were observed only in one case, using zero unification method to evaluate the overall phenomenon.

5 Conclusions

Undoubtedly, such factors as the type and dosage of the collecting reagent constitute very important features affecting the efficiency of copper ore flotation. This is due to the lack of natural hydrophobicity of copper minerals, which, under natural conditions, are unable to produce durable aggregates of air-bubbles to be floated into a froth.

The results of the sandstone copper ore flotation test were analyzed for four variables: copper content in concentrate, copper content in tailings, copper recovery in concentrate, copper recovery in tailings. By analyzing the raw experimental data, we could see the apparent differences between the values of β, ε, η in the individual products of the partition which provided the basis for the preliminary conclusions. Despite the differences in the dosage and type of collector, the υ remained at a similar level. Minor fluctuations of the metal content in the tailings significantly affect the efficiency of the whole copper ore enrichment. A complete evaluation of the variables was performed using three taxonomic methods (d_j, w_j, z_j) to compare the results for the variable dosages and types of collectors. Based on the taxonomic measures, it was found that the best enrichment effects were obtained with 150 g/Mg sodium ethyl xanthate aqueous solution at extended flotation time (30 min). The results obtained in three ways gave consistent conclusions, despite various computational techniques and approaches to the analysis of the phenomenon.

Acknowledgements

The paper is the effect of the scientific project no. 15.11.100.021.

References

1. C. Aldrich, *Exploratory analysis of metallurgical process data with neural networks and related methods*. Process Metallurgy 12. Elsevier Science B.V., Netherlands (2002)
2. T. Borys, *Elements of quality theory*. PWN, Warsaw (1980)
3. J. Drzymała, *Mineral processing*. Foundations of theory and practice of minerallurgy. Oficyna Wydawnicza PWr, Wroclaw (2007)
4. M. Duchnowska, E. Kasińska-Pilut, A. Bakalarz, K. Konieczny, P. Kowalczuk, A. Łuszczkiewicz, *Multivariate statistical analysis of upgrading copper ore in ZWR Polkowice*. CUPRUM, 2(75), 97–108 (2015)
5. D. Foszcz, *Results of determining the optimal results of multicomponent copper ores beneficiation*. Mineral Resources Management, Cracow (2013)
6. D.W. Ginsberg, W.J. Whiten, *Cluster analysis for mineral processing applications*. Trans. Instn. Min. Metall., 100, 139–146 (1991)
7. D.W. Ginsberg, W.J. Whiten, *Expert system development using ellipsoid-based clustering*. Minerals Engineering, 6(1), 31–40 (1993)
8. M. Jarocka, *The choice of a formula of the data normalization in the comparative analysis of multivariate objects*. Economics and Management, 1, 113–126 (2015)
9. J. Kurkiewicz, J. Pociecha, K. Zając, *Methods of multidimensional comparative analysis in demographic development research*. SGH ISiD, Monograph 336, Warsaw (1991)
10. M. Łuniewska, W. Tarczyński, *Methods of multidimensional comparative analysis on the capital market*. PWN, Warsaw (2006)
11. T. Niedoba, *Multidimensional characteristics of random variables in description of grained materials and their separation processes*. Mineral Resources Management, Cracow (2013)
12. M. Nieć, A. Piestrzyński, Monograph KGHM PM SA II. 2.19 *Mineralisation*. KGHM CUPRUM Sp. z o.o. CBR, Wroclaw (2007)
13. B. Pawełek, *Methods of normalization of variables in comparative studies of complex economic phenomena*. UE, Cracow (2008)
14. A. Potulska, *An influence of fine grinding operations on flotation effectiveness of domestic copper ores*. Doctoral thesis; Politechnika Wrocławska, Wydział Geoinżynierii, Górnictwa i Geologii, Wrocław (2008)
15. D. Saramak, A. Krawczykowska, A. Młynarczykowska, *Effects of high pressure ore grinding on the efficiency of flotation operations*. AMS, 59(3), 730–741 (2014)

Evaluation of Gas Phase Dispersion in Flotation under Predetermined Hydrodynamic Conditions

Anna Młynarczykowska[1,*], *Konrad* Oleksik[1], and *Klaudia* Tupek-Murowany[1]

[1] AGH University Science and Technology, Faculty of Mining and Geoengineering, A. Mickiewicza Av. 30, 30-059 Krakow, Poland

Abstract. Results of various investigations shows the relationship between the flotation parameters and gas distribution in a flotation cell. The size of gas bubbles is a random variable with a specific distribution. The analysis of this distribution is useful to make mathematical description of the flotation process. The flotation process depends on many variable factors. These are mainly occurrences like collision of single particle with gas bubble, adhesion of particle to the surface of bubble and detachment process. These factors are characterized by randomness. Because of that it is only possible to talk about the probability of occurence of one of these events which directly affects the speed of the process, thus a constant speed of flotation process. Probability of the bubble-particle collision in the flotation chamber with mechanical pulp agitation depends on the surface tension of the solution, air consumption, degree of pul aeration, energy dissipation and average feed particle size. Appropriate identification and description of the parameters of the dispersion of gas bubbles helps to complete the analysis of the flotation process in a specific physicochemical conditions and hydrodynamic for any raw material. The article presents the results of measurements and analysis of the gas phase dispersion by the size distribution of air bubbles in a flotation chamber under fixed hydrodynamic conditions. The tests were carried out in the Laboratory of Instrumental Methods in Department of Environmental Engineering and Mineral Processing, Faculty of Mining and Geoengineerin, AGH Univeristy of Science and Technology in Krakow. **Key words** – cell flotation, air bubbles distribution, bubble size distribution, empirical model.

1 Introduction

In the process of flotation air bubbles play an essential role which allow mineral particles to rise towards the flotation concentrate, and their size, quantity, and quality determine the final enrichment effect. Due to the fact that the mineralization of air bubbles is a prerequisite for the flotation act, from a practical point of view, the process of stable particle-gas bubbles aggregates formation in the enrichment chamber is a crucial research problem. Because the flotation process takes place under turbulent conditions, the results of the enrichment depend on a number of random factors [1]. The major factors include the impact of a mineral particle with the air bubble, their persistent adhesion, as well as the possible detachment process that results in the disintegration of persistent aggregates of particle-gas bubble [2]. Because

* Corresponding author: mindziu@agh.edu.pl

those events occur at random we may discuss the probability of such an occurrence which influences the flotation velocity.

There are numerous models which connect the flotation constant speed with the phenomena inside the enrichment chamber. One of the models developed by Yoon and Luttrell [3] is the probability of a particle colliding with a gas bubble, which is dependent on the quotient of their size and represented by the formula [4]:

$$P_c = A\left(\frac{D_p}{D_b}\right)^2 \tag{1}$$

A – constant, dependent on the Reynolds number value for the bubble,
D_p – particle diameter,
D_b – bubble diameter.

Bear in mind the probability of such a collision will be a random variable due to the fact that the size of the objects are also random variables [4]. Another model developed by Varbanov, Forssberg and Hallin [5] is the correlation between the flotation reaction speed and the angle of wetting the hydrophobic graig, grain and bubble diameter, as well as gas flow rate in the flotation chamber, and the chamber cross-sectional area, which has been represented by the formula:

$$k = \frac{3\,Q\,R_p\,(1-\cos\theta)}{4\pi\,S\,R_b^{\,2}} \tag{2}$$

where:
θ – equilibrium contact angle of the particle,
Q – gas flow rate in the flotation chamber [m^3/s],
S – flotation chamber cross-sectional area,
R_b – gas bubble radius,
R_p – particle radius [4].

Thus, it can be seen that the key factor influencing the achieved effects of enrichment is the size of particles and gas bubbles, and thus the mineralization of air bubbles, which depends on a number of hydrodynamic and physicochemical parameters prevailing in the enrichment work space. Therefore, it is necessary to analyze the dispersion properties of the gas phase, namely: the distribution of the size of the resulting gas bubbles [6]. If the gas bubble would be too small and completely mineralized, its buoyant force would be much less than the gravity of the resulting aggregate, which would lead to its drowning [4]. Because the size of the resulting air bubbles depends on the surface tension of the solution, which varies depending on the type and amount of added surfactant [7] as well as the characteristics of the flotation machine such as the amount of air being dispensed per unit of time, and rotor speed. Brozek and Mlynarczykowska [4] have defined a formula for calculating the minimum size of the bubble:

$$D_{b\min} = \frac{\pi(\rho_s - \rho_c)\,D_p}{\rho_c} \tag{3}$$

where:
D_b – bubble diameter,
ρ_s – density of particle under flotation,

ρ_c – liquid density,

D_p – particle diameter.

Thus the minimum bubble size depends on the size and density of the floated mineral particles. Nevertheless, it should be noted that during the process the particle-bubble aggregate is under constant central resistant force, which will result in the calculated value of the minimum value of the gas bubble size should be slightly higher [4].

Determining the appropriate size of the air bubble for a given material allows you to select the appropriate operating conditions of the floatation machine and facilitate the interpretation of the enrichment process under certain hydrodynamic and physicochemical conditions [7]. The likelihood of attaching the flotation particles to the air bubble increases as the dimensions of the air bubble increase. Hence, it can be concluded that the degree of aeration of the flotation suspension is most advantageous when the optimal saturation of the suspension with air bubbles is obtained, with maximum dispersion [8]. As a result, the size of air bubbles and airflow affect the effectiveness of the flotation process [9–11]. Smaller gas bubbles reduce the probability of detaching the solid phase from the gas phase and increase the probability of collisions and adhesions, while larger gas bubbles ensure that the aggregates of the particle-gas bubbles ascend [11].

The size of the fed particles and the gas bubbles generated in the flotation chamber are distributed in a certain type, which are useful for analyzing the flotation process. The distribution of these quantities and the state of parameters depend on the following factors: rotor speed, amount of air supplied per time unit, method of air delivery to the flotation chamber (depending on the construction / type of the flotation machine) and surface tension of the flotation solution. Therefore, the size distribution of air bubbles depends on how they are formed and under turbulent conditions is log-normal. This is confirmed by studies in which the log-normal distribution best described the obtained results of bubble size measurements [12].

Grau and Heiskanen [13] found that the best distribution of gas bubbles was the "upper-limit" distribution, which is a modification of the log-normal distribution. In contrast, Tavlarides and Stamatoudis [14] have shown that log-normal distribution seems to be the most common when describing the distribution of drops size in liquid-liquid dispersion.

The paper presents the results of analyzes of gas phase dispersion in the flotation chamber according to the distribution of air bubbles at specified hydrodynamic conditions. Based on the obtained results Feret and Sauter's diameters were calculated and the histograms along with the three types of distributions, namely Logarithmic, Weibull and Rayleigh, were plotted.

2 Measurement methodology

Measurements were made at the Instrument Methods Laboratory of the Department of Environmental Engineering and Material Processing of the Faculty of Mining and Geoengineering of the AGH University of Science and Technology in Krakow, using the FlotMin test stand [6, 7].The flotation machine has the ability to automatically control the airflow, rotor speed, maintain constant fluid level and the speed of the scraper, remotely using a specialized Fix software. The study used a cuboid flotation chamber with a symmetrical structure [6]

After filling the working chamber of the flotation machine with water so that the liquid level fills the "Luba Tube" system and determining certain parameters of the flotation machine operation in the form of rotor speed and airflow, measurements were made for the determined sectors of the working chamber of the flotation machine (0, A, B, C). Once the gas streams had flowed from the tubes, the air bubbles were broken apart by the rotor blades of the flotation machine. Then the resulting air bubbles were directed from the selected flotation chamber area using the "Luba Tube" system to the chamber where their image was recorded using the MOTICAM 2300 camera. Since the working chamber is symmetrical (Fig. 2), it is sufficient to test the air bubbles in one of the quarters, with the possibility of referencing the results to the entire space of the flotation chamber.

The image was recorded using the camera and divided into single frames from which the best quality pictures were selected. Selected images were subjected to computer image analysis to determine their size (i.a. Feret's bubble diameter, bubble circumference, bubble surface area). Professional software for qualitative and quantitative image analysis was used for this purpose. The analysis was performed for images at 5 seconds interval.

Thanks to the recording system operation optimization, a quality image suitable for further analysis was obtained. Figure 1 shows: on the left is a photo selected for further analysis, on the right is the same photo with the image analysis already done with the SigmaScan Pro image analysis software.

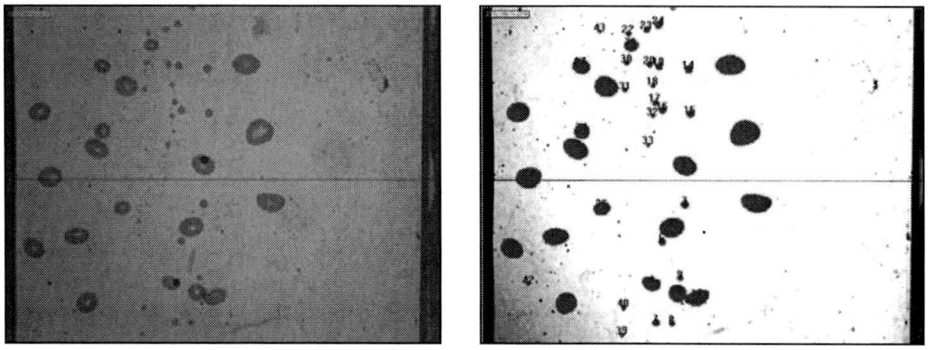

Fig. 1. Sample photo selected for analysis in an image analysis program and the same photo after processing.

3 Results

On the basis of the available literature and own analysis of the collected experimental data for mathematical description together with the graphical presentation (Figs. 4–6) of the results obtained using the STATISTICA program, three types of distributions were selected:

- Log-normal,
- Weibull,
- Rayleigh.

The experiment was carried out for two operating parameters of the flotation machine. All of the tests changed the airflow rate (24 m^3/h and 240 m^3/h) and the rotor speed was constant (1000 rpm). However, in case of measurements performed for cell 0, in addition to the airflow rate, the rotor speed (1000 rpm and 1350 rpm) has also been changed so the results obtained from this cell are more numerous and slightly different compared to the other research areas.

Using image analysis software based on two-dimensional images, it calculates the Feret's diameter from the formula:

$$d = \sqrt{\frac{4A}{\pi}} \tag{4}$$

where: A – predicted surface of the bubble in a two-dimensional plane.

Based on experimental data, Sauter's mean diameter of gas bubbles was calculated using the formula:

$$d_{32} = \frac{\sum d^3}{\sum d^2} \tag{5}$$

Sauter's mean diameter, also referred to as the volume-surface diameter, is expressed in terms of the volume-to-surface ratio, which allows the average diameter of a set of objects of different sizes to be calculated [15]. Therefore, it is an ideal tool for determining the average size of air bubbles dispersed in the flotation chamber.

Table 1 shows the mean Feret's diameter (d_0) database computed using the SigmaScan Pro software and the statistical data for various cell flotation sectors and the operating parameters, as well as the calculated Sauter mean diameter (d_{32}).

Table 1. The results in each section

Measurements number	Std. dev. of Feret's diameter	Max. Feret's diameter d_{max} [mm]	Min. Feret's diameter d_{min} [mm]	Mean Feret's diameter d_0 [mm]	Mean Sauter's diameter d_{32} [mm]	Rotor speed (ω) [rpm]	Gas flow-rate [dm³/h]	Section
1449	0,8223	4,7555	0,2393	1,0014	2,3596	1000	24	"0"
650	1,0504	5,543	0,3696	2,3028	3,1397	1350	240	
1901	0,7476	4,4689	0,2816	1,553	2,1993	1000	24	"A"
2008	0,7404	6,0885	0,6011	2,8139	3,1933	1000	240	
860	0,5984	4,2718	0,474	1,6728	2,1231	1000	24	"B"
1406	0,7459	5,1487	0,7683	2,6557	3,0633	1000	240	
890	0,8464	4,7615	0,3983	1,8435	2,6038	1000	24	"C"
1262	0,8670	6,3311	0,9934	3,2678	3,6930	1000	240	

Histograms of Feret diameters for air bubbles together with the adjustment of the log-normal, Rayleigh and Weibull probability distribution for the flotation with:

- Fig. 2. Section 0 gas flow-rate 24 [dm³/h], rotor speed 1000 [rpm];
- Fig. 3. Section 0 gas flow-rate 240 [dm³/h], rotor speed 1350 [rpm];
- Fig. 4. Section A gas flow-rate 24 [dm³/h], rotor speed 1000 [rpm];
- Fig. 5. Section A gas flow-rate 240 [dm³/h], rotor speed 1000 [rpm];
- Fig. 6. Section B gas flow-rate 24 [dm³/h], rotor speed 1000 [rpm];
- Fig. 7. Section B gas flow-rate 240 [dm³/h], rotor speed 1000 [rpm];
- Fig. 8. Section C gas flow-rate 24 [dm³/h], rotor speed 1000 [rpm];
- Fig. 9. Section C gas flow-rate 240 [dm³/h], rotor speed 1000 [rpm].

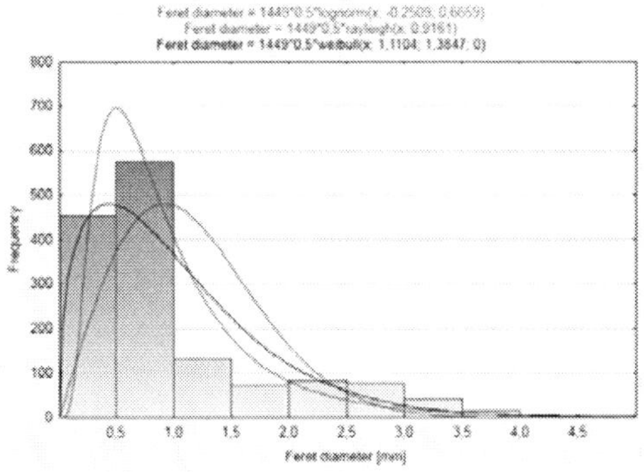

Fig. 2. Section 0 gas flow-rate 24 [dm³/h], rotor speed 1000 [rpm]

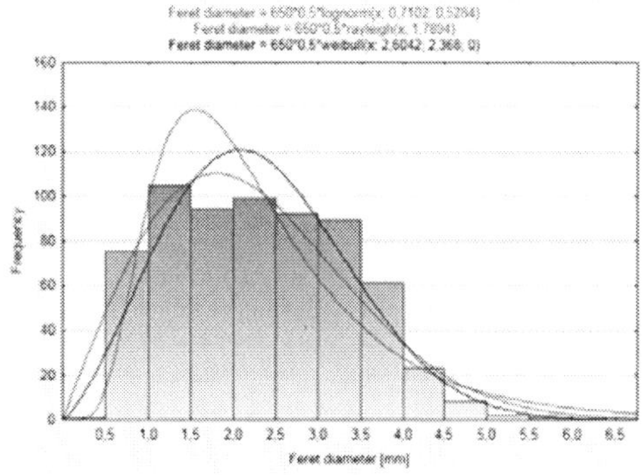

Fig. 3. Section 0 gas flow-rate 240 [dm³/h], rotor speed 1350 [rpm]

https://doi.org/10.1051/e3sconf/20183501011

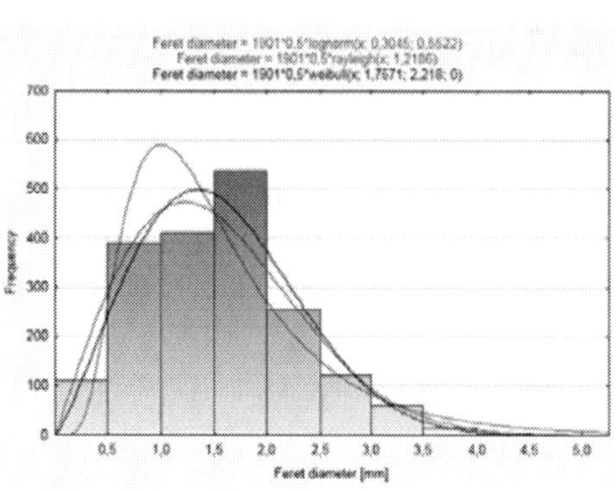

Fig. 4. Section A gas flow-rate 24 [dm^3/h], rotor speed 1000 [rpm]

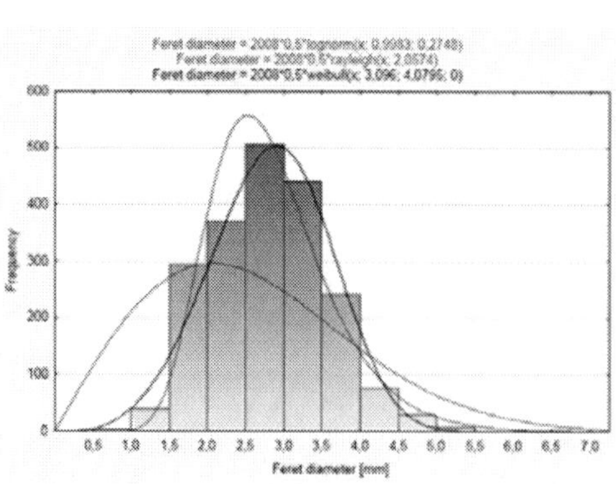

Fig. 5. Section A gas flow-rate 240 [dm^3/h], rotor speed 1000 [rpm]

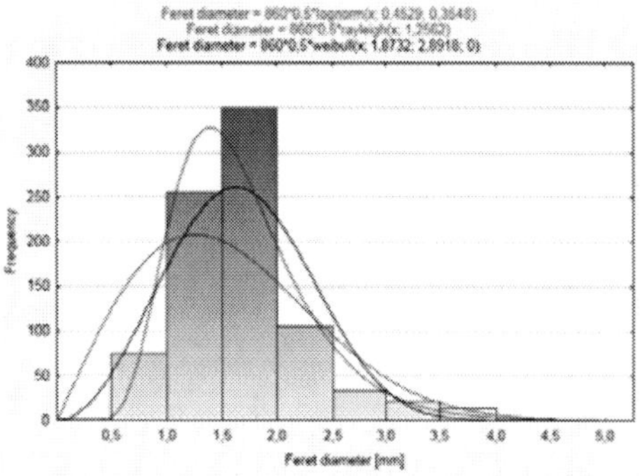

Fig. 6. Section B gas flow-rate 24 [dm^3/h], rotor speed 1000 [rpm]

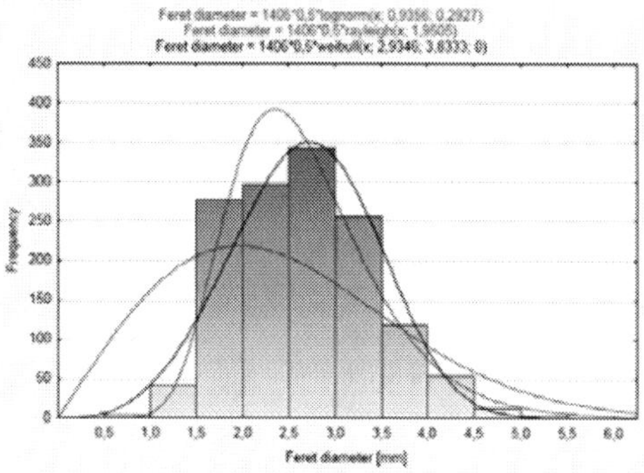

Fig. 7. Section B gas flow-rate 240 [dm^3/h], rotor speed 1000 [rpm]

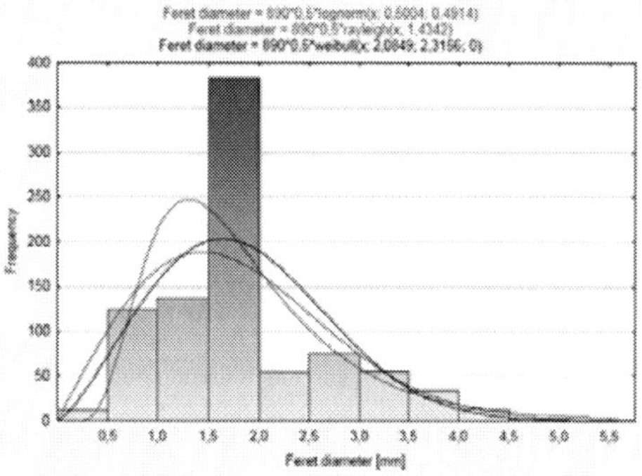

Fig. 8. Section C gas flow-rate 24 [dm^3/h], rotor speed 1000 [rpm]

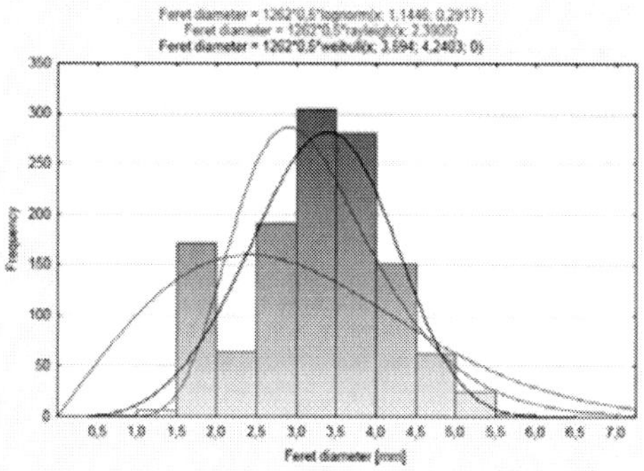

Fig. 9. Section C gas flow-rate 240 [dm^3/h], rotor speed 1000 [rpm]

4 Summary

By analyzing the distribution of determined values for individual sectors, it may be claimed that the bubbles mean Feret's diameter at minimum aeration of 24 dm^3/h of the working chamber is in the range of 1.0014 mm to 1.8435 mm and at maximum aeration of 240 dm^3/h is higher and in the range of 2.3028 mm to 3.2678 mm. Under the same conditions, the largest population was observed for cell A and the smallest for cell 0.

At the maximum aeration of the chamber (240 dm^3/h) the maximum number of bubbles reaches the size of Feret's diameter from 1.5 to 4.0 mm, and it can be concluded that the air bubbles produced are stable and have a stable shape similar to a ball. With a minimum aeration of 24 dm^3/h, the vast majority of bubbles are within the Feret's diameter range of 0.5 mm to 3 mm. The widest range of Feret's diameter values is among the bubbles in sector 0, which, combined with a small population, indicates that a small, poorly dispersed portion of air has entered the flotation chamber.

Taking into account the values of the designated Sauter's and Feret's mean diameters for individual sectors of the flotation chamber, a similar increase in the air bubble diameter size along with increasing aeration can be noted. In the case of minimal aeration of the work chamber of 24 dm^3/h, Sauter's mean diameter is in the range of 2.1231 mm to 2.6038 mm, while with aeration of 240 dm^3/h from 3.0633 mm to 3.6930 mm. Thus, it is noted that in both cases of the enrichment chamber aeration the smallest Sauter's mean diameter was obtained for cell B and the largest for cell C, where the population value varied slightly.

Charts 2 to 9 present histograms along with 3 distributions (logarithm-normal, Weibull, Rayleigh) selected from available literature and own analysis. Experimental distributions were verified with empirical distributions and their compatibility with the Chi-squared test was contrasted using the STATISTICA program. Graphs comparing experimental and empirical distributions and in the probability-probability system were also compiled.

Due to the large number of verifications carried out in the article, only two such graphs for cell A were shown as examples (the chamber aeration was 240 dm^3/h, rotor speed 1.000 rpm) – Figures 10 and 11.

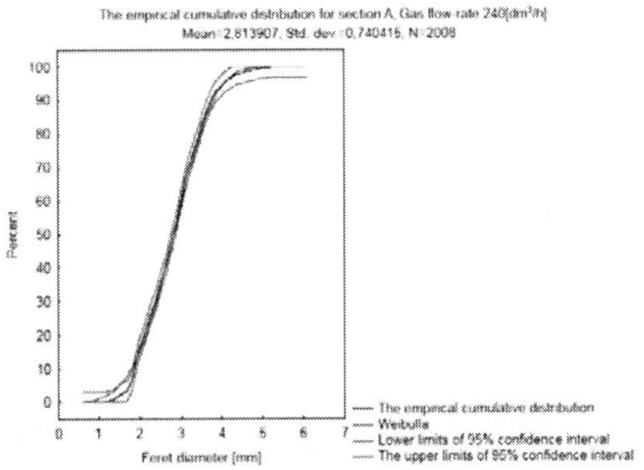

Fig. 10. The empirical cumulative distribution for section A, gas flow-rate 240 [dm^3/h]

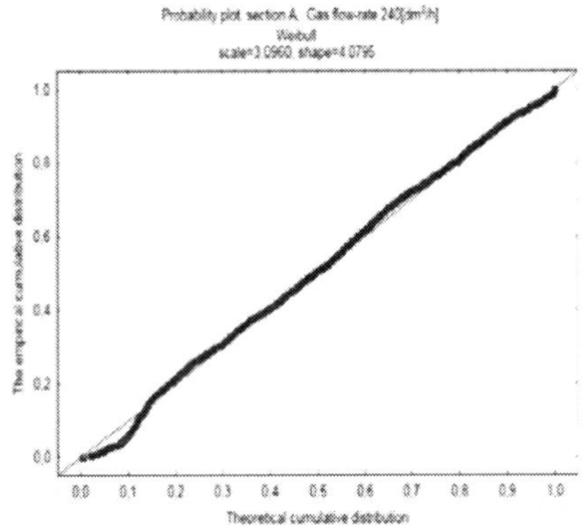

Fig. 11. Probability plot for section A, gas flow-rate 240 [dm^3/h]

The results of the verification and the tests are ambiguous, because in 4 cases (Figs. 2, 6–8) the logarithmic distribution fits to the real distributions best, and in other 4 cases (Figs. 3–5, 9). Weibull distribution seems to be a match. However, for the extreme sectors 0 and C, we can observe that the best fit for the small aeration of 24 dm^3/h is with the logarithmic distribution, while for high aeration of 240 dm^3/h Weibull distribution seems to be most matching. In addition, it is worth noting that differences in matching of these distributions are insignificant.

5 Conclusion

The performed analysis of the results and verification of the analyzed distributions confirms some of the earlier reports [12–14] of a good reflection of the experimental data by log-normal distribution and Weibull distribution.

On the basis of the results obtained, it can be stated that:

- the increase in rotor speed results in a decrease of gas bubbles dimensions and at the same time increases their amount,
- at increased rotor speed, increasing the aeration results in an increase in the size of the gas bubbles,
- aeration intensity influences the shape of the resulting air bubbles (the greater the airflow, the more irregular the shape of the air bubbles formed).

Acknowledgments

This work was done as a part of the University of Science and Technology Research Program No. 11.11.100.276. This article is an extension of the presentation at the MEC 2016, 25–28 September 2016, Swieradow-Zdroj, Poland

References

1. D. Saramak, A. Mlynarczykowska, A. Krawczykowska, Archives of Metallurgy and Materials, **59(3)**, 951–955 (2014)
2. S. Blaschke. *Przeróbka mechaniczna kopalin, Część II.* Wydawnictwo Śląsk, Katowice (1984)
3. R.H. Yoon, G.H. Luttrell, Miner. Process. Extr. Metall. Rev. **5**, 101–122, 751–756 (1989)
4. M. Brozek, A. Mlynarczykowska, Arch. Min. Sci., **57**, 3, 729–740, (2012)
5. R. Varbanov, E. Forssberg, M. Hallin, Int. J. Miner. Process., **37**, 27–43 (1993)
6. A. Mlynarczykowska., A. Nyrek, K. Oleksik, IM, **1(35)**, 181–188 (2015)
7. A. Mlynarczykowska, K. Oleksik, IM, **18, nr 1**, 257–262 (2017)
8. S. Blaschke. *Procesy technologiczne w przeróbce kopalin użytecznych.* Wydawnictwo „AGH", Kraków (1987)
9. F.F. Lins, R. Adamian, *Minerals Engineering*, 6, 267–277 (1993)
10. H.J. Schulze, *Coagulation and flocculation theory and applications*, B. Dobias (Ed.). **Chapter 7.** *Flotation as a heterocoagulation process: Possibilities of calculating the probability of flotation.* Marcel Dekker Inc., New York (1993)
11. D. Tao, Sep. Sci. Technol. **39(4)**, 741–760 (2004)
12. L. Vinnett, F. Contreras, J. Yianatos, 2011. *Gas dispersion pattern in mechanical cells.* Mineral Engineering, 26, 80–85 (2012)
13. R.A. GRAU, K. Heiskanen, Minerals Engineering, **18**, 1164–1172 (2005)
14. L. Tavlarides, M. Stamatoudis, Advances in Chemical Engineering **11**, 199–273 (1981)
15. P.B. Kowalczuk, J. Drzymała, *Particulate Science and Technology*, **vol. 34**, issue 6, 645–647 (2016)

Neotectonic and Geomorphological Characteristics of the Ma Pi Leng Area (Ha Giang Province, North Vietnam): Implications for Improving the Worth of Vietnamese Geological Heritage

Thom Van Bui[1,*], *Thinh* Huy Nguyen[1], *Tung* Duc Nguyen[1]

[1] Institute of Geological Sciences, Vietnam Academy of Science and Technology, 84 Chua Lang Str., Dong Da, Hanoi, Vietnam; buivanthom@gmail.com

Abstract. Ma Pi Leng area is the core part of Dong Van karst plateau geopark, and is a region of high diversity of geology and tectonics in Vietnam. This region has experienced strong uplift motions that have created three grades of regional relief: 1400–1600 m; 1100–1200 m and 500–600 m. This geomorphology is in turn strongly eroded by the stream systems and divided into different structural blocks controlled by the faults. During the neotectonic stage, several modes of faulting have occurred in Ma Pi Leng such as strike-slip, normal, extension, riverse faults. The NW–SE fault system had a decisive role in determining the structural pattern of the region. The tectonic fractures, at the largest scale, are hundreds of meters high and up to a kilometer long. Tectonic activities, together with exogenic processes, have made regional relief strongly diverse with strips of horsts and grabens as well as ladder-steps-shaped escarpments, inverse topography, high cliffs and deep canyons. Besides that there are also the caves at varying levels of height, chains of sinkholes, and pyramidal mountainous peaks. Ma Pi Leng area is truly majestic and worthy of tourist attractions both in terms of scientific value and natural landscape. **Key words** – Ma Pi Leng, neotetonics, active fault.

1 Introduction

The Ma Pi Leng area (MPL) of Vietnam is situated within 105°17′22″ and 105°29′13″ longitude, 23°10′05″ and 23°20′22″ latitude (Fig. 1). It covers a part of the Dong Van and Meo Vac districts, and is the central part of Dong Van plateau in the northeast of Ha Giang province. Generally Dong Van, Meo Vac districts, and the MPL area in particular, have a special tectonic position. Lying in the southern edge part of the South China block, tectonic activity is quite stable in the central part, but its edge part is significantly disordered [5]. On the other hand, the MPL area is connected with the northeast edge of Indosinian moutainous belt by the Red River Fault Zone, and so the MPL is also affected by common tectonic

* Corresponding author: buivanthom@gmail.com

processes. Because of these reasons, the MPL area has a specific geomorphological landscape and is very diverse in topographical, geological and tectonic form [1, 2]. However, it is still not paid due attention regarding its geomorphological and tectonic activities. Moreover, the MPL has been considered as a global geopark, and therefore it is extremely necessary to elucidate its tectonic characteristics for the purpose of improving the worth of geological heritage, and value for tourism and scientific activities.

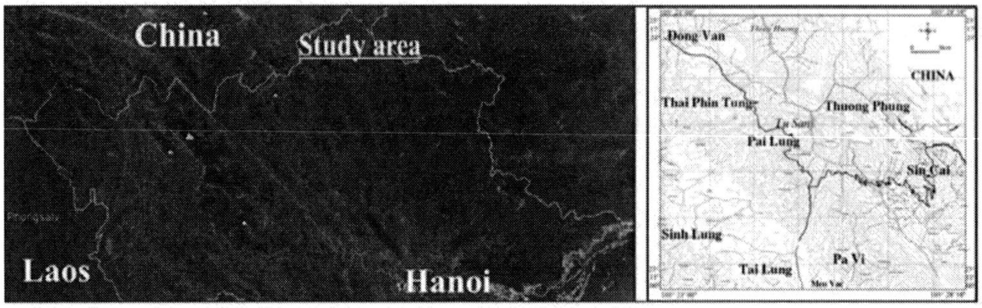

Fig. 1. Sketch-map showing the location of the study area

This paper presents new results on tectonic activities and their worth on geological heritage in the MPL.

2 Materials and methods

The present investigation has been conducted based on observations at 115 geological exposures in order to document the geomorphological, geological and tectonic data, with field work conducted in 2016. A number of important features have been systematically documented, including: geological and geomorphological anomalies (deformation of geological strata, change of river and spring channels, watershed line, alluvial fan, terraces, battures along faults and their distribution) and tectonic structural patterns. Based on the method of analysis of tectonic physics [9], stress field and characters of fault activity were reconstructed. In addition, our field-based studies were supported by powerful satellite images with high solution (landsat and Spots), maps of DEM (scale 1: 100000 and 1: 50000), and geological maps (scale 1:100000 and 1: 200000).

3 Results

3.1 Geomorphological characteristics

3.1.1 Category of relief

The MPL area includes three main topographical types: *Erosion-solution mountainous relief* in massive form developes on limestones, observed in the northwest of the study area, represented by caves, sinkholes, and solution grooves on its surface; *Erosion-solution mountainous relief* in linear-massive form develops on terrigenous rocks and thinly-bedded limestones, encountered in the central part of the study area. It is very common, presented as verticle-sloped cliffs, and small-scale caves; *Erosion-denudation mountainous relief* with lenearment-form is distributed in the NE part of the study area, and created in large ranges with the NW–SE direction. It is typical by linear and cuesta topography with peaks being cone-shaped, pyramid-shaped in form.

3.1.2 Form of topographic surfaces

- *Planation surfaces*: The MPL area has three levels of topography corresponding to three different planation surfaces: 1400–1600 m; 1100–1200 m and 500–600 m. Planation surfaces which are developmental on limestone rocks are composed of mountainous peaks with roughly similar elevation. The other with flat surfaces are narrow in width and the NW–SE elongated trends are developed by terrigenous rocks.
- *Caves*: These are concentrated in the SW part of our study area (Figs. 2A, B, C). There are two levels of cave at altitudes of 1000–1200 m and 700–800 m. They are presently dry and inactive. Caves that are located at altitudes of 500–600 m are currently developing and their surface contains active water streams.

Fig. 2. A – Cave at 600 m with active water streams; B – Dry cave at 800 m; C – Dry cave at 1200 m

- *Sinkholes:* They have circular and ellipsoidal shapes. Their bottoms with small-sizes are cavities. They are distributed in large ranges from 200 m to several kilometers according to the NE–SW, sublongitude and sublatitude direction to creat a karstic valley.
- *Tectonic cliffs*: They are built not only by limestones rocks, but also by the terrigenous rocks. Cliffs which are developed in limestones are of several meters to tens of meters in height and tens of meters to hundreds of meters in length, creating a series elongated with a NE–SW, NW–SE, sublongitudinal and sublatitudinal directions. The Nho Que river valley is the most typical example of this morphology, and nearly 800 m in height and 400 m in length.
- *Rock desert:* The whole area has been strongly uplifted, together with a dense development of fissures and faults, and differences in the altitude of erosional basement; therefore, it is completely exposed by bedrocks in the dry seasons.
- *Residual mountains:* In the MPL, topographical form clearly represents geological structures and tectonic activity. Cuesta relief is displayed by the monoclinal *attitude* of associated rock beds, while the others form roof-shaped, cone-shaped, ladder-step and pyramid-shaped forms, reflecting the presence of tectonic faults.
- *River terraces:* The second-order terraces are encountered in the Meo Vac town and Pa Vi commune. They are hills with flat peaks, 5–7 m in height, composed of polimictic conglomerates interbeded with sandstones and clayey sandstones. The first-order terraces are distributed in Meo Vac, Dong Van towns and along sides of the Nho Que river. They are situated at 2–3 m in height, and exist in the form of small and narrow plains along sides of rivers and their tributaries. They are built by sand and clayey sand.
- *Alluvial battures:* These are distributed at a height of 0.5–1 m along the Dong Van spring-sides and the Nho Que river-sides. They are built by gravel and coarse-grained sand.

3.2 Neotectonic characteristics

3.2.1 Characteristics of neotectonic structure

We have identified 4 Neotectonic structural zones in the MPL and adjacent area (Fig. 3):

- A very strongly uplifted zone with amplitude 1400–1600 m is located in the NE part of the study area in the NW–SE oriented range (Fig. 3). Its basement is built by PZ_1 rocks (Cambrian-Ordovician period) including claystones, limestones and *quartzites* [4]. During the Pliocene-Quaternary period, this structure was desconstructed strongly by faults. Currently, only residual mountains remain at elevations of 1400–1600 m and dry caves at 1300 m.
- A strongly uplifted zone with amplitude 1100–1300 m is located in the SE part of the study area, oriented the NW–SE direction (Fig. 3). Its basement is composed of PZ_3 rocks (Carboniferous-Permian age) including thickly bedded and massive limestones [4]. During the Pliocene-Quanatery, this zone was strongly uplifted. Currently, it is presented by a large range of surfaces and dry caves at an elevation if 900–1200 m with openings of 10–20 m wide and 15–30 m high (Fig. 2C).
- A medium uplifted zone with amplitude 450–600 m is located in the central part of the study area (Fig. 3). The basement is built by PZ_2 rocks (Devonian age), including claystones, siltstones, clayish limestones and thinly-bedded limestones [4]. During the Pliocene-Quaternary, this structure was moderately uplifted, resulting in a massive-linear form. Currently, its surface is at 450–800 m. Solution-leaching and erosion-denudation are precise processes thet created the NW–SE oriented cliffs and many active caves at around 450 m in height.
- Spreading-formed structure is evidenced in the Dong Van and Meovac valleys (Fig. 3). The Dong Van valley forms an acute triangle surrounded by ranges of high mountains and filled with Quaternary deposits. It was formed by lateral-slip movement of the Dong Van-Thuong Phung Fault. The Meo Vac valley is of a graben geometry and elongated in the roughly an N–S direction. Its cliffs are built by limestones. This valley is created by lateral-slip movement of the Thai Pin Tung-Pa Vi Fault.

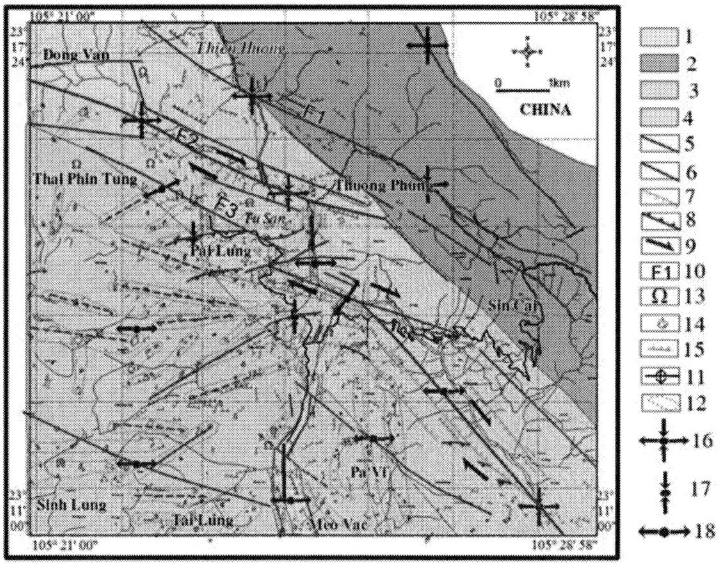

Fig. 3. Map of distribution of Neotectonic faults in the MPL and adjacent area: 1 – Structure of medium uplift in massive form with amplitude 500–800 m; 2 – Structure of strong uplift in massive form with amplitude 1000–1300 m; 3 – Structure of very strong uplift in arch-line form with amplitude 1400–1600 m; 4 – Spreading-formed structure; 5 – Main fault; 6 – Subsidiary fault; 7 – Normal fault; 8 – Inverse fault; 9 – Direction of movement of fault; 10 – Name of fault; 11 – Spreading axis; 12 – Karst valley; 13 – Karst cave; 14 – Sinkhole; 15 – Tectonic cliff; 16 – Lateral stress; 17 – Compression stress; 18 – Spreading-formed tress

3.2.2 Characteristics of tectonic fault

In the study area, main faults are orientated the NW–SE direction, playing an important role in controlling Neotectonic structures. The other faults are developed in the sublongitudinal, NE–SW and sublatitudinal directions (Fig. 3).

- The Thien Huong-Sin Cai Fault (F1) is oriented the NW–SE direction (130°). It is inclined to the NW direction and coincides with fissile surfaces of calcareous-terrigenous rocks. It shows evidence of dextral strike-slip faults (Fig. 4A). Its NE side is uplifted more strongly than the SW one. Sub-faults cut through ranges of mountains to create step-formed terrains, which decrease in height in the SW direction. Each topographic step is seperated by a vertical-sloped rock cliff. In the Sin Cai village, there is a compression zone approximately 10–15 cm wide, and which contains clay reflecting the young activity of the F1 (Fig. 4B).

Fig. 4. Tectonic evidences in the study area: A – System of faults with the NW–SE direction (Thuong Phung village); B – Compression zone in the NW-SE direction (Sin Cai village); C – Structural pattern reflects right lateral-slip movement of the F2; D – Inverse terrain; E – The Nho Que valley; F – Shear surface and striations in the sublongitude direction (Pa Vi village); G – The Meo Vac valley on the satellite image; H – Subsidence zones with sublongitude direction; I – Movement of faults in the NE–SW direction

- The Dong Van-Thuong Phung Fault (F2) is oriented in the NW–SE direction (115–130°) (Fig. 3). On the satellite images it represents a band with approximately 1000 m wide. Its sub-faults cut the topographic surfaces to form a complex of horsts and grabens, which are around 200–300 m in width. Besides this, the fault is also represented by young firsts-order springs with strongly eroded V-shaped channels without Quaternary deposits. Along the fault zone, a chains of sinkholes has developed with vertical cliffs. Evidence for the Dong Van trough is from the strike-slip activity of the F2. The main fault combines with the roughly S–N faults to form structural patterns deflecting a right-lateral-slip mechanism of the F2 (Fig. 4C).

- Thai Phin Tung-Pa Vi fault (F3) represents segmental feature according to different trends (120°, 110°, 135°) (Fig. 3). It is recognized by a number of features, such as canyons, karst valleys, ranges of sinkholes, intensive density of fractures, limestones cliffs of several hundreds meters in height, grabens, shear surfaces, strongly broken strata and step-shaped reliefs, and land-slides. They represent a right strike-slip mechanism of this fault. Especially, an inverted relief was observed, which was created by the F3 cut through a large anticline. Currently, it is represented by the valley of the Nho Que river (Fig. 4E).
- The sub longitude oriented faults scatter in the MPL area, with length from hundreds of meters to several kilometres (Fig. 4F). They were recognized based on grabens infilling Quaternary deposits, slipped evidences and striations, presented normal separation or normal slip (Figs. 4E, G, H).
- The system of NE-SW and sub latitude oriented faults is about several kilometres long, concentrated mainly in the west part of MPL. They coincide with subsided belts. Both its sides are built by ranges of limestone mountains. On its floor appears many sinkholes. The NE-SW oriented fault represents normal sinistral strike-slip (Fig. 4I).

4 Discussion

During the Neotectonic period, uplift movement has created three main levels of relief corresponding to three planation surfaces at the different elevations: 1400–1600 m; 1100–1200 m; 500–600 m, respectively. Their ages coincide with three uplift phases, but this is still debatable. An (2008) has considered that two planation surfaces occurred in the Dongvan plateau: at elevations 1200–1400 m with late Miocene age; at 1500–1700 m with early-middle Miocene age. According to Bao (2008), the planation surfaces have been recognized at 1400–1600 m and 1200–1300 m during late Miocene; the other of 500–600 m and 800–1000 m were dated to Pliocene. However, the dating of planation surfaces has still been lacked satisfactory data.

Based on analyses of topographic sections and distribution of cave steps, three main levels of relief in the MPL were recognized, including: 1400–1600 m, 1100–1200 m, 500–600 m. By correlation between deposits in filling the Dong Van and Meo Vac valleys, formation of the levels of relief was elucidated. The valleys were filled with Quaternary deposits with 2–5 m in thickness, including alluvial and alluvial-proluvial sediments. In the Meo Vac valley, three sedimentary budgets were determined corresponding to three stages of terraces and battures forming along the Meo Vac river. Based on An's data (2008), we have agreed age of terraces and battures as the following: The second-order terrace was formed during the early-middle Pleistocene (Q_1^{1-2}); The first-order terrace occured in the late Pleistocene (Q_1^3); and battures were of the Holocene (Q_2). These ages of the terraces and battures are connected with the three topographic levels which were decribed above.

The MPL is situated in the S–W edge of the South China plate, which is linked with the Indosinian mountain belt by the Red River Fault Zone. This connection leads to activities of the faults being very complicated and affected by tectonic activities from the Indosinian terrane [5]. The faults are of small-scale because they are almost intra-plate destructions [6]. Earth-quake occasionally occurs as a result of this. During the Cenozoic, from 25 to 30 milions years ago, in Vietnam, tectonic activity is divided into two main phases [8]: Eocene-early Miocene phase and Pliocene-Quaternary phase. The evidence for first phase is given by strongly broken rock strata to form a cataclastic, brecciated zone. The second phase is represented clearly on current relief's surfaces by topographical bands, planation surfaces, cave steps and tectonic cliffs. Its stress-field has compression axis more or less in the N–S direction. The NW–SE oriented fault system is right-lateral slip, the NE–SW oriented one is left-lateral slip; and roughly S–N and E–W fault systems are of reverse slip. This is suitable with the rule of distribution of faults and geodynamic model in the MPL.

Therefore, reliefs in the MPL are clearly formed during Quaternary period, with some maybe forming earlier in the Pliocene. The Late Pliocene-Quaternary tectonic compression phase is divided into three sub-phases with different directions of uplift corresponding to the three relief's levels.

The NW–SE oriented fault system is of a large-scale, playing an important role in forming and controlling types of relief's architecture with distinct features: strongly uplifted architecture with massive form is orientated in the NW–SE direction in the NE part; and the massive-form one is in the SW part, commonly as a border between different geological formations such as PZ_1, PZ_2 and PZ_3. The activities of these faults also brought about typical architectures developed along the fault zones. For example, the Dong Van valley is formed by normal-lateral-slip mechanisms of the F2; The roughly N–S oriented valley of Nho Que river and the NW–SE oriented graben Tu San occurred due to normal-lateral slip mechanism of the F2; The graben Trang Huong is created by normal separation of the F2.

The sub-meridian faults are of a smaller scale. They are formed later, caused by the activity of NW–SE oriented faults. They are distributed in a system with parallel lineament-forms. They cut topographic surface to create two main types of relief, including ladder-step-formed terrains and graben. The NE–SW oriented faults are less-developed, but their activity is represented clearly by shear surfaces, tectonic striations, and topographic shift along both sides of faults. They are combined with the NW–SE oriented fault system to create a system of fractures and conjugate faults reflecting the sub-meridian oriented force.

5 Conclusion

Ma Pi Leng is a very diverse area in Geology-Tectonics, representing a large asymmetric syncline. It is fully reflected through old to young geological formations (Cambrian-Triassic formations) which were arranged in temporary and spatial orders. Many geological boundaries of this area were discovered in accordance with the international stratigraphic chart. During the Neotetonic period, the MPL area represented a strongly uplifted regime with unstable amplitude to create topographic orders at different elevations such as 1400–1600 m, 1100–1200 m and 500–600 m. It was strongly cut by rivers and streams leading to subdividing its area into three architectural blocks controlled by regional faults.

The faults developed in different directions with typical active features (strike-slip, normal and reverse). Tectonic fractures are approximately 100 m in height, 1000 m in length, and on the largest scales compared with other regions in Vietnam. Tectonic activities brought about a strong anomaly in terrain to form horsts, grabens, ladder-step-formed or inverted terrains, large cliffs, and deep valleys. Furthermore, there are cave steps, karst funnel-formed holes, pyramid-shaped and cone-shaped mountains. Tectonic activities, together with exogenic processes, have brought the MPL area a variety in geomorphologic landscape, deserved of an international geomorphological heritage.

References

1. L.D. An, *The Dong Van-Meo Vac plateau A precious geomophorgical heritage of karst. Dong Van stone plateau: Striking values, orientation for conservation and sustainable development Ha Giang*, 6–27 (2008)
2. D.V. Bao, *Geomorphology of the Dong Van Meo Vac area, striking values and orientations for preservation and rational use. Dong Van stone plateau: Striking values, orientation for conservation and sustainable development Ha Giang*, 81–99 (2008)
3. V.V. Chinh, *Some active faults during the Neo-tectonic period in the NE Vietnam*. Resour Geol, 1, 22–32 (1996)

4. H.X. Tinh (ed.), *Geology of the Bao Lac Map sheet at 1: 200,000 scale.* General Dept. of Geol. & Min. of Vietnam, (2001)
5. T.T. Thang, N.V. Vuong, V.V. Tich, N.D. Chinh, H.H. Hiep, *Process of deformation, geodynamic evolution of the Red River Fault Zone and their significance in interaction between the South China Plate and the Indochina Plate. The results of basic researches on The Red River Fault Zone (2001–2003).* Hanoi Scien. & Tech. Publish. House, 75–107 (2004)
6. P.T. Trinh, H.Q. Vinh, N.D. Tuc, B.T. Thao, *The modern tectonic activity of the Red River Fault Zone.* J. Earth Sci, **22**, 325–336 (2000)
7. N.T. Yem, *Regimes of tectonic stress-field during Cenozoic period in the Vietnam territory.* J. Geol., Ser. A, **236**, 1–6 (1996)
8. N.T. Yem, O.I Gusenko, L.M. Quoc, A. Motrikov, *Modern stress-field and deformation mechanism of the SE Asia crust.* Resour Geol., **2**, 8–13 (1996)
9. П. Н. Николаев, *Методика тектоно-динамическогоанализа*, Изд. Недра, Москва, **295с** (1961)

Tectono-thermal Evolution of the Lower Paleozoic Petroleum Source Rocks in the Southern Lublin Trough: Implications for Shale Gas Exploration from Maturity Modelling

Dariusz Botor[1,*]

[1] AGH University of Science and Technology, Faculty of Geology, Geophysics and Environmental Protection, A. Mickiewicza Av. 30, 30-059 Krakow, Poland

Abstract. The Lower Paleozoic basins of eastern Poland have recently been the focus of intensive exploration for shale gas. In the Lublin Basin potential unconventional play is related to Lower Silurian source rocks. In order to assess petroleum charge history of these shale gas reservoirs, 1-D maturity modeling has been performed. In the Łopiennik IG-1 well, which is the only well that penetrated Lower Paleozoic strata in the study area, the uniform vitrinite reflectance values within the Paleozoic section are interpreted as being mainly the result of higher heat flow in the Late Carboniferous to Early Permian times and ~3500 m thick overburden eroded due to the Variscan inversion. Moreover, our model has been supported by zircon helium and apatite fission track dating. The Lower Paleozoic strata in the study area reached maximum temperature in the Late Carboniferous time. Accomplished tectono-thermal model allowed establishing that petroleum generation in the Lower Silurian source rocks developed mainly in the Devonian – Carboniferous period. Whereas, during Mesozoic burial, hydrocarbon generation processes did not develop again. This has negative influence on potential durability of shale gas reservoirs. **Key words** – maturity modelling, shale gas, Lublin Trough.

1 Introduction

The Lower Paleozoic basins of eastern Poland have recently been the focus of intensive exploration for shale gas. The characteristic feature of these basins is an exceptionally high volumetric contribution of mudstone and claystone, being up to 1000–2000 m thick [1–3]. However, within this section there are a few individual shale formations (in the Upper Cambrian, Upper Ordovician and Lower Silurian) identified, which are characterized by elevated organic carbon contents (~1–2 % wt. total organic carbon, TOC) and shale gas/oil potential [4]. Kerogen is mostly low-Sulphur II type-oil prone in the Lower Paleozoic sediments of the SW slope of the East European Craton [2, 5, 6].

In the Lublin Basin potential unconventional play is exclusively related to the Lower Silurian (Wenlock) source rocks, within interval of no more than approximately 100 m [4].

* Corresponding author: botor@agh.edu.pl

https://doi.org/10.1051/e3sconf/20183502002

However, the Lublin Basin was subjected to tectonic inversion during the late Carboniferous (late Variscan) compressional tectonic deformations in Late Carboniferous [7, 8].

A characteristic long time gap in sedimentary record occurs between Jurassic and late Carboniferous or Devonian time caused mainly by erosional processes. It is related to Variscan inversion, which is not well constrained [8, 9], consequently timing of hydrocarbon generation and migration is disputable. In order to assess burial and thermal history of Lower Paleozoic shale gas reservoirs, 1-D maturity modeling has been performed in the Łopiennik IG-1 well, which is the only well that penetrated Lower Paleozoic strata in southern part of study area (Fig. 1).

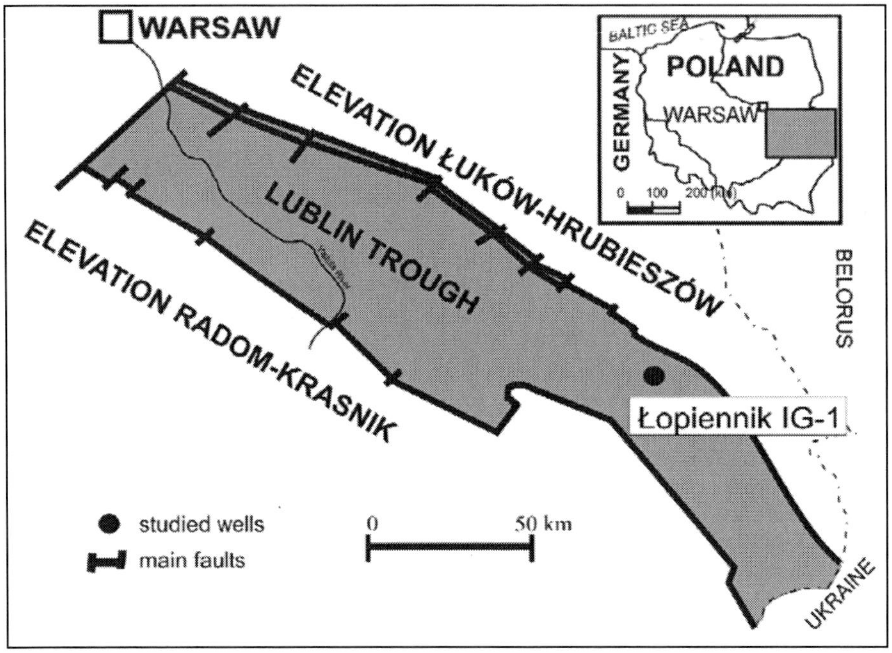

Fig. 1. Simplified map of the study area and location of the Łopiennik IG-1 well

2 Geological setting

The study area is situated at the south-western margin of the Precambrian East-European Craton at its contact with the Paleozoic Central-European units. The SW limit of the East-European Craton is conventionally defined in Poland by the Teisseyre-Tornquist Zone that represents the longest European tectonic lineament [1, 8]. Although the formation of the Lublin Trough took place between the Devonian and Early Carboniferous, its main framework is of latest Carboniferous age [7]. The Lublin Trough was tectonically inverted during late Carboniferous (Variscan) event. In the central Lublin area, the Carboniferous coal-bearing siliciclastic section begins with Visean volcanogenic rocks. The Carboniferous strata (~up to 2000 m in thickness) rest unconformably on Devonian, Silurian, Ordovician and Cambrian rocks in various parts of the basin. The pre-Carboniferous substratum includes mostly clastic Lower Paleozoic (~750–2500 m in thickness) and Lower Devonian strata (500–2300 m) as well as Middle and Upper Devonian carbonates (~700–3000 m) [1, 7–9]. Not much is known about the distribution of Lower Paleozoic strata in the Lublin Trough because it has only been penetrated by a few deep wells [1, 7, 9]. The Lower Paleozoic section is composed of the Cambrian strata, represented by deformed coarse-grained sediments (mainly conglomerates and arkoses), laterally passing into mudstones usually a few hundred meters thick; the

Ordovician carbonate and clastic complex, usually less than a hundred metres thick; and the Silurian mainly fine-grained claystone and siltstone sequence, up to 1500 m thick. The Paleozoic fill is overlain by 400–1000 m of Mesozoic (mainly Cretaceous) and Quaternary strata [1, 7–9].

3 Methods

Detailed description of applied numerical maturity modelling have been presented elsewhere [e.g. 10]. In this work, 1-D maturity modelling was carried out using PetroMod software (ver. 11) developed by Schlumberger. Integrated numerical modelling (one-dimensional, 1-D) was used to reconstruct the geological evolution of the study area and hence to derive the history of petroleum generation. The modelling procedure employed data defining the burial history, comprising the stratigraphy, thickness and lithology of the distinguished intervals as well as petrophysical parameters of rocks, the contemporary thermal regime and the present thermal maturity of organic matter. Thermal maturity has been calculated using the "Easy%Ro" algorithm [11]. Basic input parameters for modeling were taken from well data. The published present-day well temperature, porosity, density, mean vitrinite reflectance were used for verification of reliability and calibration of the calculated 1-D thermal models [12]. The thickness of the removed part of the Upper Paleozoic sediments was calculated taking into account thermal maturity, porosity and density data [12]. In the Devonian and younger sediments vitrinite reflectance has been measured, whereas in the Lower Paleozoic sediments measurements has been performed on vitrinite-like organic particle, which are not true vitrinite, but mainly graptolites [12]. Therefore, such measurements are regarded as organoclast reflectance [12]. However, recent studies has shown that such measurements are comparable to vitrinite reflectance [13].

4 Results and discussion

The Lublin Basin is characterized by very complex early to late Paleozoic geological evolution [7–9]. The burial history curves in the Łopiennik IG-1 well show generally continuous Neoproterozoic to Devonian sedimentation (Fig. 2). It was subjected to a very intense latest Devonian/ Early Carboniferous (Bretonian phase), mostly reverse faulting and block uplift related to reactivation of the Precambrian fault zones during this tectonic phase. These tectonic movements might have been genetically linked to widespread tectonic activity within the East European Craton plate farther to the east, e.g. in the Prypiat Graben [e.g. 7].

The Lublin Basin was inverted during the late Carboniferous compressional tectonic deformations [7, 8]. It can be correlated with the main phase of orogenic activity within the Variscan fold-and-thrust belt. During this tectonic phase complex system of thin-skinned compressional structures was formed [8]. A characteristic feature is the very rapid subsidence throughout the Late Carboniferous period. By the end of the Carboniferous, the vast majority of the total subsidence had occurred, and hence maximum burial and maximum temperature occurred in the latest Carboniferous just before the major tectonic inversion (Fig. 2). Paleozoic strata are covered by Jurassic and Cretaceous sediments, which were deposited in the NE flank of Mid-Polish Trough, which was finally inverted during late Cretaceous – Paleogene times (Fig. 2).

In the Łopiennik IG-1 well, the fairly uniform vitrinite reflectance values with in the Paleozoic section are interpreted as being mainly a result of higher heat flow in Late Carboniferous to Early Permian times and ~3500 m Carboniferous overburden eroded due to late Carboniferous inversion (Figs. 2–3). Although, there is no organic maturity data from the Jurassic-Cretaceous of this section, but Mesozoic profiles in adjacent area show low maturity, around 0.5% mean vitrinite reflectance. Moreover, this model has been recently supported by zircon helium and apatite fission track dating [14]. Thermal maturity development of the Wenlock source rocks is related mainly to Devonian and Carboniferous burial (Fig. 4).

General, hydrocarbon generation is controlled by organic matter quality and thermal history of the source rock. In terms of hydrocarbon generation, the critical period in the evolution of a sedimentary basin is the time of maximum burial, which occurred in the Late Carboniferous in the study area, that is documented by zircon helium and apatite fission track dating [14] (Fig. 2). The Carboniferous burial occurred in a higher than present-day heat flow regime, which can be explained by Early Carboniferous volcanism associated with Variscan tectonism, which was also pointed by many authors working in this area [e.g. 15]. This could have had a significant thermal effect for tens of million years if it was associated with deep heat sources. Therefore, in the Lower Paleozoic strata of the study area maximum temperature occurred in the Late Carboniferous period.

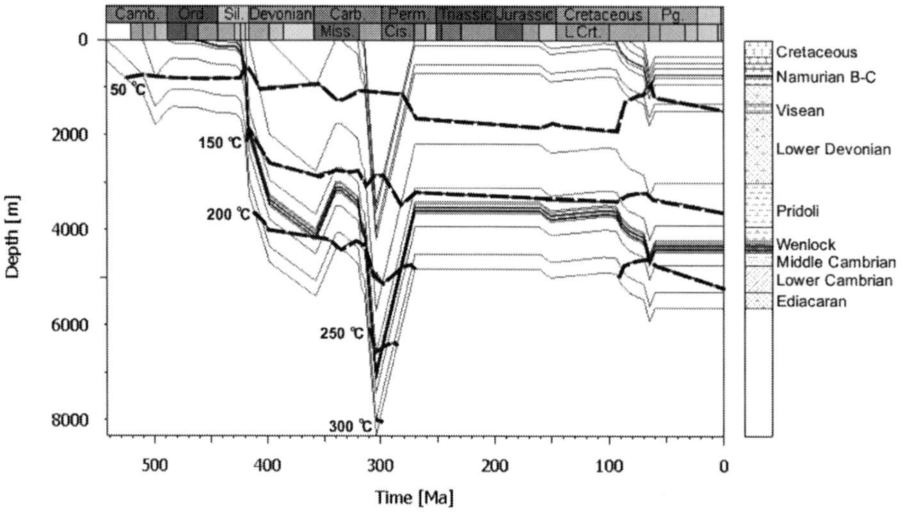

Fig. 2. Burial and thermal history model for the Łopiennik IG-1 well, assuming 3500 m of eroded Upper Carboniferous sediments and heat flow of 60 mW/m2 during maximum burial in the Late Carboniferous

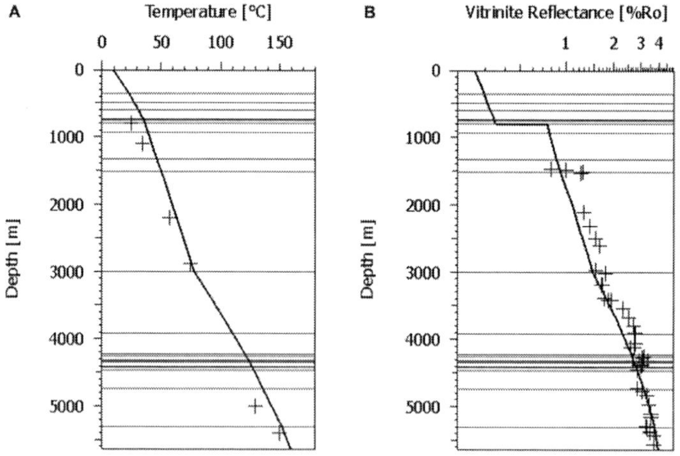

Fig. 3. Model calibration in the Łopiennik IG-1 well by means of present-day BHT temperature (A) and thermal maturity in the vitrinite reflectance scale (B)

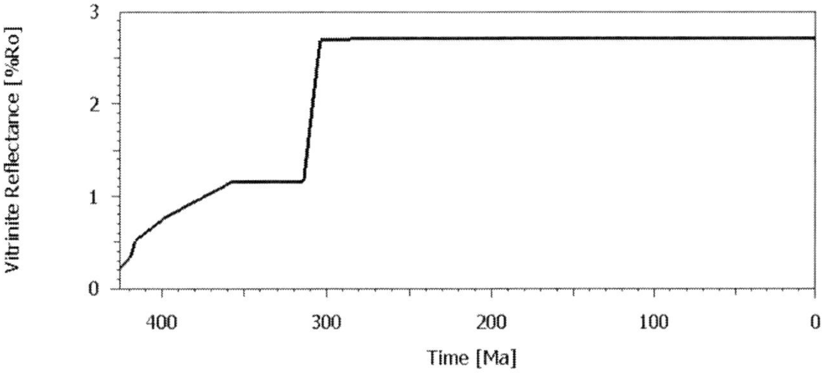

Fig. 4. Evolution of organic matter maturity in vitrinite reflectance scale evolution in the Wenlock source rocks in the Łopiennik IG-1 well

Average present-day organic carbon contents is 1.2–1.3 wt.% in the Wenlock source rocks. Taking into account geochemical data [2, 4, 6], the following kerogen parameters were assumed in modeling in the analyzed source rocks: kerogen type II, initial TOC 2.0 wt.%, and Hydrogen Index 500 mg hydrocarbons per gram TOC. Calculated kerogen transformation ratio increased significantly in Devonian period and totally reaches over 99% by the end of Late Carboniferous (Fig. 5). The hydrocarbon generation was developed in the Devonian and Late Carboniferous periods, reaching ca. 1.2 million tons per km2 of source rock. Petroleum generation was accompanied by a rapid expulsion of hydrocarbons (Fig. 6).

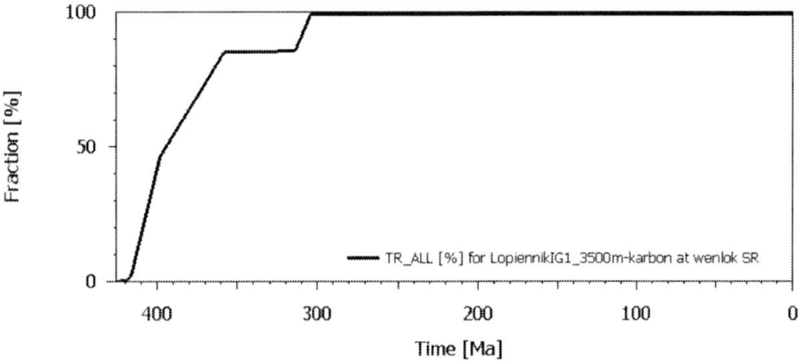

Fig. 5. Evolution of the kerogen transformation ratio in the Wenlock source rocks in the Łopiennik IG-1 well using preferred model (3500m eroded Carboniferous sediments)

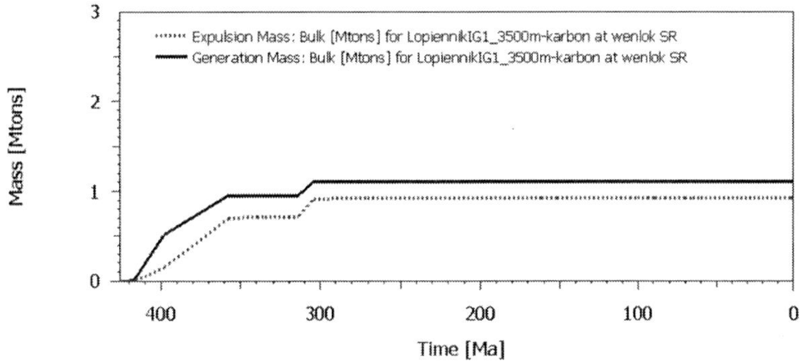

Fig. 6. Petroleum generation and expulsion in the Wenlock source rocks in Łopiennik IG-1 well

5 Conclusions

In the Lower Paleozoic strata of the study area maximum temperature occurred in the Late Carboniferous time. Accomplished tectono-thermal model allowed establishing that petroleum generation in the Lower Silurian source rocks developed in Devonian and Carboniferous periods, reaching the maximum in the Late Carboniferous time. Whereas, during Mesozoic burial, hydrocarbon generation processes did not develop again. Generated hydrocarbons were expelled rapidly from source rocks.

Acknowledgements

This research has been financially supported by the Polish National Centre for Research and Development (NCRD) grant under the BLUE GAS – Polish Shale Gas program – BG1/GAZGEOLMOD/13. I would like to express my gratitude to the reviewers (Dr. Júlia Kotulová, Dr. Yuriy Koltun and Dr. Paweł Kosakowski) for their valuable input, which significantly improved the quality of the paper.

References

1. J. Nawrocki, P. Poprawa, Geological Quarterly, **50**, 59–76 (2006)
2. P. Poprawa, Przegl. Geol., **58**, 226–249 (2010)
3. S.J. Porębski, W. Prugar, J. Zacharski, Przegl. Geol., **61**, 468–477 (2013)
4. T. Podhalańska, M.I. Waksmundzka, A. Becker, J. Roszkowska-Remin, I. Dyrka, A. Feldman-Olszewska, A. Głuszyński, I. Grotek, M. Janas, P. Karcz, G. Nowak, J. Pacześna, M. Roman, M. Sikorska-Jaworowska, M. Kuberska, A. Kozłowska, K. Sobień, Przegl. Geol., **64**, 1008–1021 (2016)
5. I. Grotek, Przegl. Geol., **64**, 1000–1004 (2016)
6. P. Karcz, M. Janas, Przegl. Geol., **64**, 995–999 (2016)
7. M., Narkiewicz, Geological Quarterly, **51**, 231–256 (2007)
8. P. Krzywiec, Ł. Gągała, S. Mazur, Ł. Słonka, M. Kufrasa, M. Malinowski, K. Pietsch, J. Golonka, Tectonophysics, **713**, doi: 10.1016/j.tecto.2017.06.008 (2017)
9. M. Narkiewicz, A. Maksym, M. Malinowski, M. Grad, A., Guterch, Z., Petecki, J. Probulski, T. Janik, M. Majdański, P., Środa, W., Czuba, E., Gaczyński, L., Jankowski, International Journal of Earth Sciences, **104**, 775–796 (2015)
10. D. Botor, D.B. Papiernik, T. Maćkowski, B. Reicher, P. Kosakowski, G. Machowski, W. Górecki, Annales Societatis Geologorum Poloniae, **83**, 353–383 (2013)
11. J.J. Sweeney, A.K. Burnham AAPG Bulletin, **74**, 1559-1570 (1990)
12. J. Pacześna, (red.) *Profile głębokich otworów wiertniczych*, Państwowego Instytutu Geologicznego - Łopiennik IG-1, **123**, (2008)
13. H.I. Petersen, N.H. Schovsbo, A.T. Nielsen, 2013. International Journal of Coal Geology **114**, 1–18
14. D. Botor, J. Golonka, I. Dunkl, A. Anczkiewicz, Tectonophysics (to be published 2017)
15. D. Botor, D.M. Kotarba, P. Kosakowski, Organic Geochemistry, **33**, 461–476 (2002)

Critical Elements in Fly Ash from the Combustion of Bituminous Coal in Major Polish Power Plants

Barbara Bielowicz[1,*], *Dariusz* Botor[1], *Jacek* Misiak[1], and *Marian* Wagner[1]

[1] AGH University of Science and Technology, Faculty of Geology, Geophysics and Environmental Protection, A. Mickiewicza Av. 30, 30-059 Krakow, Poland

Abstract. The concentration of critical elements, including such REE as Fe, Co, W, Zn, Cr, Ni, V, Mn, Ti, Ag, Ga, Ta, Sr, Li, and Cu, in the so-called fly ash obtained from the 9 Polish power plants and 1 thermal power station has been determined. The obtained values, compared with the global average concentration in bituminous coal ash and sedimentary rocks (Clarke values), have shown that the enrichment of fly ash in the specified elements takes place in only a few bituminous coal processing sites in Poland. The enrichment factor (EF) is only slightly higher (the same order of magnitude) than the Clarke values. The enrichment factor in relation to the Clarke value in the Earth's crust reached values above 10 in all of the examined ashes for the following elements: Cr, Ni, V, W, and, in some ash samples, also Cu and Zn. The obtained values are low, only slightly higher than the global average concentrations in sedimentary rocks and bituminous coal ashes. The ferromagnetic grains (microspheres) found in bituminous coal fly ashes seem to be the most economically prospective in recovery of selected critical elements. The microanalysis has shown that iron cenospheres and plerospheres in fly ash contain, in addition to enamel and iron oxides (magnetite and hematite), iron spinels enriched in Co, Cr, Cu, Mn, Ni, W, and Zn. **Key words** – Critical Elements, Rare earth elements; Coal fly ash.

1 Introduction

Bituminous coal ash is characterized by a significantly higher concentration of critical elements than ash from burning of other solid fossil fuels [1, 2, 3, and 4] and can be a potential source of critical raw materials. This is incombustible component of coal resulted in combustion processes in professional power plants and combined heat and power plants (CHP). The paper is focused on examining coal ashes from the currently exploited Polish deposits. Special attention is paid to the so-called fly ash, which are ash components collected by electrostatic precipitators. These ashes are characterized by multiphase composition, which depends mainly on the petrographic composition of coal subjected to the combustion technology.

* Corresponding author: bbiel@ agh.edu.pl

2 Research methodology

The geochemical and petrographic analysis of fly ash from Polish power plants and heat plants (14 samples) collected from ash storage sites at bituminous coal-fired power plants was carried out. The samples were collected at the following power plants: Skawina, Siersza, Połaniec, Kozienice, Jaworzno (2 samples), Ostrołęka, Poznań, Opole (3 samples), Dolna Odra (2 samples), and the Janikowo thermal power station (Fig. 1).

Fig. 1. Location of power plants (red) and thermal power stations (green) from which the analyzed ash samples were collected

The determination of chemical composition (the presence of minor, trace, and rare elements) was carried out at the Acme Labs (Canada) using inductively coupled plasma mass spectrometry (ICP-MS) A total of 53 elements, including most of the critical elements were determined.

Petrographic examination using an ZEISS reflected light optical microscope, a JEOL JXA-8230 SuperProbe Electron Probe Microanalyzer, and the WDS microanalysis of chemical composition of solids were carried out at the AGH UST – KGHM Polska Miedź SA Laboratory of Critical Elements of Critical Elements Laboratory of the Faculty of Geology, Geophysics and Environmental Protection at the AGH University of Science and Technology in Kraków. A total of 10 samples of fly ash were examined. Selected samples with the highest content of critical elements and those containing microspheres were selected for the analysis. The samples were covered with graphite before the analysis. The qualitative microanalysis of elemental composition (from boron to uranium) was performed using a wavelength-dispersive X-ray spectroscopy spectrometer. The quantitative analysis of elemental composition was performed using an X-ray diffraction (WDS) spectrometer. The images of measurement points were obtained using the secondary electrons (SE). The WDS analysis was performed only for microspheres with a current of 20 nA and an accelerating voltage of 15 kV.

3 Results and discussion

3.1 The petrographic analysis of fly ash

During industrial combustion of coal (at temperatures of up to ca. 1400°C), its mineral matter is being liquefied and simultaneously subjected to complex physicochemical changes involving

 https://doi.org/10.1051/e3sconf/20183502003

dehydroxylation of hydrated aluminosilicates and aluminum silicates, partial mullitization, carbonate decomposition, iron sulphide breakdown and reduction [5]. Rapid cooling of combustion products promotes the formation of amorphous (non-crystalline) particles, often of a spherical shape - microspheres. Gases trapped inside them can expand and create the so-called cenospheres and plerospheres [5] (Fig. 2). Minerals with a high melting point usually remain relatively unchanged [6].

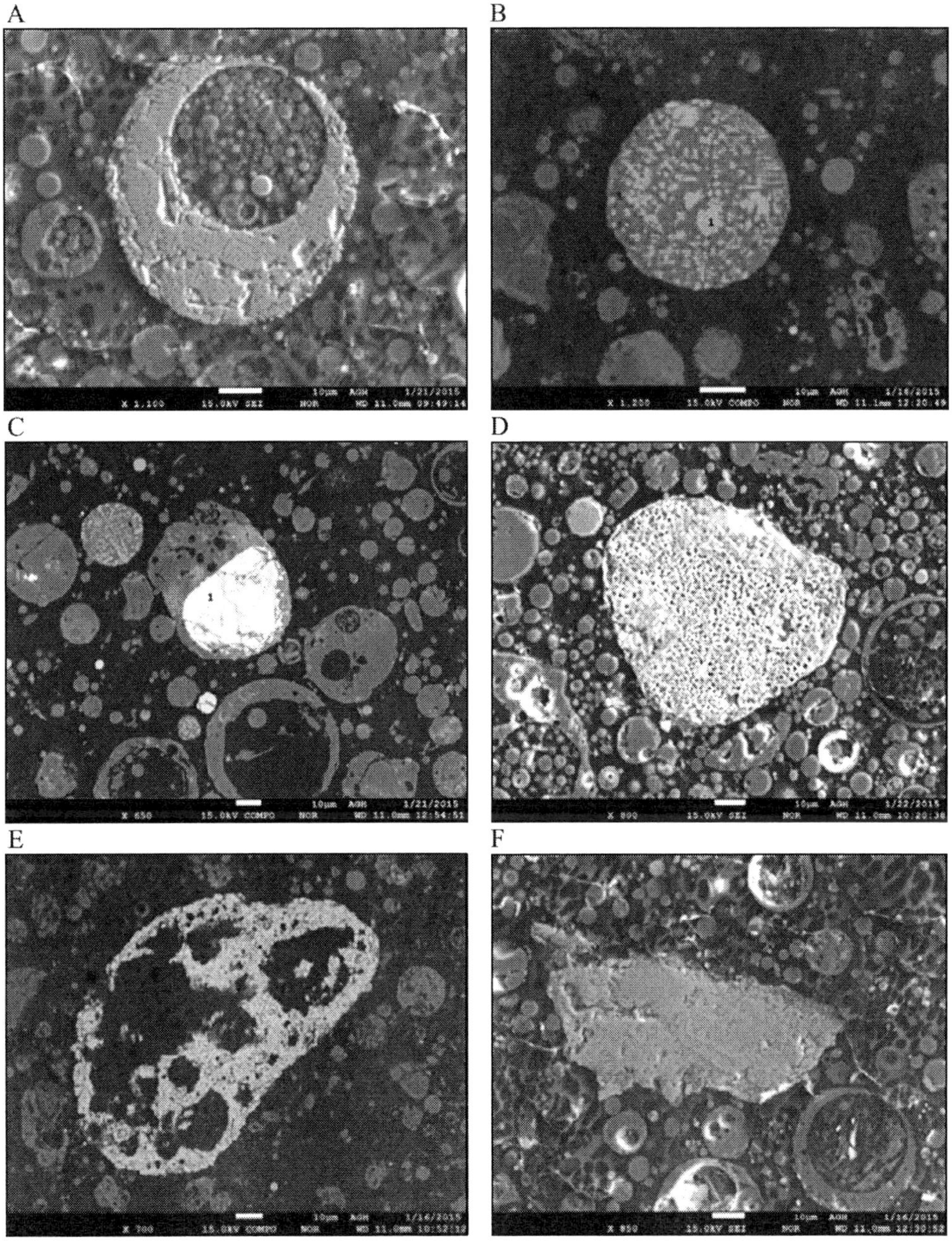

Fig. 2. A) Hematite crassisphere (Dolna Odra power plant), B) Hematite plerosphere with aluminosilicates (spinel) (Siersza power plant), C) Various forms of microspheres, the central filled with zircon (lighter) (Poznań power plant), D) Porous hematite (Opole power plant), E) Hematite crassisphere (Skawina power plant), F) Irregular iron oxide grain (Siersza power plant)

There are two types of microspheres [7]: "empty balls" in which the hollows are filled with gas and the so-called cenospheres and plerospheres, in which the empty spaces (voids) are filled with other mineral particles, most commonly in the spongy (porous) form. The term cenosphere is derived from Greek: kenos + sphere – hollow sphere, while plerosphere from Greek: plērēs + sphere – filled sphere [8].

In the examined fly ash resulting from the combustion of bituminous coals from 9 power and 1 thermal power station the occurrence of microparticles belonging to the silicates and aluminosilicates (quartz, mullite, and silicate enamel), iron oxides, carbonates, and unburned organic matter has been confirmed.

The ferromagnetic grains (microspheres), found in the following concentrations in the samples: Skawina (sample no. 2) – 0.16%, Siersza (3) – 0.31%, Połaniec (5) – 0.18%, Kozienice (6) – 0.22%, Jaworzno (7) – 0.17%, Janikowo (8) – 0.16%, Ostrołęka (10) – 0.08%, Poznań (13) – 0.29%, Opole (16) – 0.24%, and Dolna Odra (17) – 0.38%. Based on the literature, it can be stated that microspheres are promising in analyzing the occurrence of critical elements [5, 8].

Iron microspheres are highly variable in terms of both size and morphology. The size of cenospheres and plerospheres ranges from a few to several dozen micrometers (Fig. 1). Cenospheres can have very thin (tenuisphere) or thick walls (crassisphere), in which case they usually show secondary porosity within the walls. The spherical objects (plerospheres) are partially filled with porous magnetite-hematite aggregates, where lighter hematite lamellae are clearly visible against the background of the darker magnetite. Some of the plerospheres are fully filled with magnetite containing fine (lighter) hematite lamellae. In addition, plerospheres filled with enamel (of dark gray color) containing magnetite grains in different geometrical forms have also been found (Fig. 2C). In some cases, magnetite in the enamel filling the microspheres forms various dendritic-skeletal forms (Fig. 2B). In addition, magnetite-hematite aggregates filling the gaps in the enamel have also been shown in the examined samples.

3.2 The chemical analysis of ashes

3.2.1 The chemical composition of ash

The main chemical components in the examined ash include SiO_2 (30.01–52.72 wt %) and Al_2O_3 (9.97–29.02 wt %). There are also CaO (2.13–45.20 wt %), Fe_2O_3 (3.28–8.15 wt %), MgO (1.57–2.97 wt %), Na_2O (0.40–2.14 wt %), K_2O (1.32–3.5 wt %), P_2O_5 (0.16–0.83 wt %), TiO_2 (0.48–1.27 wt %), MnO (0.05–0.11 wt %), and Cr_2O_3 (0.012–0.030 wt %). Most of the examined ash samples is silica-dominated; the only exceptions are the ash samples from Janikowo (sample no. 8) and Ostrołęka (sample 10), which are calcium-dominated. The chemical composition of ash indicates the origin of coal burned in power plants (calcium-dominated ash is typical for coal from the Paralic series, while silica-dominated for coal from the limnic-fluvial series) [9].

3.2.2 Critical elements

The measured levels of critical elements were compared with the so-called Clarke value in sedimentary rocks (the global average content of a specified element in sedimentary rocks) [10] and the Clarke value in bituminous coal ash, that is the global average for this raw material [2]. The Clarke value in sedimentary rocks is a good indicator of the concentration of certain chemical element due to the genetic affiliation of coal deposits to sedimentary rocks and the apparent association of this element with the shale-like concentration pattern in the Earth's crust [11]. Coal ashes usually contain higher amounts of critical elements

 https://doi.org/10.1051/e3sconf/20183502003

when compared to coal [1, 2, 4, 12], which is a consequence of the combustion of organic matter; however, they are strongly modified by the so-called coefficient of volatility [13].

An analysis of the potential enrichment of the examined ashes was carried out by calculating the so-called Enrichment Factor (EF) as the ratio of the content of the element and the Clarke value. This coefficient is given in two variants: EF1 (mean value) and EF2 (max. value). Both factors are used for determining the content of critical elements in the examined fly ash samples in relation to the Clarke value in bituminous coal ash [2]. For a more complete analysis of the presence of critical elements in the fly ash, two other factors determining the concentration of selected elements (EF3 and EF4) referring to the Clarke value in sedimentary rocks, were used Ronov et al. [10] The EF3 value refers to the mean while the EF4 to the maximum value. The EF1 factors are low (generally lower than 1) while higher values of the EF2 factor can be observed for Be and B in the ash sample from Siersza and Mn in the ash sample from Skawina. EF4 factors higher than 1 was determined for: Cu, Li, Ga, Sb, W, Mo, Zn, Ag, and B in the sample from the Siersza power plant and for Ba from the Skawina power plant sample.

The performed analysis has confirmed that only the fly ash from coal burned in the Siersza power plant (with the exception of Ba concentration from Skawina) shows the above-Clarke values for Be, Ga, Sb, W, Mo, V, Zn, Ag, B, Cu, and Li. Although these are above-Clarke values, they are low and without economic significance, as it is believed that only EF > 10 values are economically viable. The analysis of the enrichment of fly ashes from Polish power plants and bituminous coal fired power plants in critical elements, based on enrichment factors, has unequivocally shown that the enrichment of bituminous coal and its ashes (including fly ash) in the specified elements takes place in only a few bituminous coal processing sites in Poland. These are small values with no economic significance.

3.3 The characteristic of critical elements in iron microspheres of the examined fly ash samples using the WDS analysis

The ferromagnetic grains (microspheres) found in bituminous coal fly ashes seem promising in the analysis of the occurrence of selected critical elements. The morphology of microspheres is highly variable (Fig. 2 A–F). The size of cenospheres and plerospheres ranges from a few to several dozen of micrometers. Cenospheres have walls of varying thickness (from 1 to 20 μm). Plerospheres can be filled with magnetite-hematite aggregates (Fig. 2A). In addition, plerospheres filled with enamel (of dark gray color) containing magnetite grains have also been found. In addition, magnetite-hematite aggregates filling the gaps in the enamel have also been shown in the examined samples (Fig. 2B). Locally, microsphere enamel occurs with zircon (Fig. 2C). The microanalysis has shown that ashes, in addition to enamel of the cenosphere and plerosphere, also contain iron oxides and spinels which are locally enriched in Cr, Cu, Mn, Ni, W, and Zn. In the examined samples, the iron content is definitely greater than the Clarke value of this element in the Earth's crust (Table 1). The average concentration for all points in the studied samples is 32.7%, with the Clarke value in the Earth's crust at 5.6%. In sample no. 8 (from the Janikowo power plant), the above-Clarke concentrations of Mn, with a maximum of 9867 ppm (the Clarke value in the Earth's crust and bituminous coal ash is 950 ppm and 430 ppm [2], respectively), in microspheres were recorded. In all of the analyzed samples, the Cr concentrations are up to several times higher than their Clarke values. The average concentration in the examined iron oxides is 1735 ppm, while the Clarke value of coal ash is 120 ppm. However, such enrichment may be related to the chromium oxide used during the polishing of the preparations; hence, the obtained results should be interpreted with caution. The enrichment in Cu is observed in some of the analyzed objects.

Table 1. The mean values of the critical elements determined in ferromagnetic microspheres using the WDS analysis compared with their Clarke values (all values in ppm)

Sample	Al	Ba	Cr	Cu	Fe	Mg	Mn	Ni	Si	Ti	V	W	Zn
Skawina	4606	210	227	107	336677	9793	2421	74	2932	592	25	253	213
Siersza	9378	278	13318	169	314659	1294	1047	1436	551	751	35	431	518
Połaniec	10691	220	143	527	339554	3460	432	420	2149	1187	55	407	169
Kozienice	10436	322	278	157	325130	1250	922	259	18115	1723	321	201	228
Jaworzno	4326	5	228	136	327464	1134	2334	163	21552	588	54	178	315
Janikowo	2615	183	159	98	328170	22953	2935	98	2963	116	18	274	85
Ostrołęka	2985	161	757	400	306245	17264	3281	4072	6714	197	7	82	696
Poznań	7902	462	255	56	325161	28420	2032	2298	383	628	40	206	310
Opole	7678	100	367	139	327614	28572	1527	115	464	776	74	293	196
Dolna Odra	3244	78	1220	90	343767	11152	3733	59	297	321	708	189	82
Average	6119	199	1735	181	327333	13103	2128	919	5162	638	111	252	287
Min	5	0	0	0	230041	6	0	0	70	0	0	0	0
max	36800	1030	64986	991	361468	98682	9867	19817	80584	3831	2756	1427	2981
Detection level	32	111	100	84	72	28	52	68	46	63	53	135	122
Clarke [2]	N/A	980	120	110	N/A	N/A	430	100	N/A	5300	170	8	170

The highest values (991 ppm) were measured in sample no. 10 from the Ostrołęka power plant in sample no. 5 from the Połaniec power plant, the Cu content of all tested grains was higher than the Clarke value in bituminous coal ash. In some of the analyzed samples, the Ni content was several times higher than the Clarke value in ash (100 ppm). The highest concentration of this element (19817 ppm) was recorded in sample no. 10 from the Ostrołęka power station. The average concentration of nickel for all analyzed samples is 919 ppm. Local enrichments in wolfram and zinc (their average concentration in the examined samples amounts to 252 and 287 ppm, while the Clarke value in the Earth's crust is 7.8 ppm and 170 ppm, respectively) have also been confirmed [2]. Due to their strong magnetic properties, iron microspheres in fly ash seem to be relatively easy to separate. Each year, a total of 7.5 million Mg of fly ash are captured in Polish bituminous coal-fired power plants; hence, it is theoretically possible to obtain 0.6–2.9 million tons of ferromagnetic aggregations from fly ash in order to recover the elements which concentration is higher than their Clarke value.

4 Conclusions

Fly ash samples from major Polish power plants and CHP plants were examined. The ashes contain cenospheres and plerospheres (spherical particles), in which voids are filled with other mineral particles, most commonly in the form of irregular aluminosilicate grains (mainly mullite), iron oxides, unburnt (fusained) coal particles, and spongy (porous) form. The content of iron microspheres in the examined ash ranges from 0.16 to 0.38% by volume, which has been confirmed by microscopic studies. The content of critical elements in iron microspheres contained in fly ash was examined using an electron microprobe (EDS-WDS). The content of elements ranging from boron to uranium (Ta, In, V, Ga, Rh, Nb, Pd, Pt, and Cu) has been determined according to the apparatus settings. Microspheres of all the examined ash samples contain from 23.0 to 36.1% of Fe, accompanied by numerous above-clarke concentrations of critical elements, compared to the Clarke value in the earth's crust, including barium (up to 680, the Clarke value of 425 ppm), chromium (up to 3253, the Clarke value of 102 ppm), copper (up to 991, the Clarke value of 60 ppm, manganese (up to 9867, the Clarke value of 950 ppm), nickel (up to 19817, the Clarke value of 84 ppm), vanadium (up to 2756, the Clarke value of 120), wolfram (up to 1427, the Clarke value of 1,25 ppm), and zinc (up to 2981, the Clarke value of 70 ppm).

When compared to the Clarke value in the earth's crust, it is clearly visible that the concentration of elements in the ashes is highly variable; the most interesting are elements with the enrichment factor above 10, including Cr, Ni, V, W, and, in some ash samples, Cu and Zn.

The obtained results were also compared with the average values for bituminous coal ashes, determined on a global scale and expressed in the form of specific Clarke values of bituminous coal. These Clarke values are different than the Clarke values in the earth's crust and sedimentary rocks, while take into account the geochemical affinity between the individual critical elements and the C element and generally characterize the detailed geological processes involved in the formation of coal deposits.

The obtained values are low, only slightly higher (the same order of magnitude) than the global average concentrations in sedimentary rocks and bituminous coal ashes. The ferromagnetic grains (microspheres) found in bituminous coal ashes seem promising in the analysis of the occurrence of selected critical elements. The microanalysis has shown that ashes, in addition to enamel of the cenosphere and plerosphere, also contain iron oxides and spinels. These minerals are locally enriched in Cr, Cu, Mn, Ni, W, and Zn. In the examined samples of microspheres, the iron content is definitely greater than the Clarke value in the Earth's crust. Ferromagnetic microspheres, due to their strong magnetic properties, seem to be quite easy to separate. It is estimated that about 7.5 million Mg (t) of fly ash is captured annually in Poland, so, theoretically, it would be possible to obtain from about 0.6 to 2.9 million tonnes of ferrous aggregates from fly ashes in order to recover supercritical elements.

Acknowledgements

This study was supported by Statutory Research No 11.11.140.320.

References

1. V.V. Seredin, R.B. Finkelman, Int. J. Coal. Geol. **70** 253–289 (2008)
2. M.P. Ketris, Y.E. Yudovitch, Int. J. Coal. Geol. **78** 135–148 (2009)
3. J.C. Hower, S. Dai, V.V. Seredin, L. Zhao, I.J. Kostova, L.F.O. SilvaCoal Combust Gasificat Prod. **5** 39–47 (2013)
4. R. Blissett, N. Smalley, B. Rowson, Fuel **119** 236–239 (2014)
5. I. Jonczy, J. Nowak, A. Porszke, E. Strzałkowska, *Składniki fazowe wybranych mineralnych surowców odpadowych w obrazach mikroskopowych* (Eds. Politechniki Śląskiej, Gliwice. 2012)
6. B.G. Kutchko, A.G. Kim, Fuel 85, 2537–2544 (2006)
7. V.B Fenelonov, M.S. Melgunov, V.N. Parmon, KONA Powder and Particle Journal **28** (2010)
8 G.L. Fisher, D.P.Y. Chang, M. Brummer, Science **192**, 553–557 (1976)
9. B. Roga, K. Tomków, *Chemiczna technologia węgla*, (WNT, 1971)
10. A.B. Ronov, *Sedimentary Shell of the Earth*. (Nauka, Moscow. "Science" Pub. House, 1980)
11. F. Goodarzi, Fuel **85** 273–280 (2006)
12. I. Bojakowska, D. Lech, Biul. PIG **448** 209–214 (2012)
13. M. Wagner, J. Janczyszyn, *Badania ulatniania się pierwiastków śladowych w trakcie procesu spalania węgla brunatnego z aktualnie eksploatowanych złóż węgla*, (Arch. ZGZW AGH, Inst. Fizyki AGH i ZGO AGH. 1995)

Can Satellite Remote Sensing be Applied in Geological Mapping in Tropics?

Janusz Magiera[1,*]

[1] AGH University of Science and Technology, Faculty of Geology, Geophysics and Environmental Protection, A. Mickiewicza Av. 30, 30-059 Krakow, Poland

Abstract. Remote sensing (RS) techniques are based on spectral data registered by RS scanners as energy reflected from the Earth's surface or emitted by it. In "geological" RS the reflectance (or emittence) should come from rock or sediment. The problem in tropical and subtropical areas is a dense vegetation. Spectral response from the rocks and sediments is gathered only from the gaps among the trees and shrubs. Images of high resolution are appreciated here, therefore. New generation of satellites and scanners (Digital Globe WV2, WV3 and WV4) yield imagery of spatial resolution of 2 m and up to 16 spectral bands (WV3). Images acquired by Landsat (TM, ETM+, OLI) and Sentinel 2 have good spectral resolution too (6–12 bands in visible and infrared) and, despite lower spatial resolution (10–60 m of pixel size) are useful in extracting lithological information too. Lithological RS map may reveal good precision (down to a single rock or outcrop of a meter size). Supplemented with the analysis of Digital Elevation Model and high resolution ortophotomaps (Google Maps, Bing etc.) allows for quick and cheap mapping of unsurveyed areas. **Key words** – Remote Sensing, geological mapping, tropics.

1 Peculiarities of geological Remote Sensing in tropics

Remote Sensing techniques are based on spectral data registered by RS scanner as energy reflected from the Earth's surface or emitted by it. The data can be imaging or non-imaging. In both cases it is essential that the reflectance (or emittence) comes from the objects on the Earth surface which are in focus of our interest, i.e. rock, sediment, or, to a lesser extent, soil [1, 2]. And here a problem arises in tropical, subtropical and other areas with dense vegetation: tree canopies form exceptionally effective obstacle for direct electromagnetic "communication" between geological object and scanner (the only exception is microwave or radar radiation which penetrates the vegetation cover or even top layers of soft sediments). Spectral information on the rocks and sediments can be gathered only from the gaps among the trees and shrubs and from clearings. Therefore, spatial resolution of the images is a crucial parameter: the smaller the pixel the more selective (precise) spectral information is. Two more factors control effectiveness of RS analysis: spectral and radiometric resolution. The former is defined by the number of spectral bands, the latter – by the number of bits of information per pixel (in both cases – the more the better). Temporal resolution (repeatability

* Corresponding author: magiera@geol.agh.edu.pl

of imaging of a given area), which is essential for monitoring of landuse changes, crops or natural disasters, is of marginal importance in geological RS. Overcast or atmospheric haze may substantially depreciate or totally eliminate RS images from the analysis (again with the exception for radar images). All those issues are probably a reason why RS has been so far applied in geological mapping in rather rare and specific projects [3–5].

Hard rocks are exposed clearly essentially only on the hilltops, where they often form tors. On plains they are usually covered with residual sediments or regoliths, which in tropics may attain remarkable thickness up to a few hundred meters. Mineralogy and petrography of them are more or less closely related to the parent rocks. Therefore, spectral response of both is sometimes similar. Sediment displaced e.g. on a slope reveal spectral signature quite different to the parent rock and is unsuitable for RS analysis of the bedrock lithology. It applies even more evidently to the valleys and depressions sheeted by alluvium or prolluvium.

RS study aimed at both, lithological and structural mapping in Laos (Xiangkhouang and Vientiane provinces; Fig. 1), carried out at the Faculty of Geology, Geophysics and Environment Protection of the AGH-UST a few years ago (Projects contracted by the GeoExpert, unpublished), revealed that "classical" RS techniques (classification and band ratioing of multispectral images) supported with the analysis of DEM (Digital Elevation Model) and inspection of high resolution ortophotomaps (from repositories of the Google Earth, Bing etc.) may yield highly satisfactory results. Present paper is based on results and experience gained in this study.

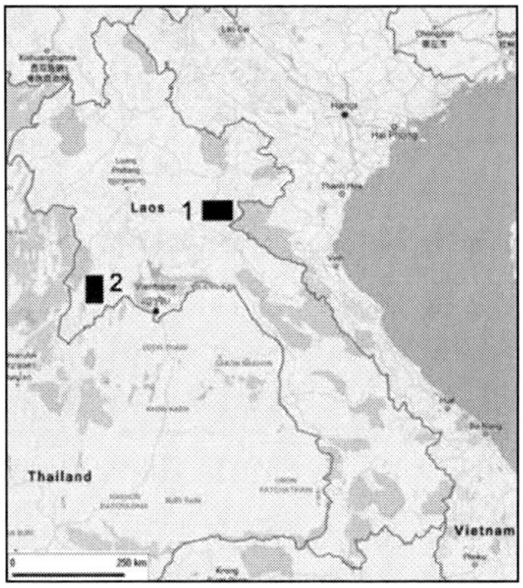

Fig. 1. Location of the study areas: 1 – Xiangkhouang, 2 – Vientiane (backdrop: Google Maps)

2 Remotely sensed data

The Landsat program (NASA) offers (free of charge) multitemporal images which have been acquired since July, 1972 [6]. This is the longest continuous satellite RS mission (carried out by three consecutive satellites and scanners). Sentinel 2 (European Space Agency) has been operating since June 2015 (Sentinel 2A) and March 2017 (identical Sentinel 2B). Images are freely available for ESA member states [7]. Both missions offer images of medium spatial resolution (10–60 m of pixel size) and good spectral inventory, comprising 6 to 12 bands. The bands range form visible blue/ultraviolet "blue-edge" to medium range infrared, plus thermal infrared and panchromatic. Particularly useful for lithological discrimination are bands around 2 µm wavelength (bands 5 and 7 of TM and ETM+, band 6: SWIR-1 and 7: SWIR-2 of OLI, bands Mid IR1 and Mid IR2 of Sentinel 2).

Radiometric resolution ranges from 8 bits (TM and ETM+) to 16 bits per pixel (OLI, Sentinel 2). This allows for quantification of the reflectance energy to, respectively, 256 and 65,536 levels. The latter figure seems, on one hand, incredible high, on the other, normally only a few thousands levels are "used" in practice.

New generation of commercial satellites and scanners World View 2 – World View 4 (Digital Globe) yields multispectral imagery of exceptionally good spatial resolution of a range of 1.24 (downsampled to 2 m for non US governmental use) to 3.7 m, very good inventory of spectral bands – up to 16 (WV3) and good spectral resolution (11 bits per pixel). In addition they offer panchromatic images of 0.36–0.41 m pixel size (downsampled to 0.5 m). WV satellites are relatively new. WV 1 operates since October 2009, while WV 2 from August 2014 [8]. Unfortunately, band of ~2 µm wavelength is missing from the WV2 scanner which imagery was used in the Vientiane study.

GDEM (Global Digital Elevation Model) data acquired by the ASTER satellite (NASA) were used for analysis of tectonic structures in the Vientiane study. Its spatial resolution declared to30 m is in reality probably around 30–100 m and vertical accuracy is a few decimeters to dozen or so meters [9].

Finally, high resolution ortophotomap (~0.5 m pixel size) acquired from the Google Satellite Maps repository was used for visual inspection and verification of the results of satellite data processing.

3 Methods of processing and analysing of RS data

Multispectral satellite RS images were processed and analyzed with rather standard methods applied normally in such a case. Good description of them can be found in, i.a., [1, 2]. In the present case they were:

- color composites created for general visual inspection and enhancement of selected objects (rock types) or phenomena (vegetation),
- band rtioing aimed at quantitative discrimination of selected objects or phenomena; vegetation index (NDVI) was used for creating vegetation mask applied later in classification,
- classification (unsupervised) aimed at producing lithological map.

GDEM was used for producing shaded relief map. It was examined in variable conditions of illumination: azimuth and angle of incident light.

Finally, results were verified by comparison to geological maps in scale 1:200,000 and 1:1,000,000.

4 Results

Spatial resolution (pixel size) of images is crucial for successful discrimination of small objects or objects exposed on small areas. The latter is the case of the tropical areas, masked by intensive and extensive vegetation. High resolution ortophotomap reveals very detailed image of the spot, while Landsat ETM+ image is quite general (Fig. 2).

Multispectral WV2 images give much better insight into the terrain than Landsat ETM+ or TM images (Fig. 3).

Spectral signatures of granitoids visible in gaps among trees and bush extracted from the WV2 images are much more different to signatures of vegetation obtained from the Landsat TM images (Fig. 4). Single pixel of the latter represents area of 30 × 30 m. It is very likely that both, rock and tree/bush might have been captured together. Therefore spectral signatures of rocks and vegetation are uniform. Well pronounced difference is visible only in band 4, smaller but probably significant occurs in band 3. Small pixels of the WV2 images (2 × 2 m) enable much more selective extraction of spectral signatures. They are significantly different in bands 6, 7 and 8, less pronounced in bands 2–5. It is obvious that discrimination of rocks and vegetation is more effective on the WV2 imagery than on Landsat images. However, processing of the latter yields surprisingly good results on the terrain with less extensive vegetation or of bigger extent, where more general results are acceptable e.g., at the initial stage of prospecting (Fig. 7).

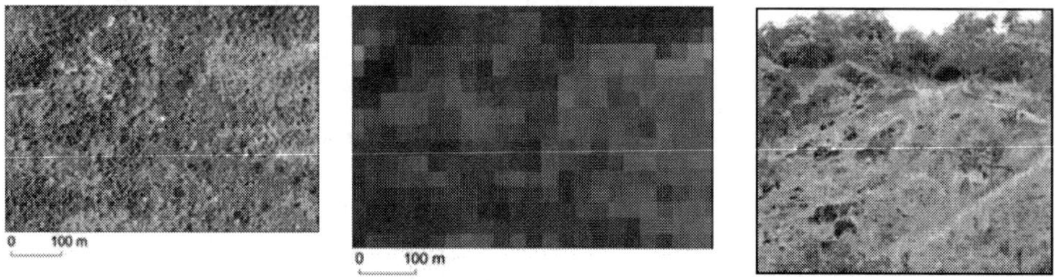

Fig. 2. Fragment of the Xiangkhouang study area shown on the ortophoto map (Google Maps; left), Landsat ETM+ "natural" colour composite RGB321 (centre) and ground photo (right)

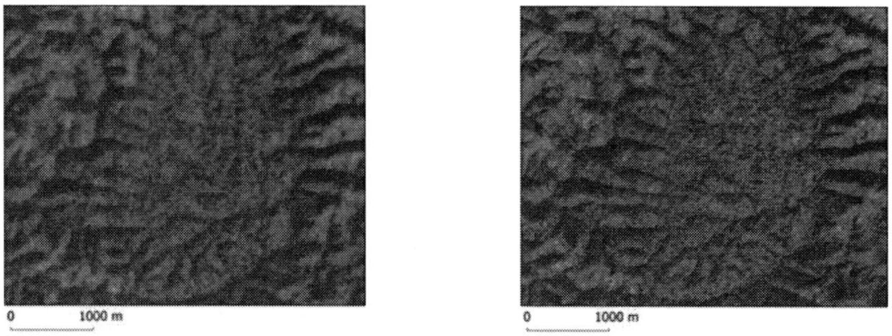

Fig. 3. Fragment of the Vientiane study area. "Natural" colour composites of Landsat TM RGB321 (left) and WV2 RGB532 (right). Intrusive Permo-Triassic granitoids form a dome. Barren rocks are represented by reddish-pink shades

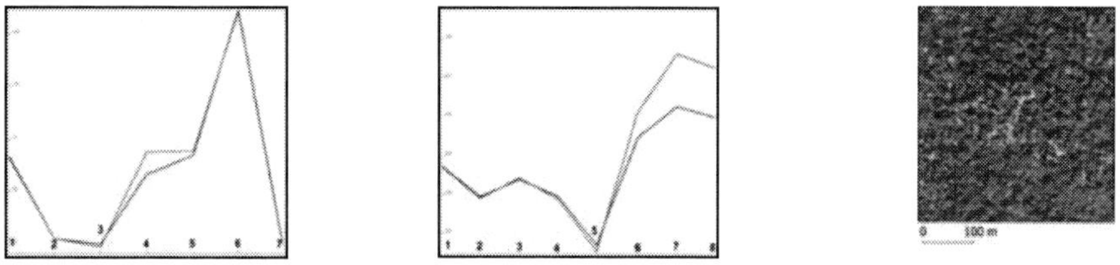

Fig. 4. Averaged spectral signatures (graphs) of granitoids (red lines) and vegetation (green lines) extracted from Landsat TM (left) and WV2 (centre) images. X – axis shows numbers of spectral bands numbers, Y – axis – DN values of reflectance.

Mixture of exposed rocks and vegetation (right) visible in close-up image of sampled area (WV2).

Vegetation which appears a particularly oppressive obstacle in scanning rocks and sediments in visible and infrared cannot be removed by any means of digital processing. However, vegetated areas can be eliminated prior to more advanced stages of the image processing to speed it considerably up. Vegetation mask, a binary (Boolean) raster map created by binarization of the vegetation index (e.g. NDVI) map, is a quite effective here. As one could expect, a share of not vegetated areas susceptible to lithological RS mapping may be quite small in tropics. In the presented fragment of the Vientiane area (Fig. 5) it is ~19%.

Fig. 5. Left: NDVI (Normalized Differential Vegetation Index). WV2 imagery. Vegetation is shown in shades of green, rocks, sediments and soil – in shades of brown, water –blue. Right: binary vegetation mask. 1 – areas with sparse or no vegetation, 0 – vegetated areas. Threshold set to NDVI = 0.45

4.1 Band ratioing

Clay, ferrous oxides, ferric iron, ferrous iron, laterite and gossan are standard band ratios applied commonly in geological mapping and prospection. They are calculated as a division of band in which reflectance of an object (e.g. iron oxides) is the highest by a band in which its reflectance is minimal. Atmospheric haze correction is sometimes added. Iron oxide ratio formula as applied to Landsat TM and ETM+ images is given below as an example:

(B3 – RMIN(,'All',I1)) / (B2 - RMIN(,'All',I2))

where: B3 and B2 – band 3 (visible red) and 2 (visible green) of Landsat TM and ETM+ scanner; RMIN(,'All',I2) – component removing atmospheric haze (in C++ syntax).

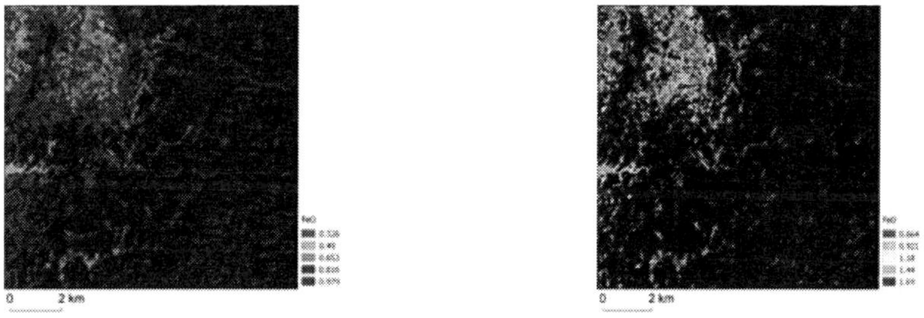

Fig. 6. Iron Oxide ratio masked with vegetation mask. Left: WV2 image, right: Landsat TM image. Relative values, not standardized

Band ratio is a quantitative measure of intensity of a phenomenon, e.g. oxidation of iron minerals in a rock and, therefore, indirectly, occurrence of the rock itself (Figs. 6 and 7). Scale of values is relative. It requires standardization in every particular case.

Image of band ratio of extensively vegetated area is more precise when high resolution WV2 images are used as an input (Fig. 6, left). Landsat images give more rough, generalized and less precise result (Fig. 6, right). However, processing of TM or ETM+ images gives satisfactory results in initial stages of prospection of bigger areas (Fig. 7).

 https://doi.org/10.1051/e3sconf/20183502004

Fig. 7. Iron Oxide ratio masked with vegetation mask. Xiangkhouang area. Landsat ETM+ image. Relative values, not standardized

4.2 Classification

Classification is definitely the most advanced tool in the RS analysis. Unsupervised classification (ISODATA – Iterative Self-Organizing Data Analysis Technique) is conducted most frequently in areas poorly recognized that suffer from the lack of reference data (geological maps, field survey, sampling etc.). This is probably the case of most areas in tropics. All limitations of application of imagery of different scanners mentioned in previous sections apply here too. Two stages are new here:

1. pre-processing stage, i.e. selection and/or decorrelation of bands to be used in classification and
2. verification (qualitative or/and quantitative) of results of the classification.

Selection of spectral bands for classification may considerable improve its results. Normally short wave visible blue bands of WV2 scanner (Coastal Blue), Landsat OLI (Coastal-Aerosol) and Sentinel 2 (Blue Edge) are omitted. They are usually biased by atmospheric haze. Decorrelation of bands removes repeating information. Two methods are applied here: decorrelation stretch and extracting principal components (PC), usually first, second and third, sometimes fourth. Classification of PCs instead of original images may result in better separability of classes and their internal consistency. However, may considerable prolong the classification process to hours or even tens of hours (depending of the image size, number of PCs and efficiency of the computer).

Figure 8 shows result of unsupervised classification of WV2 multispectral images. Bands 2 thru 8 were used and 30 classes were extracted. Classes representing vegetation and built-up areas were eliminated.

Verification of the classification results requires reference data. The best of course is a field check, which is in most cases unfortunately unavailable. Geological maps are very good reference data, especially in big scales, i.e. 1:10.000 (very rare), 1:25.000 and 1:50.000. Medium and small scale maps (1:200.000 to 1:1.000.000) allow only for general reference. (BTW: areas mapped in 1:10.000 or 1:25.000 scale essentially do not require RS mapping!). Finally, high resolution (0.5 m or less of pixel size) ortophoto maps available for observation in e.g. Google Earth service or downloadable from Google Maps repository are good reference images. They allow observation of objects on the Earth's surface down to a size of a single stone or small outcrop in nearly natural colours. However, spectral analysis, although possible (after separation of RGB or HSI components), is not reliable.

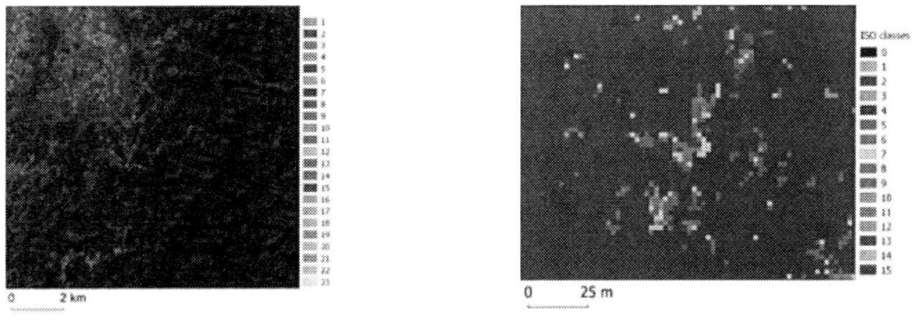

Fig. 8. Classified WV2 images masked with NDVI mask (left). Right: image zoomed and centred onto outcrops of granitoids shown in Figures 3 and 4. Barren rocks are represented on both maps by classes 9–12 shown in shades of green. Other classes represent probably rocks covered with weathering residuals and/or sparse vegetation.

Detailed geological map or results of the field mapping may be used as reference data for quantitative verification of the classified images. Cross tabulation (contingency table) is normally applied here (correlation testing cannot be conducted due to not random and not continuous character of the data). Statistical measures (overall accuracy, kappa statistics) allow for evaluation of classification results. If not satisfactory – the classification should be repeated with modified set of the data or/and with modified parameters.

4.3 Analysis of DEM

In many cases geological structure of both, hard rocks and sediments, reflects in geomorphology. Digital Elevation Model (DEM) which, further on, shows geomorphology, allows, therefore, for imaging and interpretation of geological structures. It applies to vegetated areas, including tropics, if DEM is acquired by active microwave (radar) scanning, as a large portion of microwave radiation penetrates vegetation. SRTM (Shuttle Radar Topographic Mission, NASA) and GDEM (Aster, NASA) are good sources of global DEM. Processing the latter appeared quite useful in structural interpretation of the Vientiane study area. Analysis of shaded relief model and DEM derivatives: slope angle, slope aspect, slope curvature, ruggedness, and microtopography was applied. Geological interpretation was supported by the Geological map of Cambodia, Laos and Vietnam [10]. Results are presented on a backdrop of shaded relief model (Fig. 9).

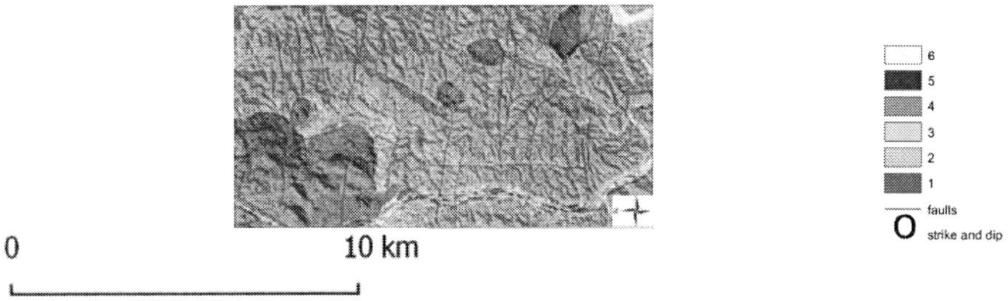

Fig. 9. Shaded DEM with interpreted lithology and structure. Illumination from NW. Items 1–6 are explained in the text

Analysis revealed 6 different types of the DEM textures.

1. Very coarse and massive texture with deep cuts and steep elevations, located in NW corner of the area shows probably thick bedded Devonian limestone and sedimentary

rocks dipping generally towards NW. They form structurally controlled mountain ridges trending SW–NE.

2. Medium coarse and moderately sculptured texture which occupies the largest portion of the area. Probably represents Carboniferous limestone, mudstone, siltstone and other thick bedded sediments. The beds dip generally towards NW and SE and form probably tight SW–NE striking folds. Structurally controlled ridges run in SSW to SW–NNE to NE and S–N directions.
3. Relatively medium grained texture that occurs in a crescent-shape along S, SE and E rim of the Devonian and intrusives (5). They may reflect Carboniferous and/or Devonian sediments changed by the intrusions. However, this texture may as well be a representation of subvulcanic intrusive rocks.
4. Fine to medium grained texture with low ridges and shallow valleys trending generally W-E represents probably Permian thin to medium bedded, relatively soft rocks. They occur along the E rim of the area and in its SE-most corner.
5. Medium to fine grained textures that stretch generally SE–NW across the area and border Devonian rocks in the NW corner. They reflect intrusive granitoids.
6. Nearly smooth texture of C-shape in SW corner of the area: Mekong river palaeo-valley.

Linear structures (faults) form 4 systems: W–E, N–S, SSW–NNE and SW–NE. Three latter seem dominating.

4 Conclusions

Intensive and extensive vegetation is the main factor controlling and limiting application of RS techniques in geological mapping of tropical regions. Therefore, good spatial resolution of imagery is crucial for detecting spectral response of rocks, sediments and soil "visible" essentially only in gaps among trees and bush and in clearings. The smaller pixel the more selective is the spectral signature gathered. Images of WV2 and WV3 scanners (Digital World) with 2 m pixel size represent actually top achievements in high resolution satellite RS imaging. Medium resolution Landsat TM, ETM+ and OLI as well as ESA Sentinel 2 scanners (10–60 m pixel size) seem not appropriate for analyzing extensively forested areas with small and rare clearings but may be good for mapping extensive areas where larger spots are dispersed with sparse or no vegetation (Fig. 6 right and 7).

Spectral and radiometric resolution are equally important in RS mapping in tropics and in any other area. Visible red and mid infrared wavelengths are important in mapping of lithology of rocks and sediments. In particular it applies to mid infrared bands around 2 μm wavelength. "Blue-edge", visible blue and yellow (e.g. band 4 of WV2) ale less informative. Radiometric resolution of 11 and 16 bits per pixel seems actually a standard in modern scanners. It allows for very precise fragmentation of even very subtle characteristics of a reflectance spectrum of an object and, thus, discrimination of quite similar objects.

Thermal infrared (emissive) images represent quite another opportunity in many fields of RS processing and analysis, i.a. in lithological mapping. It was not applied in the presented study.

Digital Elevation Model can be used in very sophisticated analysis of geomorphology related to structures of bedrock and sediments. Laser scans (LIDAR) are suitable only for not vegetated areas. Highly penetrating microwave (radar) scanning yields very useful elevation data. Its relatively low resolution (30–90 m) is not a substantial obstacle in recognizing even small scale geological structures: faults, bedding and its dip and strike (and, further on, folds), intrusive bodies, abrupt changes of lithology, as well as covers of soft sediments.

Remote Sensing supplies us with almost perfect information on lithology and structure of rocks and sediments when applied in arid areas with no or sparse vegetation (N and E Africa, Central Asia, W of both Americas, subpolar regions free of snow and ice). Used in tropical regions may yield interesting results, providing it is supplemented with interpolation of gained information to the vegetated ("hidden") fragments of the analyzed terrain. In this context we should ask here a question: is it possible to make any "classic" geological map without any interpolation of gathered observations and data over such "hidden" areas?

Acknowledgements

Publication resulted from the Faculty statute research project No. 11.11.140.320.

References

1. J.G. Liu, P.J. Mason, *Image processing and GIS for Remote Sensing* (Willey Blackwell, 2016)
2. R.P. Gupta, *Remote Sensing Geology* (Springer, 2003)
3. K. Won-In, P. Charusiri, Int. J. of Applied Earth Observ. and Geoinformation, **4**, 183–193 (2003)
4. L.Q. Hung, O. Batelaan, IGARSS 2003, **4**, 2964–2966 (2003)
5. B. Pradhan, S. Lee, Earth Science Frontiers, **14**, 143–151 (2007)
6. https://landsat.usgs.gov
7. https://sentinels.copernicus.eu/web/sentinel
8. www.digitalglobe.com
9. https://terra.nasa.gov/about/terra-instruments/aster
10. *Geological map of Cambodia, Laos and Vietnam, 1:1.000.000*, Geol. Surv. of Vietnam, Hanoi (1991)

Relative secular variations of the geomagnetic field along the Zgorzelec-Wiżajny profile, Poland

Anna Wojas[1,*], *Teresa* Grabowska[1], and *Mateusz* Mikołajczak[2]

[1] AGH University of Science and Technology, Faculty of Geology, Geophysics and Environmental Protection, al. Mickiewicza 30, 30-059 Krakow, Poland

[2] Polish Academy of Sciences, Institute of Geological Sciences, Krakow Research Centre, Senacka 1, 31-002 Krakow, Poland

Abstract. The paper presents results of the study on relative secular variations of total magnetic intensity (TMI) of the geomagnetic field along the 700 km long profile crossing the area of Poland. Surveys were carried out at annual intervals between 1966 and 2016 (50 measurement series), in 31 survey sites (secular points) separated by about 22 km. The studied profile of the SW-NE direction, called Zgorzelec-Wiżajny (Z-W), crosses large parts of the main tectonic units of Europe, namely the Palaeozoic Platform of Central and Western Europe (PLZ) and the East European Craton (EEC), connected by the Teisseyre-Tornquist Zone (TTZ). Using the original methodology of analysis of measured data, reduced to the values of geomagnetic field recorded at the Central Geophysical Observatory in Belsk, the relative secular variations of TMI with the magnetic anomalies (ΔT) and the terrestrial heat flow density (Q) were graphically presented. **Key words** – relative secular variations, geomagnetic field, Palaeozoic Platform, Teisseyre-Tornquist Zone, East European Craton.

1 Introduction

The first map of secular variations of the geomagnetic field in Poland has been presented by Kalinowski at the Congress of the Geodetic-Geophysical Union in Stockholm in 1930. The first scientific papers dealing with this issue were published in 1936, in the Geophysical Society Bulletin. After the Second World War, in 1952, Olczak presented a paper in the PIG Bulletin about secular variations of the Earth's magnetic field (D, I, H, Z) in the period of time 1900–1950. Since 1957, the Institute of Geodesy and Cartography in Warsaw has been carried out surveys of the geomagnetic field at 20 stabilized secular points (repeat stations) in Poland [1, 2]. The results with comments were compiled in the form of tables and maps (D, H, T) representing 20 year period (1970-1990) [3]. An important summary of these studies is provided in a paper by Welker and Zoltowski [4], presenting the normal field of secular variations of the geomagnetic field for the Polish territory. It is also worth to mention a paper by Welker [5] that presents changes in the geomagnetic field (D, T) in the last

* Corresponding author: awojas@agh.edu.pl

quarter of the century in Central Europe. The author emphasized the necessity of systematic measurements of the geomagnetic field in secular points.

The studies of secular changes, presented in this paper, were initiated in the 1960s by Maloszewski [6] and have been continued until present (1966–2016). The geomagnetic field surveys are performed along the 700 km profile Zgorzelec-Wiżajny (Z-W) (Fig. 1). The profile runs across fragments of two main European tectonic units, namely the Palaeozoic Platform of Central and Western Europe (PLZ) and the East European Craton (EEC) connected by the Teisseyre-Tornquist Zone (TTZ). The aim of the study is to determine the local relative secular variations of TMI and attempt to assess their causes [7].

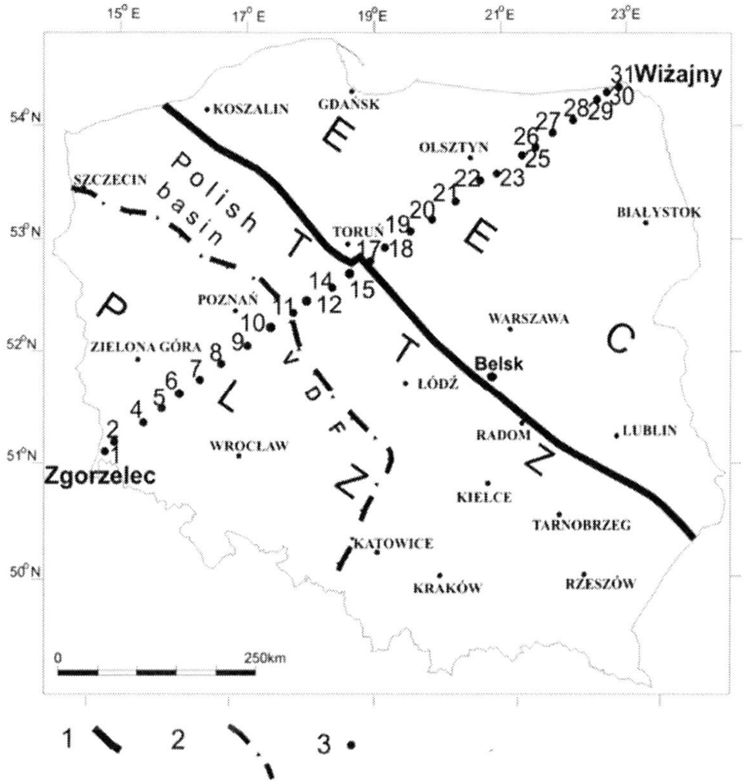

Fig. 1. Location of secular points along the Zgorzelec-Wiżajny (Z-W) profile. Explanation: 1 – Teisseyre-Tornquist Zone (TTZ) according to deep seismic surveys [8], 2 – Variscan Deformation Front (VDF) [9], 3 – secular points. Other abbreviations: EEC – the East European Craton, PLZ – the Palaeozoic Platform of Central and Western Europe

2 Methodology of survey and results processing

Annual surveys of TMI of the geomagnetic field were made in 31 GPS-determined points along the Z-W profile. Magnetic measurements were conducted with the use of proton magnetometers with 0.1 nT sensitivity. In total, 50 measurement series of TMI have been accomplished in the years 1966–2016. A series of measurements has been made at each point, called a secular point (approximately 60 records with 1 minute repetition) in order to validate the TMI value. After that, the TMI values in each secular point (n) were reduced to the values of the TMI recorded at the same time at the Central Geophysical Observatory in Belsk (formula 1). The definitive value of relative secular variation ($\Delta T_{n,t}$) in a secular point (n) in the year (t) is an average value of reduction results.

$$\Delta T_{n,t} = \frac{1}{60}\sum_{1}^{60}\left|\vec{T}_{obs,n,(\tau+\Delta\tau)}\right| - \left|\vec{T}_{Belsk,\tau}\right| \quad (1)$$

where:
τ – time of measurement in GMT (Greenwich Mean Time),
$\Delta\tau$ – time correction related to the difference in longitude between measurement point and the Belsk Observatory.

Methodology of the TMI relative variations determination and presentation along the Z-W profile was published by Maloszewski and Jankowski [10] and it is used as such until present. On this basis, the values of $\Delta T_{n,t}$ in each secular point in the 50-years time period (1966–2016) were approximated by the first-order polynomials (formula 2) and graphically illustrated by straight lines, expressing so-called trends of relative secular variations in each secular point within 50 years (Fig. 2). The next stage was graphical illustration of slope coefficient (formula 3).

$$W_n(t) = b_n t + a_n \quad (2)$$

$$b_n = \partial(\Delta T_{n,t}) \quad (3)$$

where:
b_n – slope coefficient, which corresponds to the relative secular variation of the annual geomagnetic field in an individual secular point of the profile,
a_n – mean value of relative values of the geomagnetic field in a secular point, in relation to the mean value of the geomagnetic field measured in the Belsk Observatory, during 50 years of study.

It should be mentioned that studies of relative changes of the geomagnetic field were limited to the territory of Poland and can be treated as local.

The relative secular variations shown in this paper were analysed considering the geological structure of Poland, as well as the observed magnetic anomalies (ΔT) and values of terrestrial heat flow density (Q) along the Z-W profile.

3 Magnetic anomalies along the Z-W profile against the geological setting

The magnetic anomalies (ΔT) pattern along the Z-W profile is characteristic of the central part of the Polish Lowlands. The middle part of the profile (11–17 secular points) (Fig. 1) crosses the Permian-Mesozoic Polish Basin. In its axial part (the Mid-Polish Trough), Permian-Mesozoic sediments were deposited with a thickness of up to 8 kilometres [11]. According to deep seismic surveys [12], the thickness of the entire Phanerozoic sedimentary complex in this area reaches 12–13 km. The middle part of the profile corresponds to low (negative) values of magnetic anomalies (magnetic low, Fig. 3b), that terminates from the NE against a zone of strong horizontal magnetic gradient coinciding with the TTZ and dividing Poland into two magnetic provinces.

The SW part of the Z-W profile crosses the Fore-Sudetic Monocline (section of the PLZ), where the depth to the pre-Permian basement is c. 1–4 km. In this area, a weak positive magnetic anomalies zone is present, which is probably related to Variscan metamorphic basement occurring at a depth of c. 3–6 km [13].

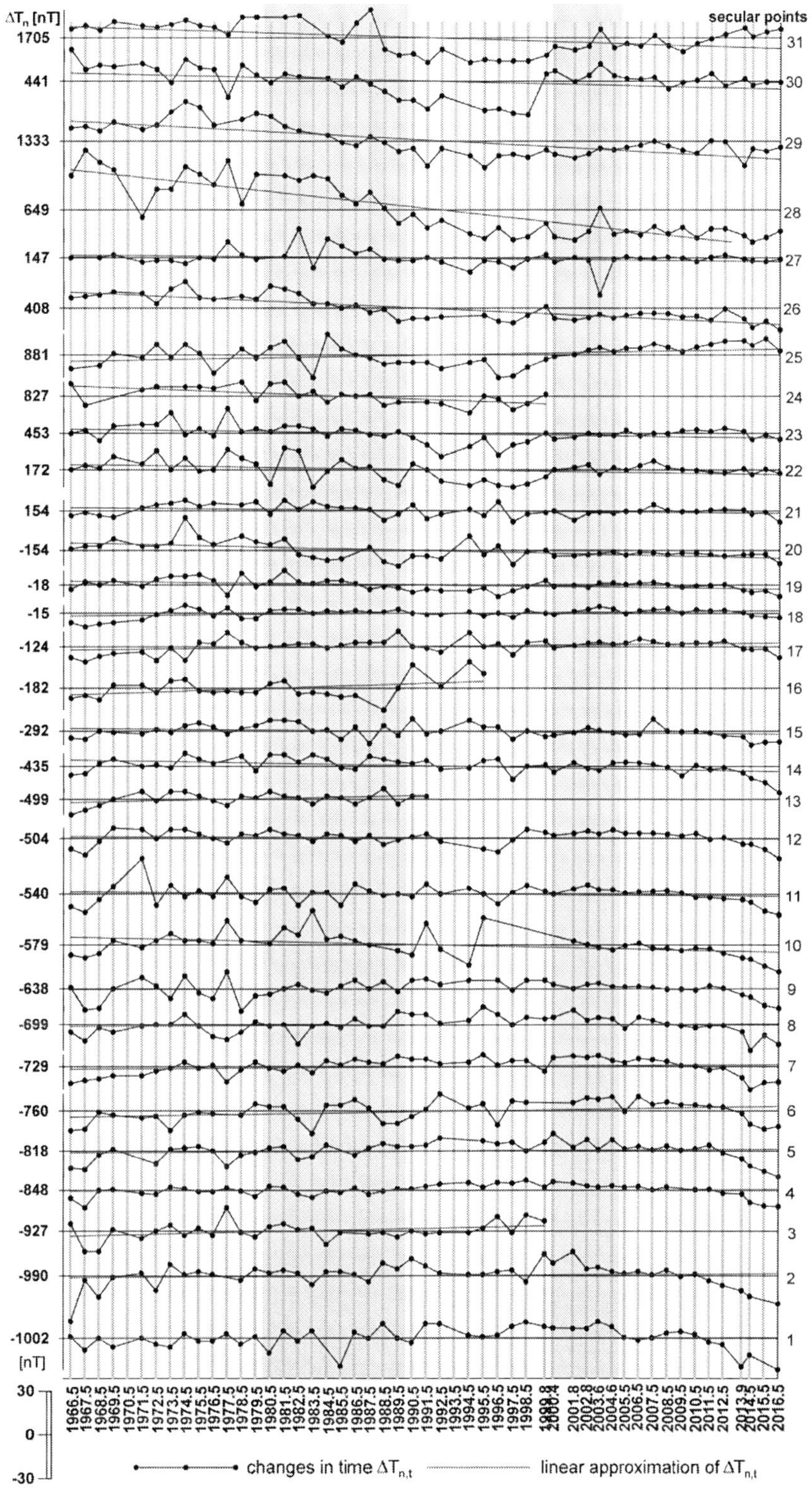

Fig. 2. Changes in time of the relative geomagnetic field values ΔTn,t in individual secular points of the Zgorzelec-Wiżajny (Z-W) profile. Explanation: ΔTn – mean values of $\Delta T_{n,t}$ in the period 1966–2016 for individual secular points, grey background – periods of time (1980–1990 and 2000–2005), when systematic changes of total magnetic intensity (TMI) in Central Europe have been observed [5]

The magnetic anomalies in the NE section of the Z-W profile reflect the geological structure of the Polish section of the EEC. The profile runs over the Dobrzyń granitoid massif, next to the Ciechanów fold zone, and its Warmia branch, which are under the cover of sedimentary rocks with NE-ward decreasing thickness (8–1 km). The metamorphic rocks in these zones are source of numerous magnetic anomalies with amplitudes in the order of hundreds of *nT*. The NE termination of the profile is located in the area with strong (over 1 000 *nT*) magnetic anomalies associated with the Mazury complex of rapakivi-like granitoids and the Suwałki anorthosite massif with the iron ores.

4 Results

Figure 2 shows the relative secular geomagnetic variations in individual secular points (1–31) of the Z-W profile in the years 1966–2016. On a vertical axis on the left side of the graph, the mean values of changes (ΔTn) for each point are given using the same scale. From the point 21 (Ciechanów zone, EEC) to the end of the Z-W profile, the ΔTn values are positive, in other words the geomagnetic field recorded in this area is stronger than in the Belsk Observatory.

Attention should be paid to the areas with the most rapid changes of the TMI secular variations. The amplitudes of changes are over 40.0 *nT* at those sites. The largest changes of the geomagnetic field are in the region of the EEC (secular point no 27, 28, 29, 30, Fig. 2) with a maximum amplitude of $\Delta T_{n,t}$ in the years 1966–2016 (61.5 *nT*) being observed in the secular point 28 (Mazury complex). Significant amplitudes of the field changes were also observed in the secular points 2 and 10, located in the PLZ area (49.0 *nT* and 42.8 *nT*, respectively).

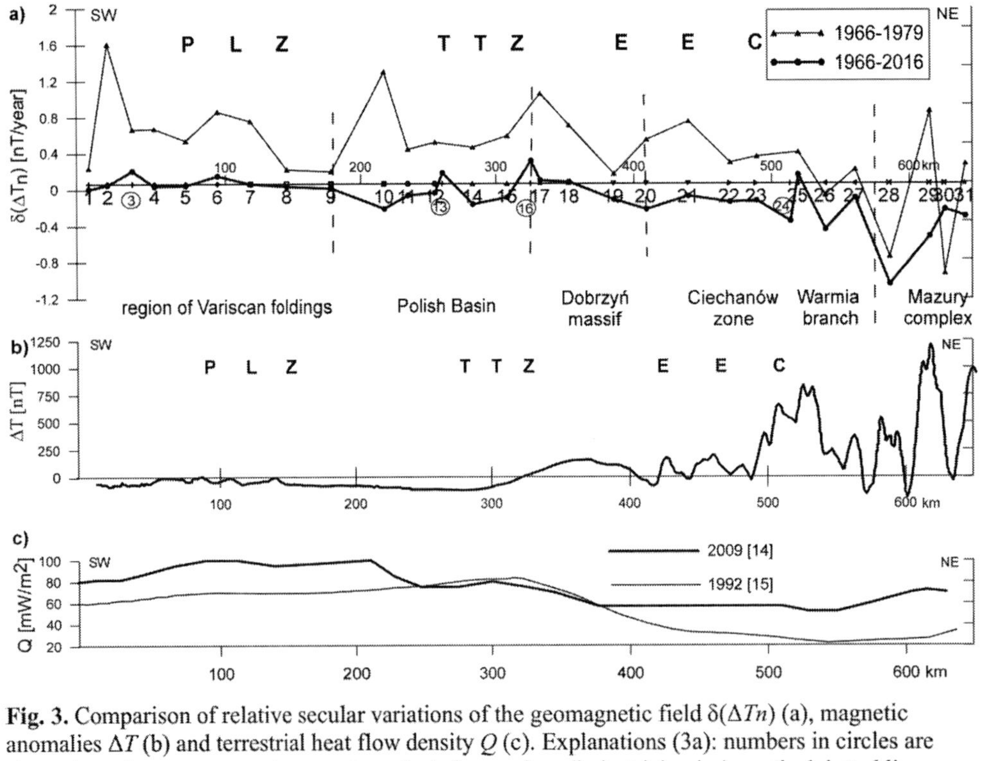

Fig. 3. Comparison of relative secular variations of the geomagnetic field $\delta(\Delta Tn)$ (a), magnetic anomalies ΔT (b) and terrestrial heat flow density Q (c). Explanations (3a): numbers in circles are sites where the measurements are not conducted nowadays (industrial noise), vertical dotted line – division into sections with different rates of change

Figure 3a presents the relative secular variations of the geomagnetic field in the two time intervals, i.e. 1966–1979 and 1966–2016 (13 and 50 measurement series, respectively). The curves obtained show trends of changes in relation to the geomagnetic field recorded in the Belsk Observatory, in the PLZ and EEC areas. The graph of the relative secular variations along the Z-W profile was compiled with the observed magnetic anomalies (ΔT) (Fig. 3b) as well as changes of the terrestrial heat flow density (Q) (Fig. 3c).

5 Discussion

In the years 1980–1990 and 2000–2005 (grey background in Fig. 2) systematic changes of the geomagnetic field were observed in Central Europe [5]. However, the years 1990–2000 were considered to be a period of disturbances in the regularity of field changes. Comparing the results of the relative values of the geomagnetic field during these three time periods in each secular point, the similar to established by E. Welker's character of field changes was noted in the few secular points of the Z-W profile. This means that probably irregular changes of the geomagnetic field, which were observed in Central Europe in the years 1990–2000, did not affect the results of the survey along the Z-W profile.

It is also interesting to note that since 2000 slight relative secular variations of the geomagnetic field are observed along the Z-W profile (Fig. 2). At the beginning of this period a lesser rate of field changes dynamics was noted by Welker [5], who defined this period as a time of return to regular changes (2000–2005). It should be emphasised that the decrease in the dynamics of field changes along the Z-W profile since 2000 caused the decrease of the bn coefficient, corresponding to relative secular variations (Fig. 2).

The analysis of relative secular variations for 13 (1966–1979) and 50 (1966–2016) measurement series (Fig. 3a) along the Z-W profile, highlights the similarity of two curves. For example, in the region at the boundary of the Polish Basin and Dobrzyń massif, the increase of relative secular variations (bn coefficient) for both curves is in agreement.

The adopted methodology of data processing and a long time of the geomagnetic field observations allow us to relate the relative secular variations of the geomagnetic field to the phenomena occurring in the Earth's crust and upper mantle as well as their different physicochemical properties affecting the transmission of the slowly changing geomagnetic field generated in the outer core. The causes of secular variations can be displacements of magnetic bodies during tectonic movements (tectonomagnetic phenomena), earthquakes (seismomagnetic effects), thermal, elastic or chemical interactions associated with changes in rocks magnetization or electrical conductivity changes (telluric currents). Time changes of the geomagnetic field may also involve a change of the induced magnetization of rocks in the Earth's crust as a result of changes of the normal geomagnetic field.

The character of relative secular variations along the Z-W profile shows a lower rate of changes (relative to the Belsk Observatory) of relative secular variations in the EEC area in comparison with the PLZ (Fig. 3a). This phenomenon was found earlier [10] and its relationship to the processes occurring in the upper mantle is not excluded. The slowdown in the increase of geomagnetic field in the EEC area (up to c. –1.0 nT/year) is especially well visible in the region of rapakivi-like granitoids and the Suwałki anorthosite massif (strong magnetic anomalies, Fig. 3b). The other part of the Z-W profile, including the area of the PLZ, is characterized by temporal dynamics of relative secular variations decreasing in time, a quiet magnetic anomalies pattern and the increasing values of terrestrial heat flow density (Fig. 3c).

In Figure 3c, the terrestrial heat flow density along the Z-W profile is presented. The variation of this parameter defines a hot province over the PLZ and TTZ, and the cold domain corresponding to the EEC, the characteristics providing important, additional information to the study of relative secular variations of the geomagnetic field. The lower terrestrial heat flow density in the EEC associated with a larger thickness of the crystalline crust indicates slower heat conduction from the mantle, which may affect the recorded geomagnetic field and cause the observed slower relative secular variations in the EEC area.

6 Conclusions

The results of observations carried out for the last 50 years along the Zgorzelec-Wiżajny profile support the following conclusions:

1. The character and dynamics of the relative secular variations of the TMI of the geomagnetic field along the Z-W profile correspond to the geological division of the studied area.
2. The observations of the relative secular variations of the geomagnetic field are important in the study of the Earth's crust, especially in places where highly magnetic rocks are present. The observations may also be affected by processes occurring in the crust-upper mantle interface.
3. The EEC area is characterized by a slower rate of the geomagnetic field changes, compared to the field recorded at the Central Geophysical Observatory in Belsk. The relative secular variations in this region during the last 50 years of study amount to c. –1.0 *nT*/year. Sources of the relative secular variations in the EEC are still not recognized.
4. In the area of the Palaeozoic Platform of Central and Western Europe, the higher rate of the geomagnetic field changes is observed. It should be noted that the results of measurements in this region are closer to the value of the geomagnetic field recorded in the Belsk Observatory.
5. Long-term studies show that since 2000 the relative secular variations of the geomagnetic field along the Zgorzelec-Wiżajny profile were low. They may correspond to the regular secular variations observed in Central Europe in this time.

Acknowledgements

The studies were carried out by the Department of Geophysics, statutory research of the Faculty of Geology, Geophysics and Environmental Protection, AGH-UST, grant no 11.11.140.768. The results presented in this paper rely, among others, on the data collected at the Belsk Observatory. We thank the Polish Academy of Sciences, for supporting its operation and INTERMAGNET for promoting high standards of magnetic observatory practice (www.intermagnet.org).

Referencess

1. A. Uhrynowski, Works of the Institute of Geodesy and Cartography **24**, 2, 3–32 (1977a)
2. A. Uhrynowski, Works of the Institute of Geodesy and Cartography **24**, 3 (1977b)
3. A. Uhrynowski, Works of the Institute of Geodesy and Cartography **39**, 1, 87 (1992)
4. E. Welker, A.M. Żółtowski, Works of the Institute of Geodesy and Cartography **40**, 1, 88, 103–118 (1993)
5. E. Welker, Works of the Institute of Geodesy and Cartography **53**, 111 (2007)
6. S. Małoszewski, Scientific Papers of the University of Mining and Metallurgy (AGH) **44** (1965)
7. G. Bojdys, T. Grabowska, Acta Geophys. **55**, 2, 169–181 (2007)
8. N. Narkiewicz, A. Maksym, M. Malinowski, M. Grad, A. Guterch, Z. Petecki, J. Probulski, T. Janik, M. Majdański, P. Środa, W. Czuba, E. Gaczyński, L. Jankowski, Int. J. of Earth Sciences **104**, 775–796 (2015)
9. R. Dadlez, Z. Kowalczewski, J. Znosko, Kwart. Geol. **38**, 2, 169–190 (1994)

10. S. Małoszewski, J. Jankowski, Acta Geophys. Pol. **45**, 23–31 (1997)
11. R. Dadlez, *Mapa tektoniczna kompleksu cechsztyńsko-mezozoicznego na Niżu Polskim, 1 : 500 000.* PGI-NRI (1998)
12. M. Wilde-Piórko, M. Swieczak, M. Grad, M. Majdański, Tectonophysics **481**, 108–115 (2010)
13. A. Koblański, Scientific Papers of the University of Mining and Metallurgy (AGH) **5**, 92–93 (1989)
14. J. Szewczyk, D. Gientka, Geol. Quart., **53**, 1, 125–140 (2009)
15. M. Plewa, S. Plewa, D. Poprawa, A. Tomas, Geothermal Atlas of Europe (1992)

Modern Geodetic Measurement Techniques in Gravimetric Studies on the Example of Gypsum Karst in the Siesławice Region

Sławomir Porzucek[1,*], *Monika* Łój[1], *Karolina* Matwij[2], and *Wojciech* Matwij[2]

[1] AGH University of Science and Technology, Faculty of Geology, Geophysics and Environmental Protection, A. Mickiewicza Av. 30, 30-059 Krakow, Poland

[2] AGH University of Science and Technology, Faculty of Mining Surveying and Environmental Engineering, A. Mickiewicza Av. 30, 30-059 Krakow, Poland

Abstract. In the region of Siesławice (near Busko-Zdrój, Poland) there are unique phenomena of gypsum karst. Atmospheric factors caused numerous gypsum outcrops, canals and underground voids. The article presents the possibility of using non-invasive gravimetric surveys supplemented with geodetic measurements to illustrate karst changes occurring around the void. The use of modern geodetic measurement techniques including terrestrial and airborne laser scanning enables to generate a digital terrain model and a three-dimensional model of voids. Gravimetric field studies allowed to map the anomalies of the gravitational field of the near-surface zone. Geodetic measurement results have made it possible to accurately determine the terrain correction that supplemented the gravimetric anomaly information. Geophysical interpretation indicate the presence of weathered rocks in the near surface zone and fractures and loosened zones located surround the karst cave. **Key words** – karst, gravimetry, geodesy, terrain correction, laser scanning.

1. Introduction

Gravity and geodetic survey was carried out in "Cave in Siesławice" area. It is located in rod-like gypsum crystals and sabre-like gypsum crystals layer of Miocene age. It has a bell shape cross-section and according to the 2013 study its length is 17 m and its roof is 2 m deep below ground level [7]. It is partially filled with water (Fig. 1). The area around the profiles had a complicated topography. There were a lot sinkholes and slopes close by and on the profiles.

The aim of the study was to detect any cavities and loosened zones over and near the cave, the existence of which may threaten the stability of the cave. To make this possible it was necessary to maintain high accuracy measurements with gravimetric measurements and eliminating adverse factors affecting the measured values of gravity. These factors were the relief of the terrain and the cave itself. To eliminate gravitational influence of both factors,

* Corresponding author: porzucek@agh.edu.pl

accurate geodetic measurements were required. Therefore, the latest terrestrial laser scanning (TLS) and airborne laser scanning (ALS) techniques have been used.

Fig. 1. "Cave in Siesławice"

2. Geological settings

"Cave in Sieslawice" is located in layer of evaporates (gypsum) in Miocene series, in the Nida Gypsum cave sub-region of Busko Staszów [3, 9]. The layer of evaporates has a significant thickness (40 m) and is built of four main parts. In the footwall part there is a glassy gypsum (3.5 4 m thickness) with large crystal intergrowths. It is covered by 3 m thickness fine-crystalline gypsum. The next part is composed as rod-like gypsum crystals and sabre-like gypsum crystals. The last evaporate layer is laminated and detrital gypsum [5].

All karst caves, in this area are most often found in layer of rod-like gypsum crystals and sabre-like gypsum crystals but also in the layer of glassy gypsum. Because of the different structure of these two layers, caves of various shapes and durability are formed. Caves of glassy gypsum layer are most often trapezoidal and do not often collapse. Caves of rod-like gypsum crystals and sabre-like gypsum crystals layer have bells and lenticular shapes resulting from the scattering and collapse of their roof part [10]. In the roof part of evaporate layer the cavers are very rare, which is related to their very rapid collapse.

3. Methodology of gravimetric and geodetic measurements

The application of the gravimetric method was used due to the fact that the most contrasting physical parameter in the tested area was the bulk density. This method takes an advantage of the variability of the Earth gravitational field, depending on its structure. Every heterogeneity in the rock mass density distribution generates its own gravitational field. Such heterogeneity are, for example, various geological structures and anthropogenic forms such as faults, voids, mines shaft or drifts [4]. It means that in the first instance the gravity distribution depends on difference of rock mass bulk density, but also from the size, shape and depth of occurrence of this heterogeneity [8].

Gravimetric survey was made using the intermediate points method. The size of the study area and the number of measuring points determined the choice of one basic point. It was located in the studied area. This location provides stable conditions for conducting gravimetric cyclic (every 1 hour) to control gravimeter drift.

Gravimetric surveys was carried out using a CG-5 Autograv gravimeter by the Canadian company Scintrex. Measurement accuracy of this gravimeter is 0.005 mGal, with repeatability of 0.001 mGal. At each measuring point the observations were taken twice to avoid random error.

After field observations and their initial processing, the Bouguer anomaly values were calculated for all measuring points. Terrain relief requires the calculation of the terrain correction and the existence of a cave partially filled with water requires the calculation of the "cave correction" [1]. This meant the need for additional geodetic surveys that were carried out using new geodetic techniques.

Gravimetric studies are always supplemented by geodetic measurements. They generally refer to the spatial coordinates of the measuring points and to create a digital model of the surveyed area. In the analyzed case a number of modern measuring technologies were used, which allowed to generate the shape of both the research object and its surrounding region. Geodetic field measurements included the use of classical (tachymetry) as well as modern (terrestrial laser scanning TLS) measurement technology. These measurements have also been supplemented by a digital model of surface area derived from airborne laser scanning ALS. For this purpose, a surface model derived from a CODGiK resource at a resolution of 1 point/m^2 was taken. Tachymetric measurements included the execution of several profiles that were positioned so that their center was approximately over the void. The area of the study terrain was 500 m^2 where five gravimetric profile were made. Due to the shape of the terrain and the most accurate mapping of the gravimetric anomaly, the distance between the profile points was about 2 m (further away from the cave) and 1 m in the part above the cave. Tachymetric measurements were carried out in the local coordinate system created for the object of research. The later transformation of the local system to the Polish coordinate system "1992" allowed the data from three measurement technologies to be linked together.

4. Geodesy TLS and ALS

One of the most important tasks required to precisely design a topographic correction is a faithful and detailed mapping of the terrain. For this purpose, TLS technology has been used to achieve a dense model of terrain around a void of resolution of several thousand dots per square meter [2]. Due to the small size and high speed Faro Focus 3D scanner was applied, which makes it ideal for measurements both on the surface and underground excavations [6]. To create a model of the surface area around the void scans from 9 positions were made, and to create a 3D model of the cave, because of its small size three scan were sufficient. The height of the cave did not exceed 1m in many places, so inside of the cave scanner was only placed on the tribrach (Fig. 2). This was possible because the bottom surface of the cave was covered with a thick layer of ice.

Fig. 2. Photo of cave interior during scanner operation

To register scans usually checkboard targets and spheres were used but in difficult areas a cloud to cloud option was also applied. The registration process was carried out in Leica Cyclone 9.1. The mean absolute error for the used targets in the registration process was 4 mm, which may indicate the correctness of the process itself and its high precision. Due to the small range of the scanner, a dense measurement has covered the area over the emptiness and about 25 m around it. Correct definition of the topographic correction value required information from a larger area. The data needed for this purpose were imported from the digital model of the surface from the ALS technology (Fig. 3). Transformation the measurement data from local coordinate system into a global one made it possible to create a complete and detailed surface model with sufficient accuracy.

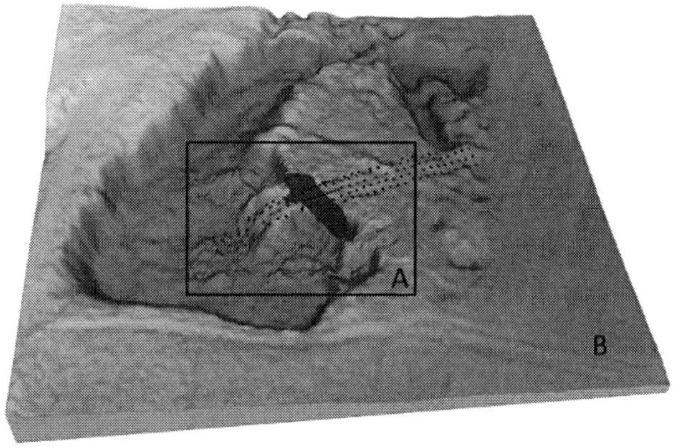

Fig. 3. Digital terrain model (DTM) based on integrated data derived from TLS (A) and ALS (B) measurements. The profile points are marked with black dots, the location of the cave is shown in a darker shade

Transformation parameters were calculated during cloud to cloud process of both TLS and ALS terrain measurements. The same transformation parameters were applied to the point cloud representing the interior of the cave. Next step was to unify the point cloud to a resolution of 25 mm and exported it to Geomagic Qualify. In this program a continuous vector model was created and on its basis cross sections with a gap of 10 cm (Fig. 4) were made.

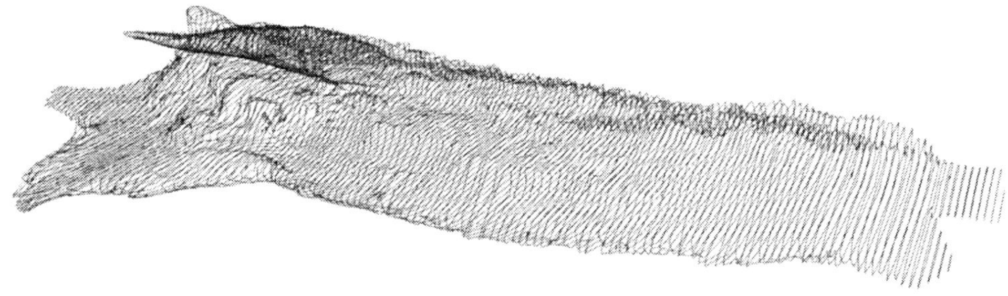

Fig. 4. Spatial model of the cave presented in cross-sections with a gap of 10 cm

5. Terrain correction

Detecting any cracks above the cave or loosened zones in the overlay requires that the gravimetric survey has the measuring error as small as possible. The gravity influence of a terrain relief is one of the significant the error, which can be eliminated by the terrain correction. By analysing the position of measuring points against the background of the relief (Fig. 3), it is easy to see that some of the points are located close to a significant depression of the terrain and even on the scarp. For this reason, it was necessary to calculate a terrain correction. For this purpose, digital numeric model was used, which was obtained from terrestrial and airborne laser scanning. A program from Geosoft's Oasis montaj package was used for calculation the terrain correction, with density equal 2.3 g·cm^3. This density was similar to the average density of gypsum. The terrain correction was calculated by collecting terrain data from a radius of 50 m around each measuring point (Fig. 3). Increasing radius did not significantly alter relative changes in correction values.

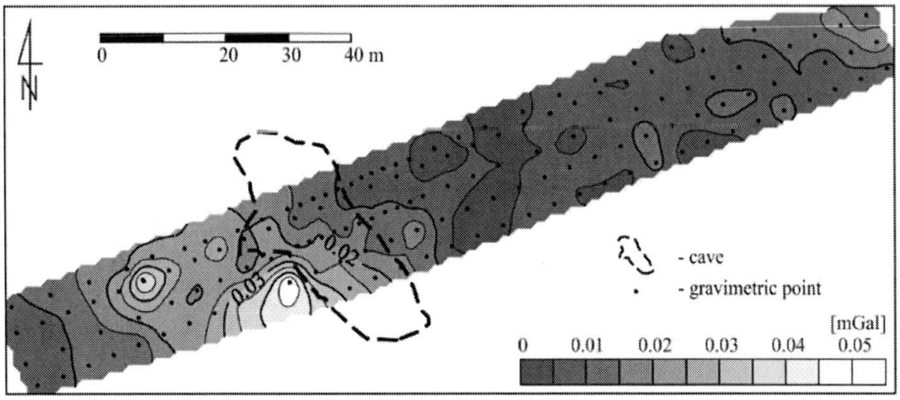

Fig. 5. Distribution of terrain correction

As expected, the terrain correction has achieved its greatest value in the region of the scarp. The maximum values were about 0.049 mGal and were significant in proportion to expected effects from possible voids, cracks and loosened zones.

6. Cave elimination

To be able to observe possible changes in the rocks around and above the cave, it was necessary to calculate the "cave correction", which gravitational influence of "Cave in Siesławice". For this reason, it was necessary to determine the geometry of the cave and its relative position to measuring points. In a survey carried out in 2013 [7] the cave geometry was determined by a tachymeter. Due to difficult measurement conditions the cave shape was very approximate. The entire cave was flooded with a depth of about 1 m, which made it impossible to enter and perform accurate observations. In the next phase of the study, the survey was made in the winter, when water was frozen. This made it possible to enter the cave. The modern geodetic laser scanning techniques were used to accurately reflect the shape of the cave (Fig. 4).

There were two parts of "cave correction": one was the correction of the empty cave (void), and the second is the correction of the water-filled cave (cavity). As show in Figure 4 the cave shape was very irregular, which made it difficult to calculate the correction. In the calculation, the method described by Gołębiowski 2016 [1], was used. In this method, the complex spatial body is divides into slices of specified thickness, which makes calculations of the correction much easier.

Based on a few measurements, the water depth at the bottom of the cave was determined. It allowed to create a model of water and to calculate the second part of the correction. Both of them, the cave and water had a bulk density much smaller than the surrounding gypsum rocks, so they reduced measured gravity value. For this reason the correction has a positive value (Fig. 6).

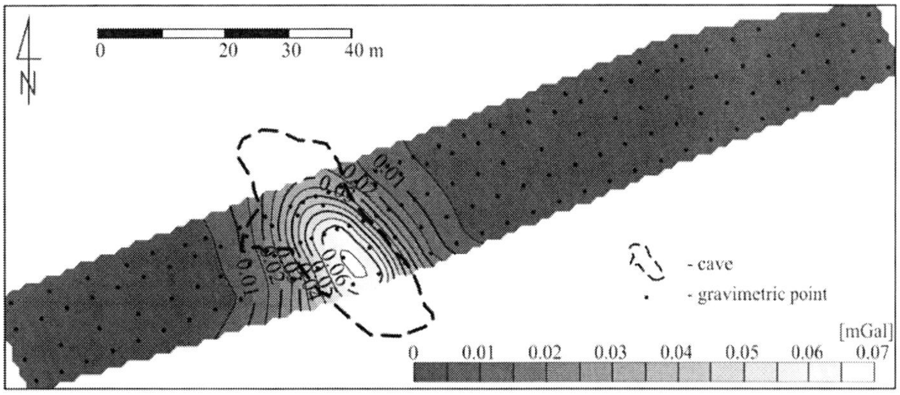

Fig. 6. Distribution of the "cave correction"

Of course, the greatest values were observed directly above the cave, which were decreasing rapidly at a distance from it. The values obtained from the calculation were significant and had a noticeable impact on the measured values. Of course, the highest correction value was obviously from the empty part of the cave (max. 0.047 mGal), while water correction was only max. 0.019 mGal. The analysis shows that the error of void correction did not exceed ± 3% and for water ± 5%.

7. Interpretation

Bouguer anomaly values were a base on the interpretation of gravimetric surveys. In the first step, they were calculated from the simple Bouguer anomaly formula, which means that only free-air correction and Bouguer correction were included. There is a distribution of this anomaly in Figure 7.

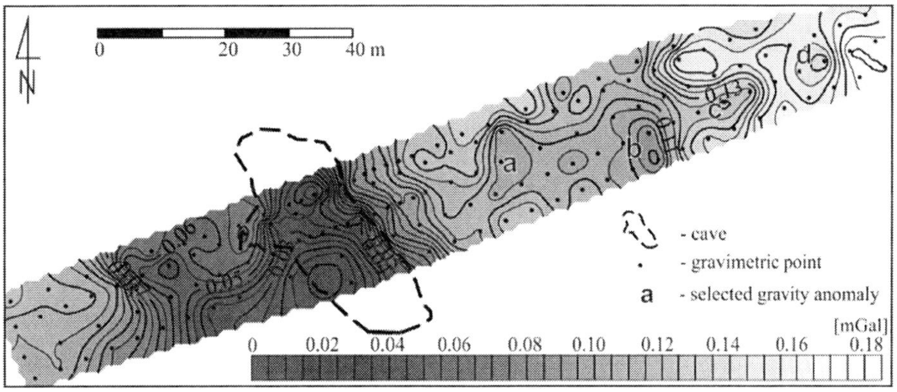

Fig. 7. Distribution of Bouguer anomalies

By analyzing the distribution, there are large variations in Bouguer anomaly values throughout the study area. This clearly demonstrates that the karst processes are responsible for the strong diversity of the bulk density of near-surface rocks. The largest changes in values are observed above the cave and on the scarp on SW from the cave. It means that a large part of these disturbances come from the cave and the scarp. There are also four minor, relatively negative anomalies (a, b, c, d) in the NE of the test area.

In the next step, a terrain correction was taken into account (Fig. 8). It has effectuated significant changes in Bouguer anomalies distribution in the scarp area, and minor changes in the rest of this distribution. There were still an area of relatively negatives anomalies in SW part of the survey area. Undoubtedly they were caused by the existence of the cave. The minor a, b, c and d anomalies were still visible but have slightly changed shape.

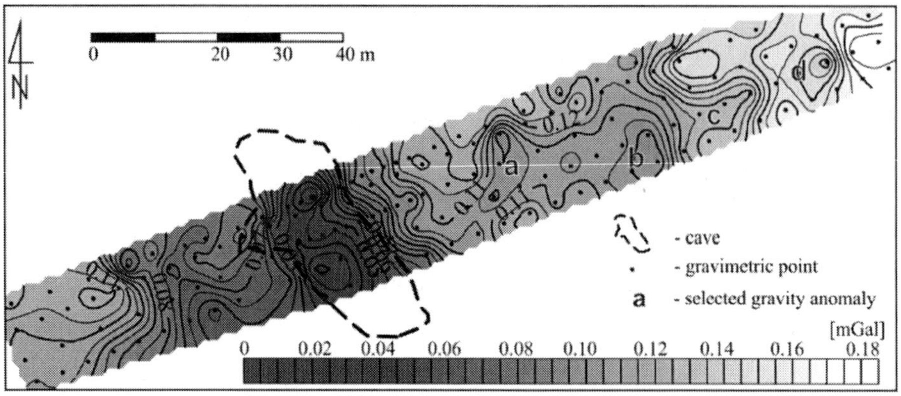

Fig. 8. Distribution of Bouguer anomalies with terrain correction

Last but not least, the correction of "Cave in Siesławice" was included. This has caused significant changes in the distribution of anomalies in the western part of the study area and practically remained unchanged in the rest of this area (Fig. 9).

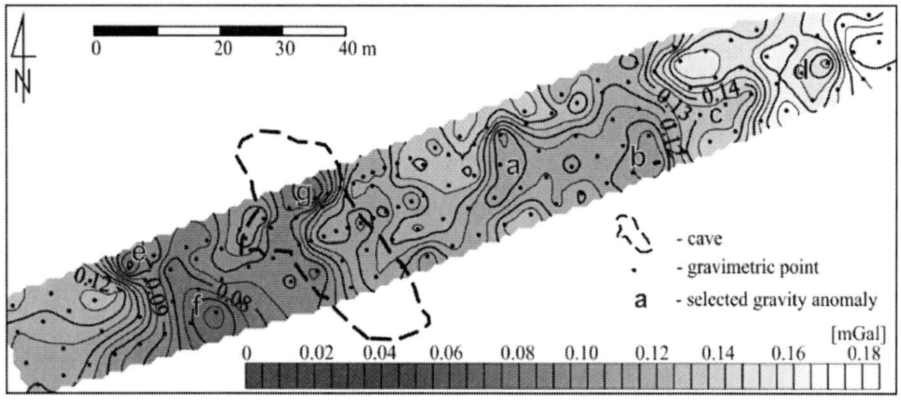

Fig. 9. Distribution of Bouguer anomalies with terrain correction and "cave correction"

Once all the corrections have been taken into account, the final interpretation of the results could be carried out. There is no doubt that the studied area was subject to strong karst phenomena. This is evident in complicated distribution of Bouguer anomalies. Undoubtedly, the a, b, c and d anomalies were derived from small, shallow voids or hollows in a gypsum filled with regolith. The a and b anomalies were located in a largest relatively negative anomaly, which may derived from a geater, deeper located void of small height.

However, the most interesting large-scale anomaly was located in the west part of the cave and partially above it. Its existence should be connected to two elements: the scarp and the cave (Figs. 3 and 9). The e and f anomalies were certainly associated with loosened zones and small voids within the scarp there were noticed fissures during the surveys. There were decreased anomaly values above the SW part of the cave, which indicates that there were small voids and loosened zones in the area. On this basis, it should be assumed that the stability of the cave in this area is at risk and threatens to collapse. In addition, the g anomaly indicates that also there is a decrease in bulk density in the northern part of the cave, which, in this case, should be associated with the existence of void. This fact makes the threat of cave collapse even more real.

8. Conclusion

Detailed gravimetric surveys require accurate geodetic measurements, and to achieve the required precision, modern measurement techniques should be used. In this work, terrestrial and airborne laser scanning (TSL and ASL) was used for the measurements. Data from these measurements have been used to create accurate digital terrain model (DTM) and spatial model of the cave. Thanks to that it was possible to eliminate their gravity influence with better accuracy.

The result of gravimetric surveys, in the form of Bouguer anomaly, showed a high concentration of karst phenomena in the studied region. Removing interfering or unnecessary information from gravimetric measurements has also shown that cave stability is at risk. It is very likely to collapse.

In conclusion, it can be said that the use of modern, precise geodetic measurements allows for better localization of even small, near-surface density changes and their accuracy significantly increases the credibility of gravimetric interpretation.

References

1. T. Gołębiowski, S. Porzucek, B. Pasierb, *Ambiguities in geophysical interpretation during fracture detection-case study from a limestone quarry (Lower Silesia Region, Poland).* Near surface geophysics, **14**, 4, pp. 371–384 (2016)
2. W. Gruszczyński, W. Matwij, P. Ćwiąkała, *Comparison of low-altitude UAV photogrammetry with terrestrial laser scanning as data-source methods for terrain covered in low vegetation.* ISPRS Journal of Photogrammetry and Remote Sensing, **126**, pp. 168–179 (2017)
3. J. Gubała, A. Kasza, J. Urban, *Jaskinie Niecki Nidziańskiej.* Wyd. PTPNoZ, Warszawa (1998)
4. Z. Fajklewicz, *Geofizyka stosowana.* Wyd. Nauk Technicznych AGH, Kraków (2008)
5. A. Kasprzyk, *Lothofacies and sedimentation of the badenian (Middle Miocen) gypsum in the northern part of Carphatian Foredeep, Southern Poland.* Annales Societatis Geologorum Poloniae, **63**, pp. 33–84 (1993)
6. T. Lipecki, W. Jaśkowski, W. Gruszczyński, K. Matwij, W. Matwij, P. Ulmaniec, *Inventory of the geometric condition of inanimate nature reserve Crystal Caves in „Wieliczka" Salt Mine.* Acta Geodaetica et Geophysica Hungarica, **51**, pp. 257–272, (2016)
7. M. Łój, T. Gołębiowski, S. Porzucek, *Geophysical surveys and modelling for recognizing of gypsum karst.* Geoinformatica Polonica, **13**, pp. 83–97 (2014)
8. M. Łój, *Microgravity monitoring discontinuous terrain deformation in a selected area of shallow coal extraction.* Conference: 14th International Multidisciplinary Scientific Geoconference (SGEM) Location: Albena, Bulgaria (2014)
9. I. Walaszczyk, *Turonian through Santonian deposit of the Central Polish Uplands, their facies development, inoceramid paleontologice and stratigraphy.* Acta Geologica Polonica, **42**, 1–2, pp. 1–122 (1992)
10. B.W.Wołoszyn, *Jaskinie Zespołu Parków Krajobrazowych Ponidzia.* Studia Ośr. Dok. Fizjogr. **18**, pp. 275–341 (1990)

Seismic Linear Noise Attenuation with Use of Radial Transform

Żaneta Szymańska-Małysa[1,*]

[1] AGH University of Science and Technology, Faculty of Geology, Geophysics and Environmental Protection, A. Mickiewicza Av. 30, 30-059 Krakow, Poland

Abstract. One of the goals of seismic data processing is to attenuate the recorded noise in order to enable correct interpretation of the image. Radial transform has been used as a very effective tool in the attenuation of various types of linear noise, both numerical and real (such as ground roll, direct waves, head waves, guided waves etc). The result of transformation from offset – time ($X - T$) domain into apparent velocity – time ($R - T$) domain is frequency separation between reflections and linear events. In this article synthetic and real seismic shot gathers were examined. One example was targeted at far offset area of dataset where reflections and noise had similar apparent velocities and frequency bands. Another example was a result of elastic modelling where linear artefacts were produced. Bandpass filtering and scaling operation executed in radial domain attenuated all discussed types of linear noise very effectively. After noise reduction all further processing steps reveal better results, especially velocity analysis, migration and stacking. In all presented cases signal-to-noise ratio was significantly increased and reflections covered previously by noise were revealed. Power spectra of filtered seismic records preserved real dynamics of reflections. **Key words** – seismic, noise attenuation, radial transform.

1 Introduction

Recording of seismic data always goes hand in hand with recording of unwanted noise. One of the goals of seismic processing is to attenuate the noise in the simplest and fastest way to enable correct interpretation of the image. The large group of unwanted energy constitutes linear noise, such as ground roll, air waves, direct waves, head waves, guided waves etc. After noise filtration all further processing steps become more reliable, especially deconvolution, velocity analysis, migration and stacking. Therefore proper signal-to-noise ratio improvement is so important.

In standard processing flow few transformations are used to treat such type of energy. Among the available tools the most popular are methods based on velocity, amplitude and / or frequency difference between reflections and noise. Author used radial transform ($R - T$ transform, apparent velocity – time transform) because presented method offered effective attenuation of all the linear noise present in data in one processing step, although it could also be used in iterative and / or adaptive mode.

* Corresponding author: zmalysa@agh.edu.pl

Radial transform was first used in migration and imaging techniques [1–2]. Attenuation of multiples was another early implementation of use of $T-R$ domain [3–4]. Although coherent noise attenuation is general application of $R-T$ transform, recently described for instance by [5–7], it has become useful in other seismic fields, as for example seismic data interpolation [8]. Filtration in radial domain has formerly been studied by the author [9–10].

2 Mathematical background

Mathematical background of radial transform was discussed in detail for example in [11–13]. Wavefield amplitudes from gather of seismic traces S with respect to radial origin (x_0, t_0) are being extracted and interpolated into coordinates of apparent velocity V and time t' (radial domain).

The $R-T$ transform is defined by:

$$R\left[S(x,t)\right]=S'(V,t') \tag{1}$$

where x is the source-receiver offset, t is two-way travel time,

$$V=(x-x_0)/(t-t_0) \text{ and } t'=t-t_0.$$

The inverse radial transform is given by:

$$R^{-1}\left[S'(V,t')\right]=S(x,t) \tag{2}$$

The result of radial transform is frequency separation between reflections and linear events. A sloping event which originates from the $R-T$ origin becomes vertical in the radial trace domain and is situated in narrow group of constant-velocity traces in the $R-T$ domain [13]. Furthermore, its temporal apparent frequency decreases significantly, while reflections, which do not pass through the $R-T$ origin, are almost unchanged in amplitude and phase. Bandpass filtering aimed at low frequency attenuation executed in radial domain attenuates such noise very effectively.

The important parameters of radial trace transform are: the interpolation method which influences the relative precision of remapping events in the $R-T$ domain [1], coordinates of the $R-T$ origin, apparent velocity range and type of transformation (radial fan transform or radial dip transform).

3 Methodology

Synthetic and real data were tested. The typical processing sequence consisted of: gather preconditioning (field statics – if necessary, normal moveut correction, top muting, gain – if necessary), forward radial transform, proper noise filtering and scaling, reverse radial transform.

Noise identification was carried out on the basis of apparent velocities and frequency bands in $X-T$ (offset-time) and $R-T$ (radial) domains. Power spectra of examined data have also been analyzed. To attenuate particular type of noise author measured its apparent velocity and examined its power spectrum in $X-T$ domain and $R-T$ domain.

Two modes of radial transform in used seismic software (Vista, Schlumberger) are available. Depending on type of noise and its strength so called 'fan' or 'dip' transform is used. The best option for the attenuation of ground roll is formation of fan of velocities, which cover the whole dataset, although this method can be used to attenuate any noise originating from the source. In this mode complete shot gather is transformed into $R - T$ domain and then processed. In radial domain one or set of bandpass filters followed by proper scaling can be used to attenuate stretched ground roll. In author's opinion this mode can be more "dangerous" because it converts the whole dataset, therefore may cause unwanted artifacts. To avoid distortion of reflections bandpass filter must be set very carefully. This results normally only in partial attenuation of noise.

Another available tool is dip radial filter which offers transformation of seismic amplitudes in limited range of apparent velocities. On the basis of apparent velocity of noise only very small portion of dataset can be processed, avoiding larger distortion of the signal. Then more 'aggresive" bandpass filter can be used. Sometimes it is difficult to attenuate all linear events in one step because of their different apparent velocities and origins. By careful velocity and frequency range identification great precision in noise separation can be achieved.

4 Results

4.1 Synthetic dataset

First examined example is a set of seismic shot gathers which were produced by elastic modelling. On the top of Figure 1 first and last shot gathers in the dataset were shown.

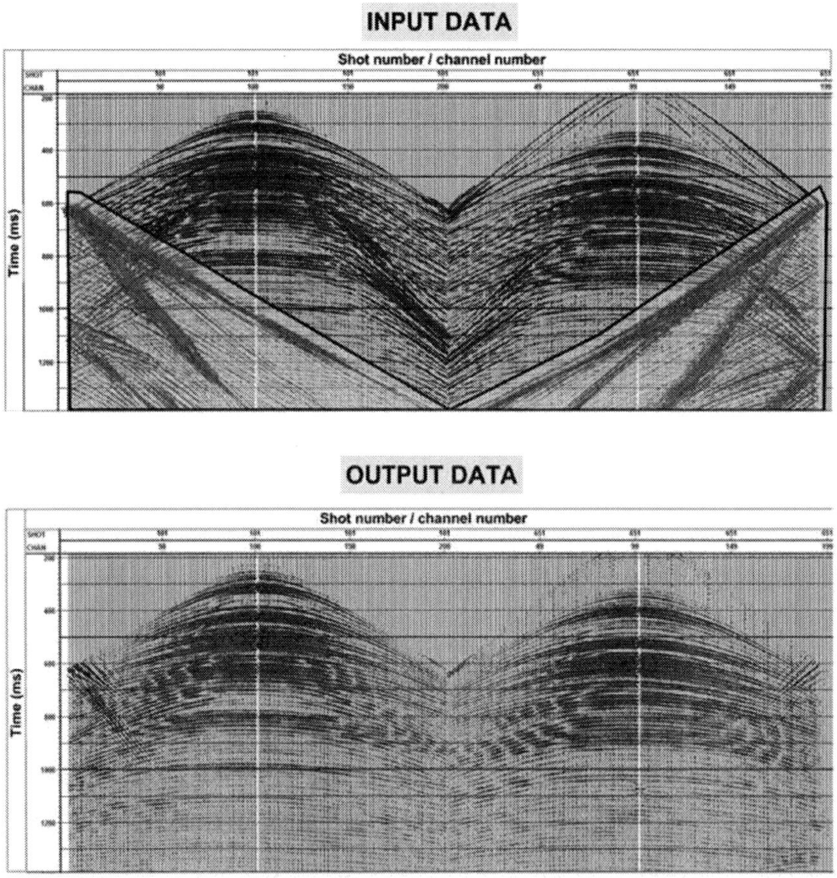

Fig. 1. Two synthetic shot records before (top picture) and after filtration in radial domain (bottom picture). Area of strong noise is marked on the top picture

The linear artifacts produced by modelling (shaded area on the top of Fig. 1) are strong compared to seismic reflections and cover large area of the dataset. This type of noise is piecewise linear with relatively broad power spectrum shifted towards low frequencies. On the bottom of Figure 1 filtrated seismic record is shown.

In presented case the optimal solution was the use of two iterations of dip transform with proper filtration and scaling in the $R-T$ domain. To limit the negative influence of the transomation narrow range of velocities was used (1600–1800 m/s, positive and negative). The shot gather 101 from Figure 1 was transformed to the radial domain where high-amplitude noise was spread all over the data and had variable curvatures (left on Fig. 2), while reflections remained hyperbolic. Two iterations were conducted: first for positive velocities, second for negative velocities. Bandpass filtering set to attenuate low frequencies was performed to limit the noise and the result in radial domain is presented on the right on Figure 2. It is obvious that bandpass filter attenuated noise very effectively and left signal almost unchanged, therefore causing relative increase of signal amplitudes.

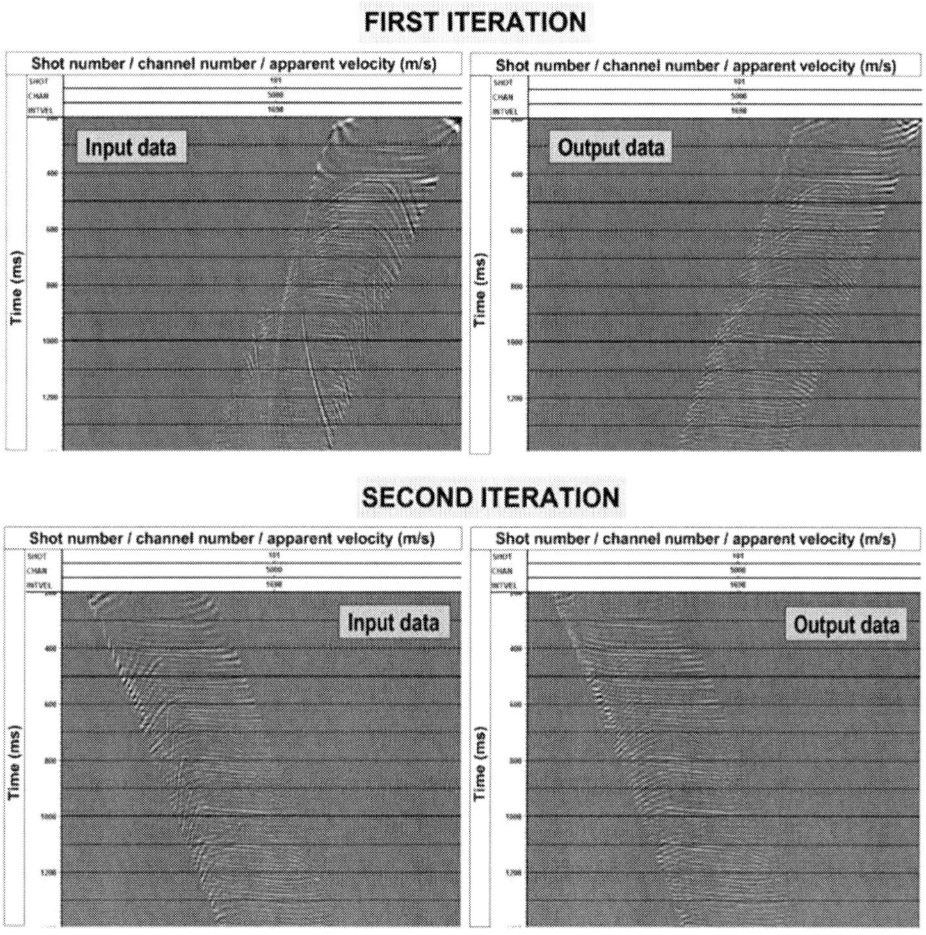

Fig. 2. Synthetic record in radial domain before and after first iteration (top picture) and after second iteration (bottom picture). Input record is shown on the left. The same record after first and second pass of bandpass filtration is shown on the right

When comparing the power spectra of input and filtered dataset (Fig. 3) it is clearly seen that proposed method left the signal almost unchanged except the area of low frequencies where noise dominated (10–30 Hz). In this zone power spectrum was slightly attenuated.

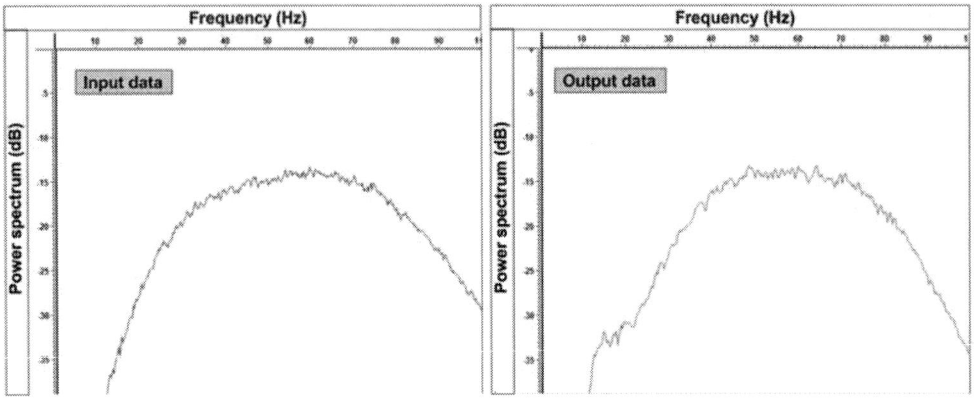

Fig. 3. Power spectrum of synthetic shot record before (left picture) and after sets of radial filters (right picture)

4.2 Real dataset

Another tests were conducted on land shot gathers acquired in Poland where very poor signal-to-noise ratio was observed. The presented real dataset was targeted on far offsets where reflections and noise had similar apparent velocities and frequency bands. At the top of the Figure 4 the far offset part of original shot record with its power spectrum is presented. The shot gather consists of reflections (very weak and almost invisible on raw data) and various types of linear noise: ground roll, direct wave, first breaks (refraction) and guided waves. Unwanted energy is here extremely strong and various filtration methods were examined to treat such kind of energy. The final results of use of radial transform were compared to results received from more common methods such as Radon (tau-p) transform and $f-k$ filter (second and third from the top on the Figure 4, respectively). The shot gathers and their power spectra were compared.

Linear Radon transform (tau-p transform) was performed to achieve noise model which was then adaptively subtracted from the shot gathers. Because transformation introduced some high-frequency noise additional bandpass filter was applied. It can be concluded from Figure 4 that such filtration caused attenuation of many of visible types of noise. Despite the fact that weak reflections became more evident (in particular reflection at 1900 ms), it was not as successful as proposed method based on radial transform.

Data transformed to the $f-k$ domain revealed the aliasing of noise caused both by too sparse receiver interval and very low apparent velocity of ground roll. Designed filter has been found ineffective in noise attenuation except ground roll case. The aliasing was the reason why data became smeared and distorted, while noise was still strong and covered reflections.

Radial transform was performed in three steps by means of both fan and dip transform. First ground roll was attenuated. Afterwards other types of noise were handled. On the bottom of the Figure 4 record after radial transform with its power spectrum is shown. From the wavefield large amount of noise vanished, consequently improving the visibility of reflections at times 1100–1200 ms, 1700 ms and 1900–2000 ms. When comparing power spectra on Figure 4 it can be seen that every filtration method attenuated large amount of low frequency energy (10–30 Hz corresponding to ground roll). Other types of noise which were spread all over the signal spectrum were effectively attenuated only by means of radial method.

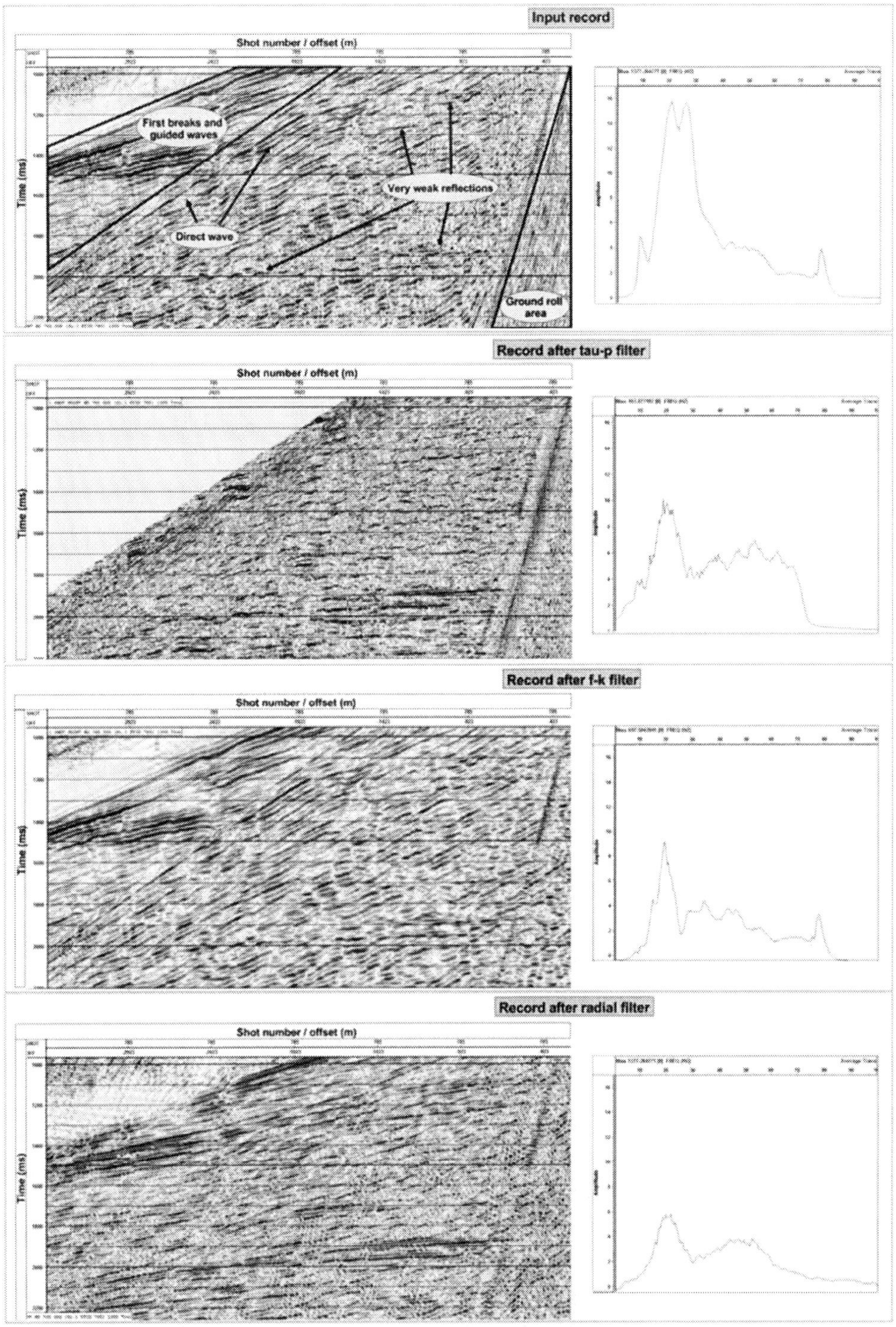

Fig. 4. Comparison of different filtration methods – shot gathers (on left) and the corresponding power spectra (on right)

One of the best evidence of importance of noise attenuation is creation of semblance functions which are used to map stacking velocity field (Fig. 5). On the left on Figure 5 raw semblance function was created. Velocity picking can not be executed with high reliability because the image is smeared. Set of filters executed in radial domain improved semblance image so much that velocity picking has become clear and easy (energy is focused to red dots on the right picture on Fig. 5).

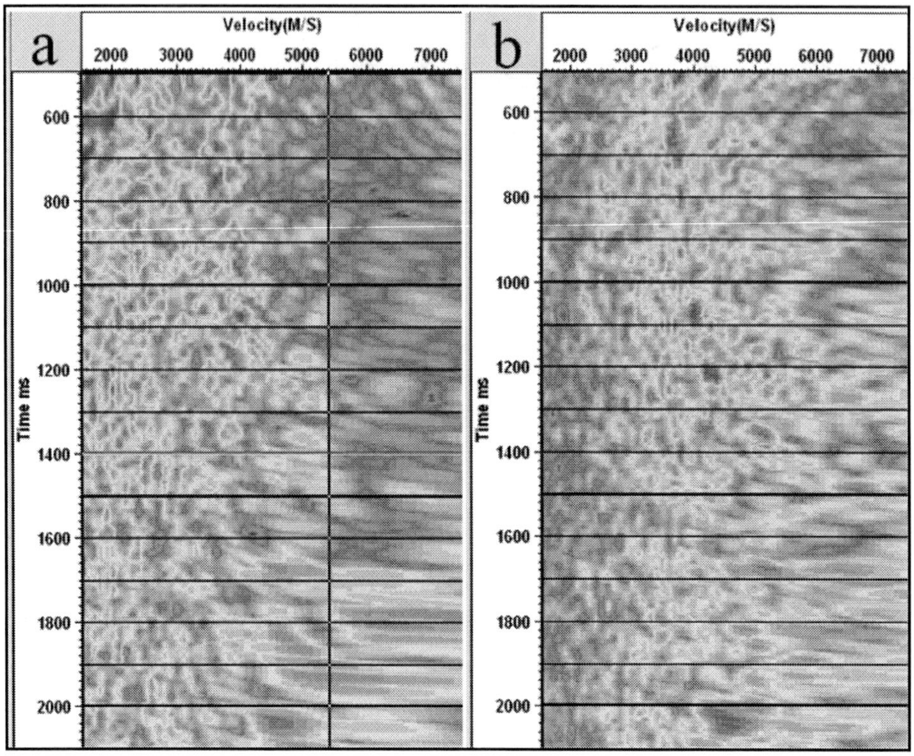

Fig. 5. Semblance function created on raw data (on the left) and the same function created on filtrated shot records (right)

5 Conclusions

Bandpass filtering in the radial domain enables effective and easy way of attenuation of various types of linear noise observed on seismic data. In all presented cases signal-to-noise ratio was significantly increased and reflections covered previously by noise were revealed. Power spectra of presented data are very little distorted and reveal real dynamics of reflections.

Radial transform performed in VISTA software (Schlumberger) presents many advantages with respect to other transforms, which mainly are: simplicity of the algorithm, ability to track every type of linear event, little distortion of reflections. Another advantages are: efficiency in transformation of irregularly spatially sampled data and ability to cascade the process to better treat particular type of noise. In author's opinion the most important advantage is the ability to attenuate wide range of generated linear noise in one step, even without additional data preconditioning or using adaptive noise attenuation algorithms.

Presented method reveals also some disadvantages. Among them the most important in author's opinion are: long computational time (the computational time increases with the amount of radial traces used), weakness in case of dispersive or non-linear events attenuation (they become smeared in radial domain and filtration is not so efficient). Because of amplitude balancing in radial domain, which generated better images, discussed results can not be used in AVO analysis. Without this special amplitude treatment results of filtration are little worse but also maintain high value.

On the basis of conducted research author concludes that among presented methods radial transform eliminated wide range of noise in relatively fast and easy way while preserving original frequencies of signal. The presented examples prove that proper noise filtration and scaling in radial domain is competitive and effective way of noise elimination, guaranteeing better image interpretation.

References

1. J.F. Claerbout, Stanford Exploration Project Report, **35**, 43–53 (1983)
2. R. Ottolini, Stanford Exploration Project Report, **20**, 97–115 (1979)
3. M.T. Taner, Geophys. Prosp. **28**(1), 30–48 (1980)
4. M.G. Lamont, B.M. Hartley, N.F. Uren, The Leading Edge **18**(1), 110–114 (1999)
5. L.J.A. Almeida, R.R. Manenti, M.J. Porsani, *Coherent Noise Attenuation using the Wavelet Transform on Radial Basis*, SEG Technical Program Expanded Abstracts 2015, 4720–4724 (2015)
6. O.F. Adizua, J.O. Obeniro, C.N. Ehirim, Int. J. Res. Innov. Earth Sci, **3**, 11–14 (2015)
7. W.L. Wang, W.Y. Yang, X.J. Wei, X. He, Appl. Geophys, **14**(1), 96–104 (2017)
8. M. Zhang, Y. Liu, *Seismic data Interpolation based on radial basis function*, International Geophysical Conference, Quingdao, China, 17–20 April 2017, 435–439 (2017)
9. Ż. Szymańska-Małysa, *Radial transform potential of coherent linear noise attenuation*, 75th EAGE conference & exhibition incorporating SPE EUROPEC 2013, London, UK, 10–13 June 2013 (2013)
10. Ż. Szymańska-Małysa, *Seismic image improvement developed on the basis of radial transform*, IX Krakow conference of young scientists 2014, Krakow, Poland, 2–4 October 2014, book of abstracts (2014)
11. D.C. Henley, *The radial trace transform: An effective domain for coherent noise attenuation and wavefield separation*, SEG Technical Program Expanded Abstracts 1999, 1204–1207 (1999)
12. D.C. Henley, *Wavefield separation and other useful applications in the radial trace domain*, SEG Technical Program Expanded Abstracts 2000, 2111–2114 (2000)
13. D.C. Henley, CSEG Recorder, **26**(2), 21–29 (2001)

The Application of Depth Migration for Processing GPR Data

Dang Hoai Trung[1,*], *Nguyen* Van Giang[2], and *Nguyen* Thanh Van[1]

[1] Faculty of Physics and Engineering Physics, VNUHCM – University of Science

[2] Vietnam Academy of Science and Technology

Abstract. Migration methods play a significant role in processing ground penetrating radar data. Beside recovering the true image of subsurface structures from the prior designed velocity model and the raw GPR data, the migration algorithm could be an effective tool in bulding real environmental velocity model. In this paper, we have proposed one technique using energy diagram extracted from migrated data as a criterion of looking for the correct velocity. Split Step Fourier migration, a depth migration, is chosen for facing the challenge where the velocity varies laterally and vertically. Some results verified on field data on Vietnam show that migrated sections with calculated velocity from energy diagram have the best quality. **Key words** – Split Step Fourier, energy diagram, depth migration.

1 Introduction

Ground Penetrating Radar is the most accurate and fast method in surveying near-surface structures. There are five main steps for processing GPR data: removing bad data, filtering noise, gaining, calculating electromagnetic propagation velocity and migration. In seismic method, migration is usually the last step for collapsing diffraction hyperbolae on unmigrated sections to points, thus moving reflection events to their proper locations, creating a true image of subsurface structures [2, 7]. Nowadays, in GPR, migration can become a middle step of processing data for estimating true velocity.

There are two kinds of migration methods: time migration (using root mean square (RMS) velocity) and depth migration (using interval velocity). RMS velocity is the average velocity that are taken into account by considering the influence of the upper layer's instantaneous velocity; whereas the interval velocity only reflects the practical velocity of one layer. In our research, depth migration is chosen for producing highly detailed images.

2 Split Step Fourier Migration

Phase shift is a migration method considered vertically varying velocities and based on wavefield extrapolation [1]. Phase shift is performed in the frequency wave number domain, so a constant velocity must be used for each depth interval. To solve this problem, phase

* Corresponding author: dhtrung@hcmus.edu.vn

shift plus interpolation is introduced. In this method, image for each depth interval are extracted from migrations of different constant velocities and the results are combined to form last migrated section [5].

The SSF migration algorithm is an alternative to above methods. It is based on a modification to phase shift migration that makes it possible to accommodate lateral variations in the velocity for each migration interval. The lateral changes are considered as a perturbation; therefore, only one additional spatial Fourier transform is required for each depth extrapolation. We can split the laterally varying velocity field into a constant term and a small perturbation term [1, 4]:

$$v(x,z)=v_0\left(z\right)+\delta v(x,z) \tag{1}$$

Firstly, we use $v_0(z)$ to propagate wavefields in the (ω, k_z) domain [1]:

$$P^{*}=P\left(z,k_x,\omega\right)\exp\left[\pm i\sqrt{\left(\frac{\omega}{v_0(z)}\right)^2-k_x^2}\mathrm{d}z\right] \tag{2}$$

Then, P^* is transformed back to the (ω, x) domain, and a phase correction is applied to it to account for the lateral velocity variation [1]:

$$P\left(z+\mathrm{d}z,k_x,\omega\right)=\hat{P}^{*}\exp\left[\pm i\left(\frac{\omega}{v(x,z)}-\frac{\omega}{v_0(z)}\right)\mathrm{d}z\right] \tag{3}$$

The accuracy of SSF can be improved by making the propagator symmetric. That is, this term can be achieved by splitting the phase shift term into two identical parts and applying them before and after the wave propagation in the (ω, k_x) domain [1]:

$$\pm i\left(\frac{\omega}{v(x,z)}-\frac{\omega}{v_0(z)}\right)\mathrm{d}z=\pm i\left(2\cdot\frac{1}{2}\right)\left(\frac{\omega}{v(x,z)}-\frac{\omega}{v_0(z)}\right)\mathrm{d}z \tag{4}$$

However, for strong lateral variation of the velocity field, the perturbation theory fails, and more than one reference velocity must be used for SSF.

3 Field data

In this part, we illustrate the accuracy of SSF migration with some real data.

3.1 Underground Construction in Ho Chi Minh City street

The GPR data were collected by Detector Duo (IDS – Italy) with 700 MHz shielded antenna on a street of district 3, Ho Chi Minh City. In this GPR section after filtering noise (Fig. 1), we can see two hyperbole signals at 2.4 m and 5.4 m of distance. Note that black show negative values, while white is positive ones in Figure 2. The GPR machine emits the wavelet pulse with the strongest positive value in its early time. The reversed polarity of the reflection from the top of those two objects show high conductive anomalies. According to the prior information, this line crosses two subsurface metal water pipes.

The GPR section in figure 1 will be processed by finite-difference migration with the RMS velocity range of 0.06 to 0.15 m/ns (its jump valued 0.001 m/ns). We estimate a set of the total energy values extracted from migrated results with different velocities for one specific zone. This zone is illustrated as a dashed rectangular in Figure 1. Each total energy value is summation of all energy points of migrated section from one velocity. Therefore, we can achieve a graph expressing relationship between total energy values and velocities [3]. Note that energy is square of amplitude.

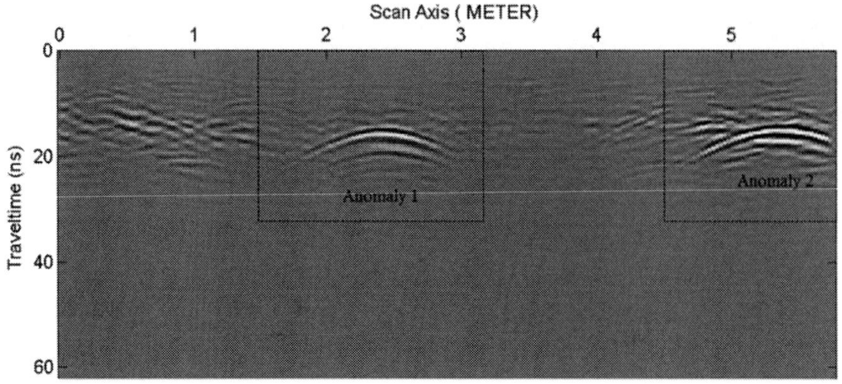

Fig. 1. GPR section after processing

The result shows that the calculating signal zone around the first hyperbole has maximum energy value at the velocity of 0.09 m/ns (Fig. 2a). Similarly, we also estimate the energy diagram of the second hyperbole. The results in Figure 3b show that the EM wave velocity in the environment above anomaly 2 is equal to 0.093 m/ns.

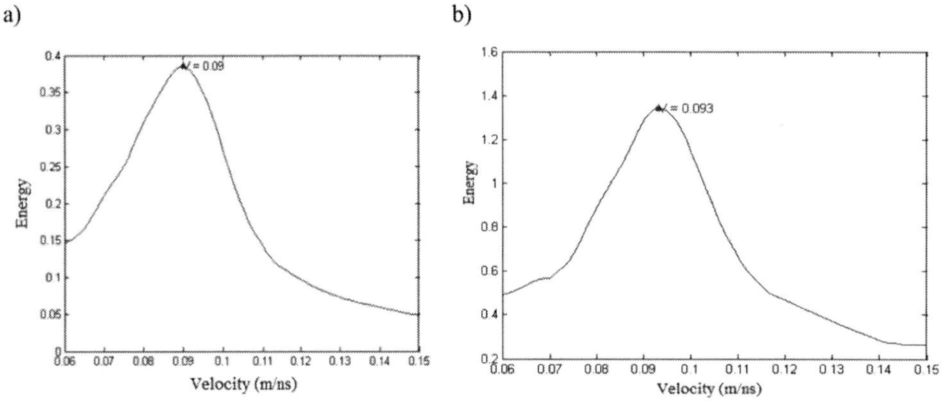

Fig. 2. Energy diagram of zone around: a) the first hyperbole; b) the second hyperbole

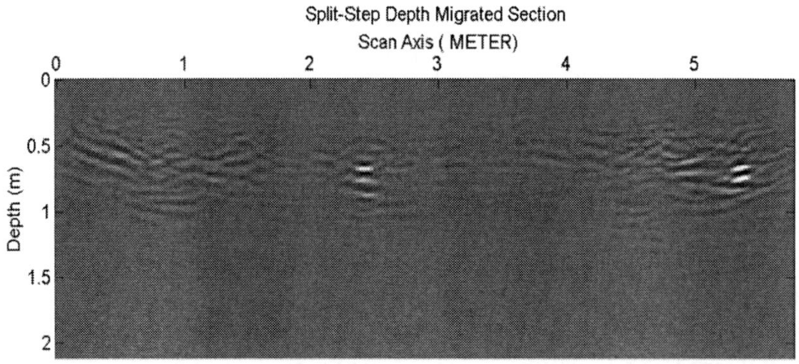

Fig. 3. The GPR migrated section by SSF

Generally, the use of maximum energy standard in order to optimize the migration problem has shown the variations in velocity laterally in this survey area. Therefore, SSF will be applied to migrate the GPR section (Fig. 3) [6]. On the migrated section, the two signals are clear and convergent. Consequently, the propagation velocities to the top of two hyperbolae are true. The depths and sizes of the two pipes based on the calculated velocity are $d_1 = 0.65$ m; $\Phi_1 = 142$ mm and $d_2 = 0.68$ m; $\Phi_2 = 142$ mm, respectively. This result has proven the accuracy of research method.

3.2 Surveying geology in Khanh Hoa province

The GPR data were collected by Zond 12e (Radar System – Latvia) with 150 MHz unshielded antenna at the seashore of Cam Lam district, Khanh Hoa province. The purpose of this survey is to determine the depth of a shell deposit layer. This is an organogenic structure that has been created from bay alluvial process. There are 5 longitudinal lines and 15 transverse lines at this survey area.

Based on data from two electrical resistivity imaging sections and a geological borehole, there three layers in the survey site:

- The first layer has rather low resistivity values which range from 120 to 160 Ωm. This layer can be interpreted as fine sand with its thickness of 3 m. It has been formatted from whitey grey, whitey yellow fine sand. It is in wet and loose state.
- The second layer has low resistivity values which range from 70 to 120 Ωm. This layer is also fine sand the same as the above layer, but it is in wet and medium dense to dense state.
- The last layer resistivity values are lower than 70 Ωm. This layer is muddy clay which consists of bluish grey, whitey grey, organic muddy clay. It is high plasticity and in water saturated and very soft state.

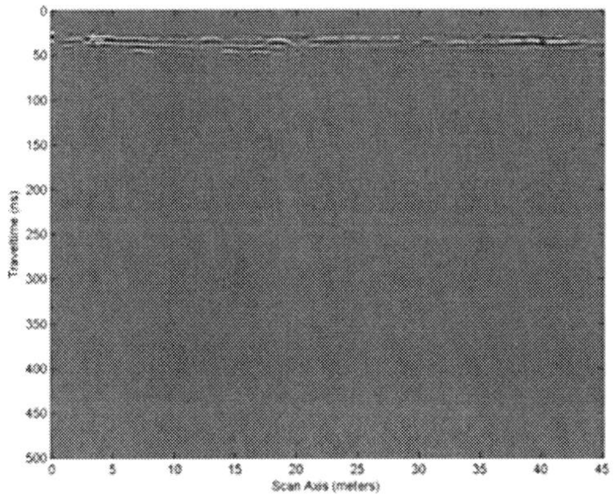

Fig. 4. The raw GPR section of line 5

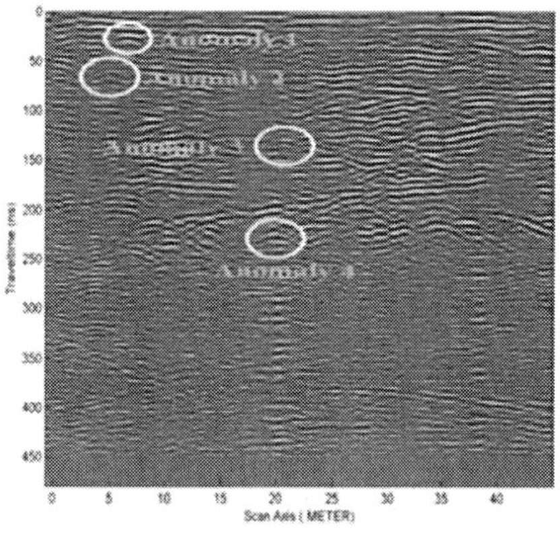

Fig. 5. The GPR section of line 5 after processing

We use the transverse line 5 and 14 to show the essential and accuracy of SSF migration. Figure 4 is a raw section of line 5 and Figure 5 is the section after processing. We can find a lot of hyperbolae in this image at $x_1 = 14$ m, $t_1 = 22$ ns; $x_2 = 5.2$ m, $t_2 = 52$ ns; $x_3 = 20.6$ m, $t_3 = 126$ ns and $x_4 = 19.6$ m, $t_4 = 222$ ns. Several hyperbolae at the depth of 200 to 250 ns can be reflection signals from the shell deposits. From 250 to 450 ns, the signal intensity decreases quickly due to low resistivity environment.

This GPR section will be processed by finite-difference migration with the RMS velocity range from 0.07 to 0.16 m/ns (its jump valued 0.001 m/ns) for signal zone around the first anomaly. The energy diagram (Fig. 6a) shows that the velocity at the apex of the hyperbola 1 (RMS velocity) is 0.141 m/ns (highest energy).

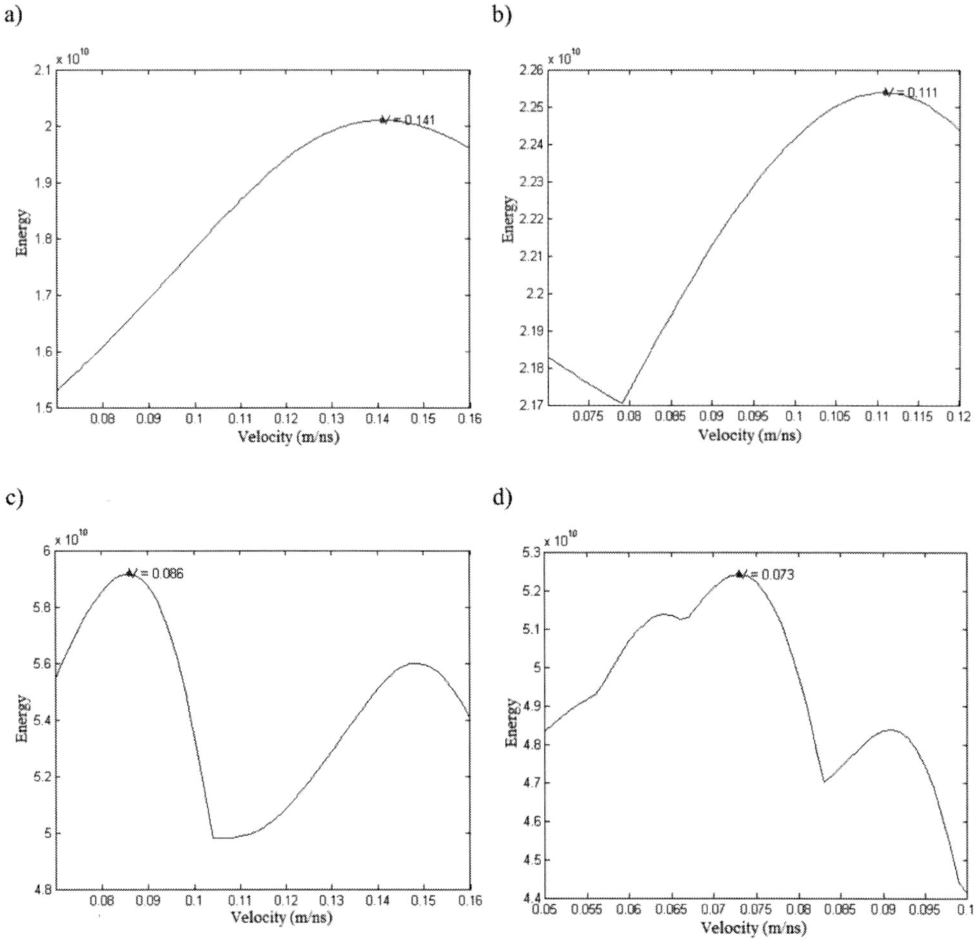

Fig. 6. The energy diagram of signal zone around the anomaly a) 1; b) 2; c) 3; d) 4

Similarly, the RMS velocity at the apex of the anomalies 2, 3 and 4 can be estimated at 0.111, 0.086 and 0.073 m/ns, respectively, from energy diagrams (Figs. 6b, c, d). To build the velocity model of this site, we must have the interval velocities derived from RMS velocities. The relevant formula of interval velocity and RMS velocity is shown as below [7]:

$$v_{\text{int}} = \sqrt{\frac{v_{rms2}^2 \tau_2 - v_{rms1}^2 \tau_1}{\tau_2 - \tau_1}} \qquad (5)$$

We calculate the interval velocities by using the above equation and build the velocity model of this survey area as shown below (Fig. 7). We can see that the velocity varies strongly in vertical direction due to the change of water content of the coastline (according to geological borehole).

The SSF migrated section 5 using the above velocity model is shown in Figure 8. Evidently, all hyperbolae are convergent and the geological boundaries are clear and visible (see white arrows in Fig. 8). The time axis is changed to depth axis; thus, we can determine the depth of geological boundaries. The shell deposit layer can be recognized at the depth of 6–8 m (see white lines in Fig. 8).

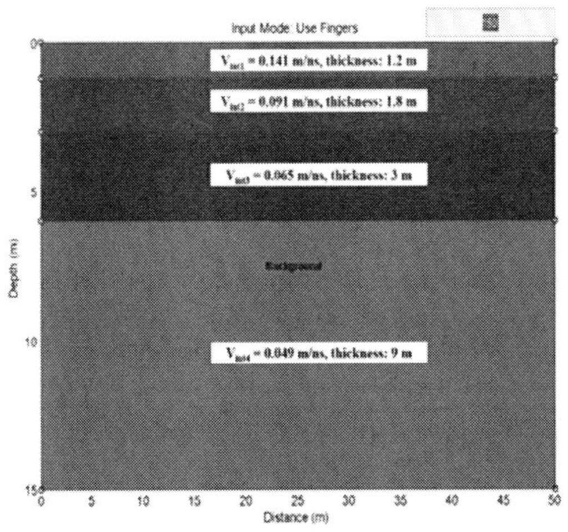

Fig. 7. The velocity model

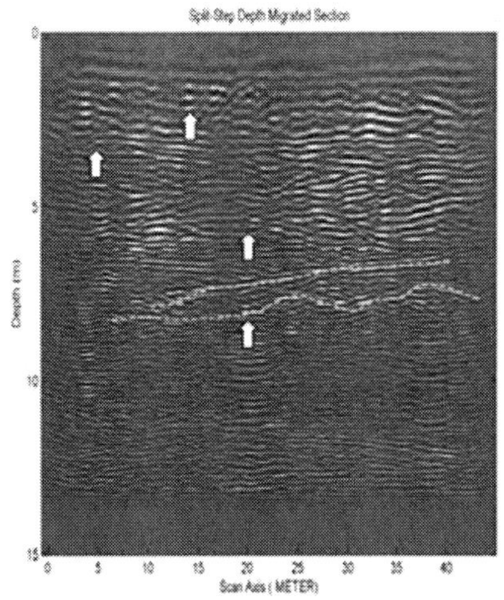

Fig. 8. The GPR migrated section 5

Similarly, we perform all above steps for line 14. The GPR sections before and after migration are shown in Figures 9a, b. We also recognize that all diffraction hyperbolae are collapsed, so the shell deposit boundary becomes easy to see (tracing positions of white arrows in Fig. 9b).

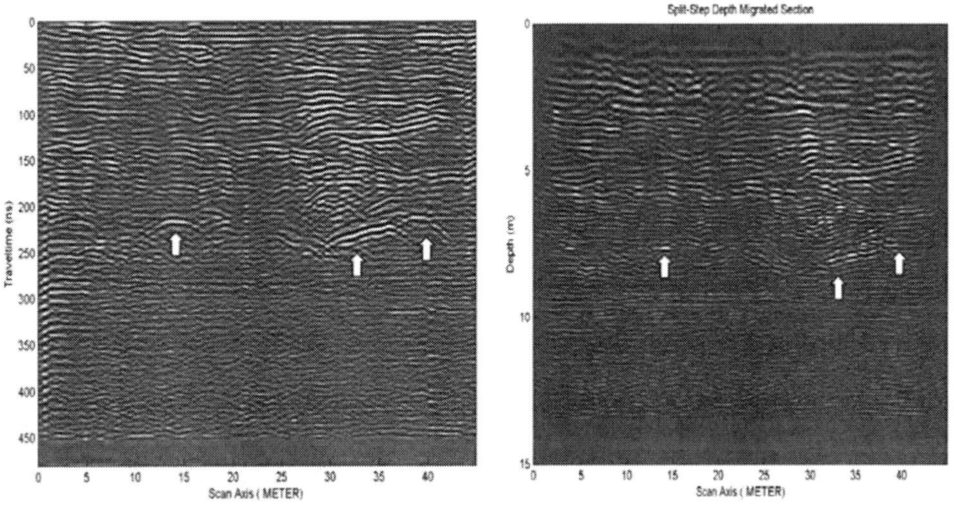

Fig. 9. The GPR section 14: a) before migration; b) after migration

4. Conclusion

We have presented the application of migration method to interpret GPR data and to build velocity model. Energy diagram is a good criterion to optimize the accuracy of determining propagation velocity by a migration algorithm. Split Step Fourier, depth migration, could solve the environment issue with velocity varying laterally and vertically. The migrated section is clear and reflect the depth of the apex of objects. Therefore, we can calculate exactly the size and the location of objects (example 1) and the depth of shell deposit layer (example 2) in migrated image.

However, the processing steps are complicated. Firstly, we must use time migration to calculate RMS velocity from energy diagram. Secondly, we need estimate the interval velocity from RMS velocity and build velocity model. Finally, SSF migration method is used to process GPR section. Therefore, this research should be expanded in the way of using depth migration to calculate energy diagram of each layer. The velocity that has highest energy can be an interval velocity of this layer. Thus, the processing steps become easier and decrease the error of determining interval velocity.

Acknowledgments

We would like to thank Cuong Van Anh Le from University of Science – VNUHCM Vietnam for his support.

References

1. B. Han, *A comparision of four depth migration methods.* SEG Technical Program Expanded Abstracts, pp. 1104–1107, doi:10.1190/1.1820080 (1998)
2. J. Gazdag and P. Sguazzero, *Migration of Seismic Data.* Proceedings of the IEEE, 72(10), pp. 1302–1315 (1984)
3. V.T. Nguyen, V.A.C. Le, T.V. Nguyen, H.T. Dang, M.T. Vo and N.N.L. Vo, *Energy Analysis in Semiautomatic and Automatic Velocity Estimation for Ground Penetrating Radar Data in Urban Areas: Case Study in Ho Chi Minh City, Vietnam.* Advances and Applications in Geospatial Technology and Earth Resources, Springer International Publishing, ISBN: 978-3-319-68239-6 (2017)
4. A.R. Sena, P.L. Stoffa and M.K. Sen, *Split-step Fourier migration of GPR data in lossy media.* Geophysics, Vol 71 (4), pp. 77–91, doi: 10.1190/1.2217157 (2006)
5. P.L. Stoffa, J.T. Fokkema, R.M.D.L. Freire and W.P. Kessinger, *Split-Step Fourier Migration.* Geophysics, Vol 55, No. 4, pp. 410–421, doi: 10.1190/1.1442850 (1990)
6. A.Tzanis, 2006. MATGPR: *A freeware MATLAB package for the analysis of common-offset GPR data.* Geophysical Research Abstracts, Vol. 8, 09488, 2006
7. O. Yilmaz, *Seismic Data Analysis, Chapter 4.* Society of Exploration Geophysicist, Electronic Edition, ISBN 978-0-931830-46-4, pp. 476–491 (1987)

Application of Geodetic Techniques for Antenna Positioning in a Ground Penetrating Radar Method

Ewelina Mazurkiewicz[1], *Łukasz* Ortyl[2,*], and *Jerzy* Karczewski[1]

[1] AGH University of Science and Technology, Faculty of Geology, Geophysics and Environmental Protection, A. Mickiewicza Av. 30, 30-059 Krakow, Poland

[2] AGH University of Science and Technology, Faculty of Mining Surveying and Environmental Engineering, A. Mickiewicza Av. 30, 30-059 Krakow, Poland

Abstract. The accuracy of determining the location of detectable subsurface objects is related to the accuracy of the position of georadar traces in a given profile, which in turn depends on the precise assessment of the distance covered by an antenna. During georadar measurements the distance covered by an antenna can be determined with a variety of methods. Recording traces at fixed time intervals is the simplest of them. A method which allows for more precise location of georadar traces is recording them at fixed distance intervals, which can be performed with the use of distance triggers (such as a measuring wheel or a hip chain). The search for methods eliminating these discrepancies can be based on the measurement of spatial coordinates of georadar traces conducted with the use of modern geodetic techniques for 3-D location. These techniques include above all a GNSS satellite system and electronic tachymeters. Application of the above mentioned methods increases the accuracy of space location of georadar traces. The article presents the results of georadar measurements performed with the use of geodetic techniques in the test area of Mydlniki in Krakow. A satellite receiver Leica system 1200 and a electronic tachymeter Leica 1102 TCRA were integrated with the georadar equipment. The accuracy of locating chosen subsurface structures was compared. **Key words** – Ground Penetrating Radar, geodetic positioning, robotic tachymeter, GNSS system.

1 Introduction

Geodetic measurement techniques have been successfully applied in many fields of science, for instance geological engineering, geophysics or archaeology, mainly to determine precise coordinates of a characteristic points in a given area (such as the initial and final points of the measuring profiles, a location of a discovered anomaly etc.) [1]. One of the geophysical methods in which the above-mentioned techniques can be employed is the georadar method. A comprehensive introduction to the georadar method can be found in [2–4].

Modern georadar system make it possible to gather huge amounts of data and to conduct surveys of "global character" with profiles of substantial length covering a vast area.

* Corresponding author: ortyl@agh.edu.pl

Measurements can be conducted with high speed (with georadar antennas mounted on a train or a car). Georadar data obtained in this manner provide additional geospatial information, which, when reduced to a coordinate system, can be combined with other data (gathered for instance in the Geographical Information System) and used to complement thematic maps with information regarding subsurface objects and structures. Typical determination of georadar antenna positions is based solely on measurement performed with the use of an odometer in the form of a wheel or an uncoiling chain. The accuracy of pinpointing the exact location of a recorded trace/route in space is, in this case, very low.

Spatial interpretation of the distribution of the objects registered with the use of georadar methods based only on the data recorded by a georadar, is a significant geometric simplification. A natural and diverse lay of the land over a vast area usually proves impossible to conduct rectilinear and parallel profiling. Georadar surveys are often carried out in the areas where some denivelations occur and the awareness of height differences in the land shaping is vital when the accuracy of positioned objects and the interpretation of the obtained georadar results, are concerned.

Georadar research makes it possible to mark initial and final points of the profiles in the given test field as well as the points where profiles change their direction, and then to measure them with geodetic methods taking into account specific land denivelations. However, such a two-stage survey process is very time consuming, both in terms of gathering geodetic information as well as subsequent post processing of the obtained georadar data. In order to improve the accuracy of the results of georadar profiling, the survey should be based on geodetic techniques using satellite (GNSS) and tachymetric measurements. Georadar data can be gathered simultaneously when a position of the antenna is being determined and recorded, primarily by means of geodetic GNSS satellite techniques or if their application is not possible, by robotic tachymeters of the TCA type capable of automatic target recognition (ATR) and tracking an EDM prism. However, then both devices must be physically integrated with one another, which means that a GNSS receiver or an EDM prism should be mounted directly on a georadar antenna or in its vicinity with a given shift interval.

The accuracy of determining the location with the use of geodetic techniques will depend on the environmental factors and capabilities of a given measuring equipment. Differences in test results will be particularly noticeable while performing GNSS measurements. It should be noted that GNSS measurement can be carried out using the navigational or differential mode.

In the navigational mode the position of the receiver is determined directly on the basis of the signal transmitted by satellites. Due to various factors impairing the signal propagation between a satellite and a receiver, the end result is vitiated with a measurement error (the impact of troposphere, ionosphere and clock errors). Handheld GNSS receivers usually operating in a navigational and occasionally differential mode, use a single frequency, which allows them to determine a given position with the accuracy of no more than 0.5–0.25 m. Therefore, such handheld receivers could be used to indicate an approximate location of georadar measurements

In the differential mode the position of the moving receiver is determined in relation to the Continuously Operating Base Station (a reference one) or to the receivers of the Active Geodetic Network situated over a given point whose coordinates are known in any arbitrary reference system. Thus, in this case far greater accuracy can be achieved.

GNSS receivers for geodetic applications operate in a differential mode using two frequencies, which allows to determine the position of a given object by means of the Real-Time Kinematic method with the accuracy of ca. 0.02 m (X,Y) ca. 0.03-0.05 m (Z). As georadar measurement surveys are of dynamic character, application of the RTK method would be the most optimal solution while using a set of GPR-GNSS equipment.

Similar accuracy can be expected while using electronic homing tachymeters for conducting measurement in a dynamic mode. Lehmann and Green [5] cited the value of ca. 0.04 m at the prism speed of 1 m/s.

In both cases it is worth remembering that it is the accuracy determined by stable positioning of the receiver over a given point, which lasts a couple of seconds. However, when georadar measurements are concerned, the accuracy achieved is lower due to uncoordinated movements of georadar antennas.

An additional factor influencing the positioning accuracy of georadar traces is the frequency estimation for a new location computed by a GNSS receiver or a tachymeter.

Modern GNSS geodetic receivers determine a given position with the frequency of 5–20 Hz (some models even with 100 Hz), while tachymeters with the frequency of up to 10 Hz. Thus, a new position will appear every 0.1 s in the case 10 Hz frequency, and every 0.05 s if the frequency is 20 Hz.

The article presents the results of georadar measurements performed in the test area of Mydlniki in Krakow. The survey was carried out with the use of georadars of the Swedish company Mala GeoScience (ProEx System and RAMAC/GPR CU II) fitted with shielded antennas of the frequency 100 and 250 MHz and working with various distance triggers (such as a measurement wheel or a hip chain). A satellite receiver Leica system 1200 and a homing tachymeter Leica 1102 TCRA were integrated with the georadar equipment. The accuracy of locating chosen subsurface structures with the use of different positioning methods was compared and evaluated.

2 Research site description

The research site is located in Krakow in the district of Bronowice on the terrain that rises slightly in the eastward direction. The test profile is 200 m long. Figure 1 shows a schematic geological cross-section of the test area.

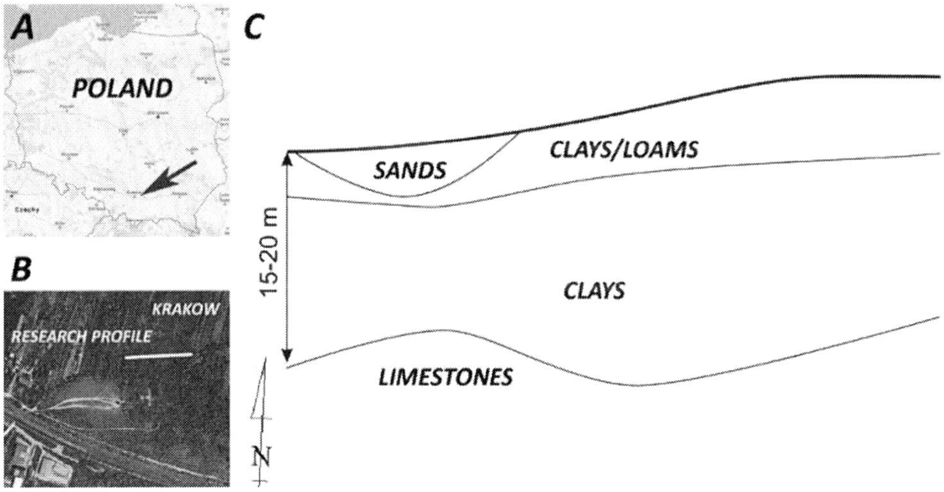

Fig. 1. Localization of the test area in Mydlniki (A, B) and cross-section of probable geological structure of test area the test area in Mydlniki (C by Grzegorz Pacanowski, modified)

According to Kondracki's physico-geographical regionalisation of Poland [6] the test area of Mydlniki is situated in the Kraków-Częstochowa Upland within the Lesser Poland Upland which in turn can be included in the province of Polish Uplands.

Compact structures in the form of clays and loam (very high conductivity ca. 80 mS/m) with occasional sand layers of low thickness are found directly in the surface/substratum of the investigated area. Miocene silt structures or Jurassic limestones are likely to be situated underneath [7]. However, no opening in the rock bed confirming such geological structure can be found in the immediate vicinity of the test area. The assumption is based on a geological cross-section located at a distance of 2 km in the eastern section of the research site.

3 Description of the equipment

The measurements were performed in the test area of Mydlniki with the use of georadars RAMAC/GPR CUII and ProEx System of the Swedish company Mala Geoscience. More detailed description of both models can be found among others in [8, 9]. Antennas of different construction (shielded, unshielded and Rough Terrain Antenna – RTA) as well as different frequencies (50, 100, 200, 250 and 500 MHz) were used in the conducted survey. This article, due to its size, focuses only on the analysis of the results recorded with the antennas of 100 MHz frequency (see Table 1).

Georadars of Mala Geoscience company enable electromagnetic impulses to be emitted at fixed time or distance intervals, or to be triggered remotely by means of a keyboard. The tests are most often conducted at fixed distance intervals – traces are registered after the antenna travels a given distance. Devices which make such measurements possible are a hip chain and a measurement wheel. The principle they both work on is the same –both of them contain a roller and a device which counts the number of its rotations, which is equivalent to a given unit of distance. A survey conducted with distance triggers makes the operator independent of changes in the antennas' moving speed; moreover, it is more accurate from the measurements carried out at fixed time intervals and much faster than the tests triggered remotely with a keyboard.

However, the measurement of that type is not devoid of flaws. While using a hip chain possible errors of antennas' positioning might result from: pulling of a chain, its slipping over a measurement roller, getting stuck in natural obstacles, etc. Errors in measurement conducted with the use of a measurement wheel might be due to its skidding over the surface of an investigated area. When used on an uneven terrain the wheel erroneously overstates the reading of a covered distance. Geodetic tools should be used in order to eliminate such inaccuracies. Various ways of GPR antennas positioning are shown in Figure 2.

Fig. 2. Positioning systems for GPR antennas: a hip chain (A), a measurement wheel (B) and a robotic tachymeter, a measurement wheel and a GNSS receiver (C)

GPR antenna positioning was conducted with the use of satellite measurements and Leica GPS 1200 Smart Rover receiver operating in the RTK mode as well as the implementation of differential correlations from a single reference station. A robotic tachymeter Leica TCRA 1102+ was also utilized in the process. The integration of the geodetic instruments with the georadar was carried out by means of a measuring system specially developed by the Geodesy Department at the Faculty of Mining Surveying and Environmental Engineering of the AGH

University of Science and Technology [10–12]. The receiver and tachymeter were configured to determine and send a new position to the RS232 port using the NMEA protocol with the frequency of 5 Hz for the receiver and 0.66 Hz for the tachymeter.

4 Measurement methodology

A series of georadar profiles was determined in the test area of Mydlniki with the use of various georadars, different types of antennas and measurement parameters. The following article analyses only selected readings, their list can be seen in Table 1.

Table 1. List of analyzed profiles

GPR model	Type of antenna	Frequency, [MHz]	Distance between traces, [m]	Profile number	Type of GPR positioning	Type of geodetic positioning
ProEx	unshielded	100	0.1	prof004	hip chain	none
RAMAC/GPR CUII	shielded	100	0.1	prof001	measurement wheel	tachymeter
RAMAC/GPR CUII	shielded	100	0.1	prof002	measurement wheel	GNSS

As the table demonstrates, the survey was carried out with georadar distance triggers-a hip chain and a measurement wheel as well as geodetic positioning techniques – the GPS system and a tachymeter. The next part of this article analyses the accuracy of antenna positioning when the above- mentioned techniques are applied.

5 Results

The total length of the measuring profile counted along the surface of the investigated area and determined on the basis of the measurements of direct profile points marked every 2 m, was 187.98 m. However, when measured with a hip chain, the profile appeared to be 187.98 m and 182.90 m when the measurements were conducted with the use of a wheel and a tachymeter. The tests involving the application of the wheel and a GNSS receiver recorded the profile length as mere 167.63 m, which indicates the appearance of a significant error, even though the surveys were carried out with the identically calibrated wheel.

It is difficult to notice this data discrepancy while comparing recorded echograms. If positioning was conducted only with the use of a measurement wheel, without any information concerning the total length of a given profile, locating the anomaly placement in the echogram would be significantly imprecise. Therefore, it is vital to implement calibration correlations as the ratio of the profile total length to the profile length as provided by the georadar's gauge. Unfortunately, such a solution has its weaknesses as it does not exclude errors which are non-linear in character (such as a wheel getting stuck at some part of a given profile).

Additional geodetic procedures must be carried out to control the length of the profiles measured along the surface of the investigated area, especially when profiles' topography change. Height differences increase the total length of a profile with regard to its projection (values of the initial and final profile point are insufficient for positioning amendments). The solution based on geodetic positioning systems eliminates this problem.

The echogram (Fig. 3) shows the position of the anomaly between 147 and 168 m on the profile 1 (Fig. 3B), which in the profile Prof002 is situated between 135 and 153 m (Fig. 3C).

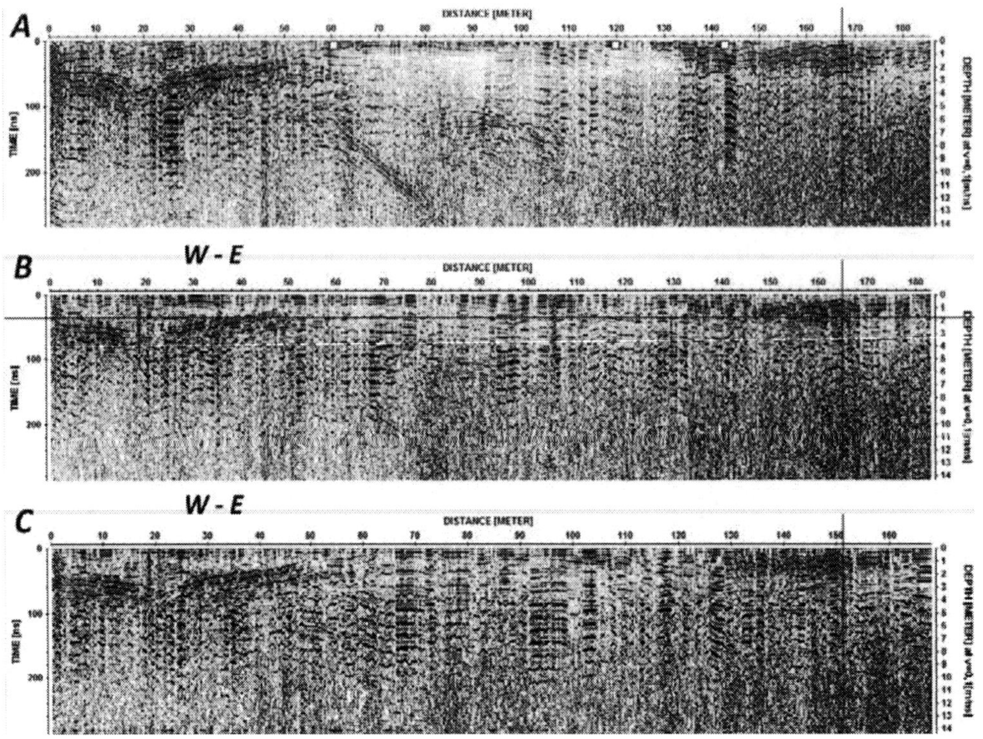

Fig. 3. Prof004 recorded with a hip chain (A), Prof001 recorded with a measurement wheel and robotic tachymeter (B), Prof002 recorded with a measurement wheel and GNSS receiver (C).

Figure 4A shows a comparison of approximated values of geodetic coordinates for both of them in the radargram about 147 and 135 meter of the profiles 1 and 2 respectively.

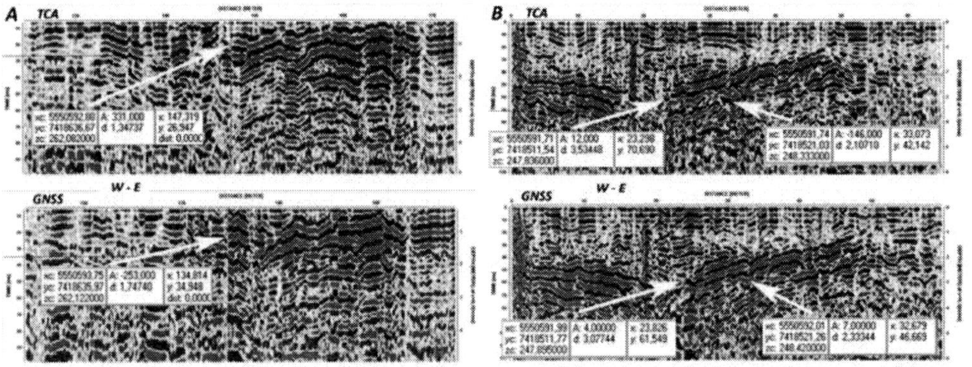

Fig. 4. Summary of geodetic coordinates recorded by robotic tachymeter and GNSS receiver for various anomalies areas: end profile area (A), initial profile area (B)

The results demonstrate correlation at the level of 1m without marking any control points, which confirms the efficiency of the positioning carried out with the use of the geodetic techniques. However, when a part of an echogram regarding 168 and 153 m of the profiles was analyzed, coordinate discrepancies increased to about 3m. Consequently, the analysis of the obtained geodetic data was conducted.

Figure 5 depicts changes in the accuracy of the geodetic data for the TCA (Fig. 5A) and GNSS (Fig. 5B) by means of evaluating the diagram of plane coordinates and heights/altitudes determined by both of the devices during the survey.

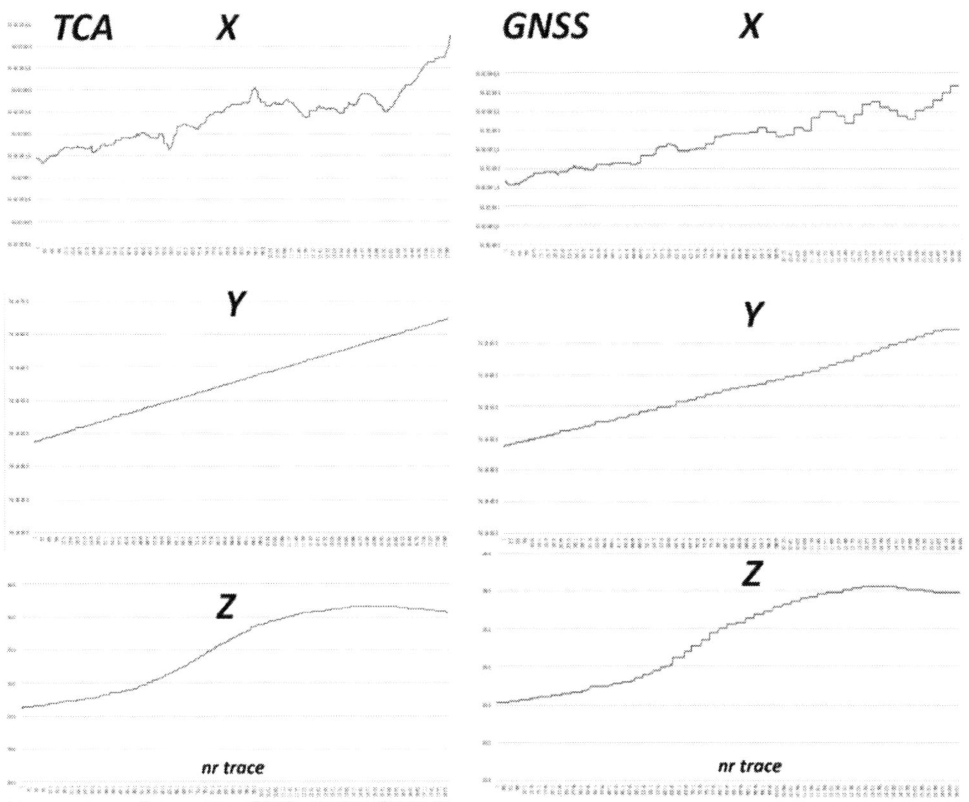

Fig. 5. Summary of geodetic coordinates obtained by robotic tachymeter and GNNS receiver on analyzed measurement profiles

In the case of GNSS receiver, a clear decline in accuracy of the determined positions starts at 1/3 of the profile (stepped character of plane coordinates and altitude data), which differs from the values recorded by the TCA along the whole profile. The area where the GNSS and TCA data correspond is located at 1/3 of the profile. Figure 4B demonstrates correlation of spatial positioning results at the level of 0.1–0.3 m.

6 Conclusions

While conducting georadar research certain limitations of various positioning techniques regarding the process of subsequent placement of the recorded anomalies, should be taken into account.

In the analysed case (Fig. 3A) the measurement conducted with the use of a hip chain proved to be reliable as far as the profile length was concerned, but at the same time insufficient to determine spatial position of the anomaly. Consequently, performing additional geodetic procedures would be required, especially in order to determine the altitude coordinate.

The profile recorded by means of the measurement wheel (Fig. 3C) turned out to be erroneous even though it seemed to be visually counterintuitive as may be the case when a hip chained is used.

The article pointed out the advantages of positioning with the use of the geodetic systems. Even though they decrease the accuracy of the wheel, they situate an anomaly correctly.

However, it has been demonstrated that even such data should not be exempt from criticism. Verification of the obtained information especially when gathered by means of the GNSS is highly recommended. During the survey carried out in the test field of Mydlniki in perfect visibility, the GNSS system failed for some unexplainable reasons decreasing the positioning accuracy in the final part of the echograms (from 60 m).

In the area where the data from the GNSS system was recorded with high accuracy, such a problem does not occur. In such conditions high positioning correlation (0.1–0.3 m) to the results obtained by means of the TCA method, which is highly reliable but at the same more demanding, can be observed.

Acknowledgements

The research has been funded from the appropriations received from the NCBR (the National Centre for Research and Development) and GDDKiA (the General Director for National Roads and Motorways) and as part of the project and the competition for the RID (Development of Road Innovation) Project "Modern methods of subsoil identification in road engineering" with the acronym OT1-1E/PIG-AGH-PW.

References

1. S. Porzucek, W. Matwij, 17th International Multidisciplinary Scientific Geoconference: Science and Technologies in Geology, Exploration and Mining, Conf. Proc. **17**, 57–64 (2017)
2. H. M. Jol, *Ground penetrating radar: theory and applications* (Elsevier 2009)
3. J. Karczewski, Ł. Ortyl, M. Pasternak, *Zarys metody georadarowej* (UWND AGH, Kraków, 2011)
4. M. Łyskowski, E. Mazurek, J. Ziętek, Journal of Cave and Karst Studies, **76**, 184–190 (2014)
5. F. Lehmann, A.G. Green, Geophysics, **64**, 719–731 (1999)
6. J. Kondracki, *Geografia Regionalna Polski* (Wyd. Naukowe PWN, Warszawa, 2002)
7. J. Rutkowski, *Szczegółowa mapa geologiczna Polski w skali 1:50000, arkusz 973-Kraków* (1989)
8. A. Benedetto, L. Pajewski., ed., *Civil Engineering Applications of Ground Penetrating Radar* (Springer, 2015)
9. M. Pasternak (red.), *Radarowa penetracja gruntu* (Wyd. Komunikacji i Łączności, Warszawa, 2015)
10. J. Gocał, Ł. Ortyl, M. Sołtys, in: Geodesy, photogrammetry and monitoring of environment, 5–14 (2006)
11. Ł. Ortyl, A. Bałut, [in:] Geodesy, photogrammetry and monitoring of environment, 27–44 (2006)
12. Ł. Ortyl, T. Owerko, Zeszyty Naukowe Politechniki Śląskiej, **278**, 335–352 (2007)

Detection of Unknown Crypts under the Floor in the Holy Trinity Church (Dominican Monastery) in Krakow, Poland

Anna Strzępowicz[1,*], *Mikołaj* Łyskowski[2], *Jerzy* Ziętek[1], and *Sylwia* Tomecka-Suchoń[1]

[1] AGH University of Science and Technology, Faculty of Geology, Geophysics and Environmental Protection, A. Mickiewicza Av. 30, 30-059 Krakow, Poland

[2] AGH University of Science and Technology, University Computer Centre, A. Mickiewicza Av. 30, 30-059 Krakow, Poland

Abstract. The GPR surveying method belongs to non-invasive and quick geophysical methods, applied also in archaeological prospection. It allows for detecting archaeological artefacts buried under historical layers, and also those which can be found within buildings of historical value. Most commonly, just as in this particular case, it is used in churches, where other non-invasive localisation methods cannot be applied. In a majority of cases, surveys bring about highly positive results, enabling the site and size of a specific object to be indicated. A good example are the results obtained from the measurements carried out in the Basilica of Holy Trinity, belonging to the Dominican Monastery in Krakow. They allowed for confirming the location of the already existing crypts and for indicating so-far unidentified objects.
Key words – GPR, crypts, Basilica of Holy Trinity, Krakow.

1 Introduction

For many years, ground penetrating radar (GPR) surveys serve as a tool for investigation of historical complexes such as: churches, basilicas or ancient settlements. In most cases it is closely related to the specifics of place – e.g. churches are place of worship, and also a place of burial for the monks, people of merit or with a high social status. Among numerous historic buildings which can be found in Krakow, the Dominican Monastery is one of the oldest ones. The origins of this architectural object date back to the 13th century, i.e. the period of economic and cultural growth in Krakow [1]. Inter alia, the facilities belonging to the Dominican Monastery complex were taken into account when the water supply and sewage disposal system of the city, future medieval capital of Poland, was created. Despite the importance of the Church at that time, there were disputes between the city and the Dominicans which would continue until the 17th century. They referred to the most crucial issue – access to clean drinking water, as well as to water that would drive water mills [2]. The Basilica of Holy Trinity, belonging to the Dominican Monastery, originally had a form of a three-nave hall in the redbrick Gothic style. At the turn of the 14th and 15th centuries it

* Corresponding author: tomecka@agh.edu.pl

was converted into the basilica style. Until half of the 19th century, one of the most characteristic elements of the exterior of the temple was a stone bell tower, separated from the main building and erecting in front of its facade at the end of Stolarska Street. However, after the fire of 1850, the only remains of the church bell tower were walls blackened by smoke. They were demolished in the course of reconstruction works, and in 1876 a neo-Gothic porch was added to the facade of the temple. It sheltered a 14th century Gothic portal of the main entrance, which was renovated in 1893 [3].

Investigations, initiated by administration employee and archaeologists, were aimed at checking if as then undiscovered crypts, other burial rites or other important archaeological objects could exist under the floor of this historic church. These works were supposed to be carried out before the collision-free installation of a heating system or other elements of the infrastructure included in the design.

The GPR method, selected for this purpose, is one for the basic tools for this kind of research [4, 5] and it was the main reason for its implementation in this case study. Application of this survey type is very popular in Poland. The main advantage of the GPR method is its totally non-destructive nature which allows to perform measurements even within the most buildings, especially those with cultural and historical biggest value. Moreover, considering the characteristics of the site and physical working space (which was limited), the method chosen by the authors was more than adequate.

2 Method

The GPR method belongs to non-invasive, high resolution, and fast application geophysical survey. Sometimes in undergoing measurements, one is capable of identifying locations of anomalous areas at the very early stage such as data collection. This method has applications in various fields, one of them is archaeology. The basic measuring equipment consists of a central unit and two antennae: a transmitting and receiving one. The first one, the transmitting antenna, emits short pulses of electromagnetic waves, which are subject to reflection, refraction and attenuation when propagating. The most important phenomenon from the point of view of the GPR method is the reflection of the wave on the interface of two media characterised by different physical parameters, which occurs in the case of the contrast between the dielectric constants of these media. The value of the dielectric constant has a major impact on the velocity of electromagnetic wave propagation in a medium. The reflected wave is recorded by the receiving antenna. The obtained wave image reflects the structure of the medium. The depth range of the GPR method usually does not exceed the distance of several meters, and the depth of penetration heavily depends on the structure of the examined medium, the frequency of the emitted electromagnetic wave and the humidity of the soil. It also depends on the pre-set recording parameters. It is generally assumed that for antennae of 500 MHz frequency, the depth range is 6 m on average. It is also important to take the resolution parameter into account, as it increases with an increase of the antenna frequency, which deteriorates the penetration capability of the antenna [6, 7].

It has to be emphasised that the quality of obtained results strongly depends on the optimal selection of measurement parameters such as: the sampling frequency of the input signal, the number of samples, the stacking and the distance between particular traces [8].

The equipment used in the measurements was a Mala GeoScience GPR with a 500 MHz shielded antenna. Distance on the profile was measured with a calibrated mechanism calculating the number of revolutions into meters.

3 Raw data processing

In total, there were 62 measured profiles. The entire investigated area was divided into sub-areas: "Main transept", "Right nave", "Left nave", "Crosswise", "Presbytery", "Vestry", "Altar" and "Behind altar".

In order to adjust the range of the depth scale, the velocity of wave propagation of 0.1 m/ns was adopted for the purposes of the time-depth conversion. The depth scale was assumed on the basis of specified propagation velocity of the electromagnetic wave in the studied medium [9]. The value of the velocity changes form 0.3 m/ns (in the air) to 0.03 m/ns (in the fresh water) [7]. It was impossible to perform the WARR (Wide-Angle Reflection and Refraction) velocity profile. The permissible error in adjusting the velocity is 0.01 m/ns.

Prior to the presentation of the results, the raw data were processed, so they were subjected to an increase in the signal-to-noise ratio. The data used in this process were subjected, inter alia, to [10]:

- static adjustments, the aim of which is to move the first penetration to 0 on the depth scale (move starttime), or to cut off the lower part of information which includes the noise (time cut);
- procedures bringing the average value of the amplitude to zero (subtract-DC-shift);
- eliminate low frequency signal disturbance (subtract-mean (dewow));
- frequency filtering aimed at the elimination of the noise (bandpassbutterworth);
- enhancement, increasing the value of the amplitude weakened due to the electromagnetic wave attenuation in the investigated medium (gain function);
- smoothening, the aim of which is eliminating single noises by means of the averaging of the amplitude value, e.g. in the time-distance window (median *xy*-filter);
- remove horizontal ringing from signal traces (background removal).

All above processing procedures allowed for acquiring very good quality data for interpretation.

4 Results and discussion

From among all the profiles, the profile 5 from the areas labelled as "Left Nave" and "Right Nave" and the profile 9 from the "Altar" area were selected for presentation. The profile across the area called "Vestry" seemed to be interesting for the authors, therefore the profiles 4 and 16 from the area will also be presented. Also interesting seems to be profile 19 in the "Vestry" area, on which anomalies form construction elements can be observed.

The interpretation of the results is presented on the enclosed map (Fig. 1), on which the sites where significant anomalies were spotted are marked. The areas marked with the letter "C" are connected with the occurrence of crypts under the church floor. The remaining sites do not allow for assuming any definitive conclusions.

On the below figure (Fig. 2), in the "Left Nave" area, on the profiles 2 to 7 there is a clearly distinguished anomalous zone, probably resulting from the presence of a crypt of a larger size. At the ends of the profiles (from the distance of approx. 30–32 running meters – means that this in the distance from the beginning of the profile), an anomaly at a larger depth (approx. 2–2.5 m) can be seen. The possibility that it is caused by a crypt cannot be excluded. The perpendicular profile 3 – "Crosswise" area, shows strong anomalous zone between 14 and 20 running meters, correlating with that seen on profile 5. Similarly, in the "Right Nave" area, on the profiles 6, 5, 1, 2 and 3 significant anomalies can be observed, which is most probably caused by the existence of crypts under the floor. Anomalies between 2 and 6 running meters on "Crosswise" profile 10 correlates with that seen on profile 5 in the "Right Nave" subarea. Also anomalies which can be seen at the ends of the profiles 5 and 6 may be due to the existence of a crypt. The presence of a crypt on the profile 4 is confirmed by the "Crosswise" profiles 1, 2 and 3. What was innovative here was the instantaneous verification of the anomalies by archaeologists by means of drilling small-diameter boreholes in the floor.

The main aim of conducting the perpendicular profiles was to confirm the occurrence of the earlier discovered anomalies. They confirmed all the anomalous areas that had been indicated before. As it has already been mentioned, they allowed for the identification of other crypts as well. It was at the beginning of the profiles 1, 2 and 3 that such a site was indicated at the threshold of the chapel.

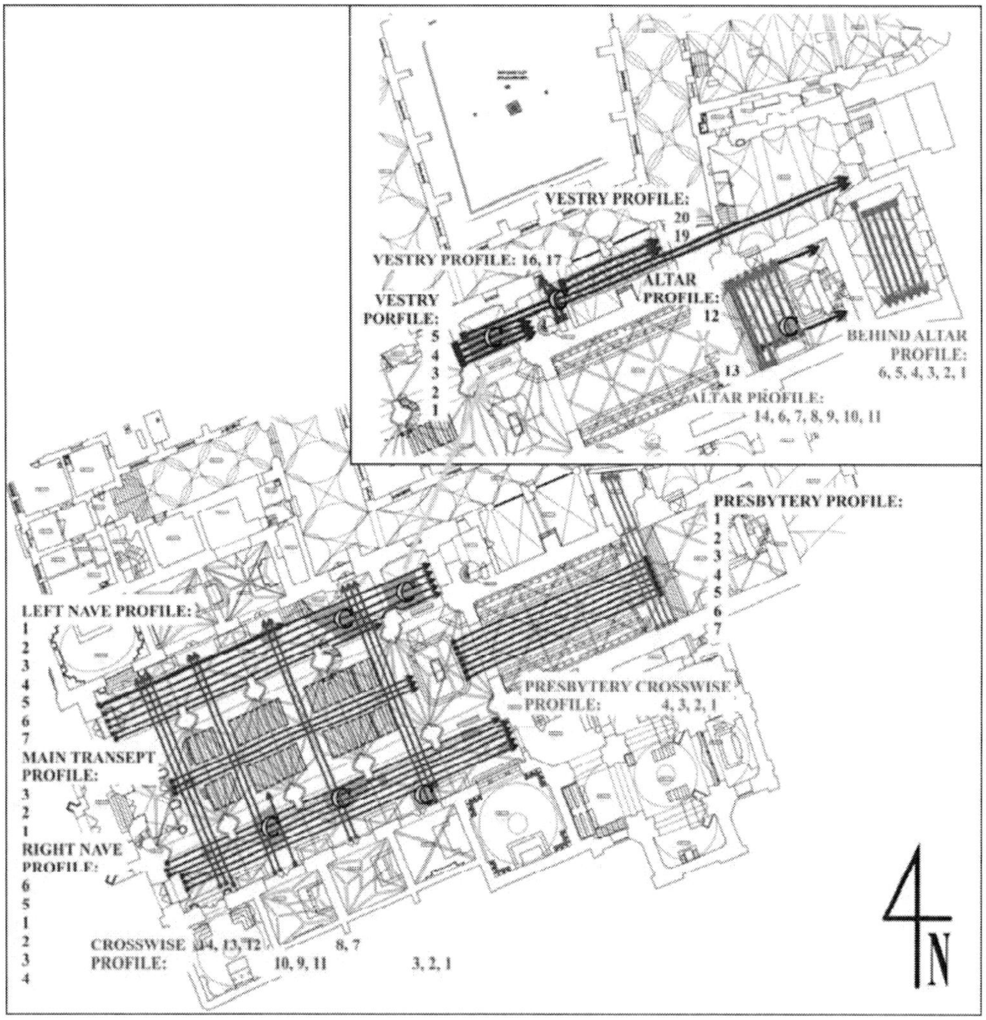

Fig. 1. Schematic map of the Basilica of Holy Trinity (non scaled) with drawn GPR profiles and anomalous area marked (C – crypts)

 https://doi.org/10.1051/e3sconf/20183503006

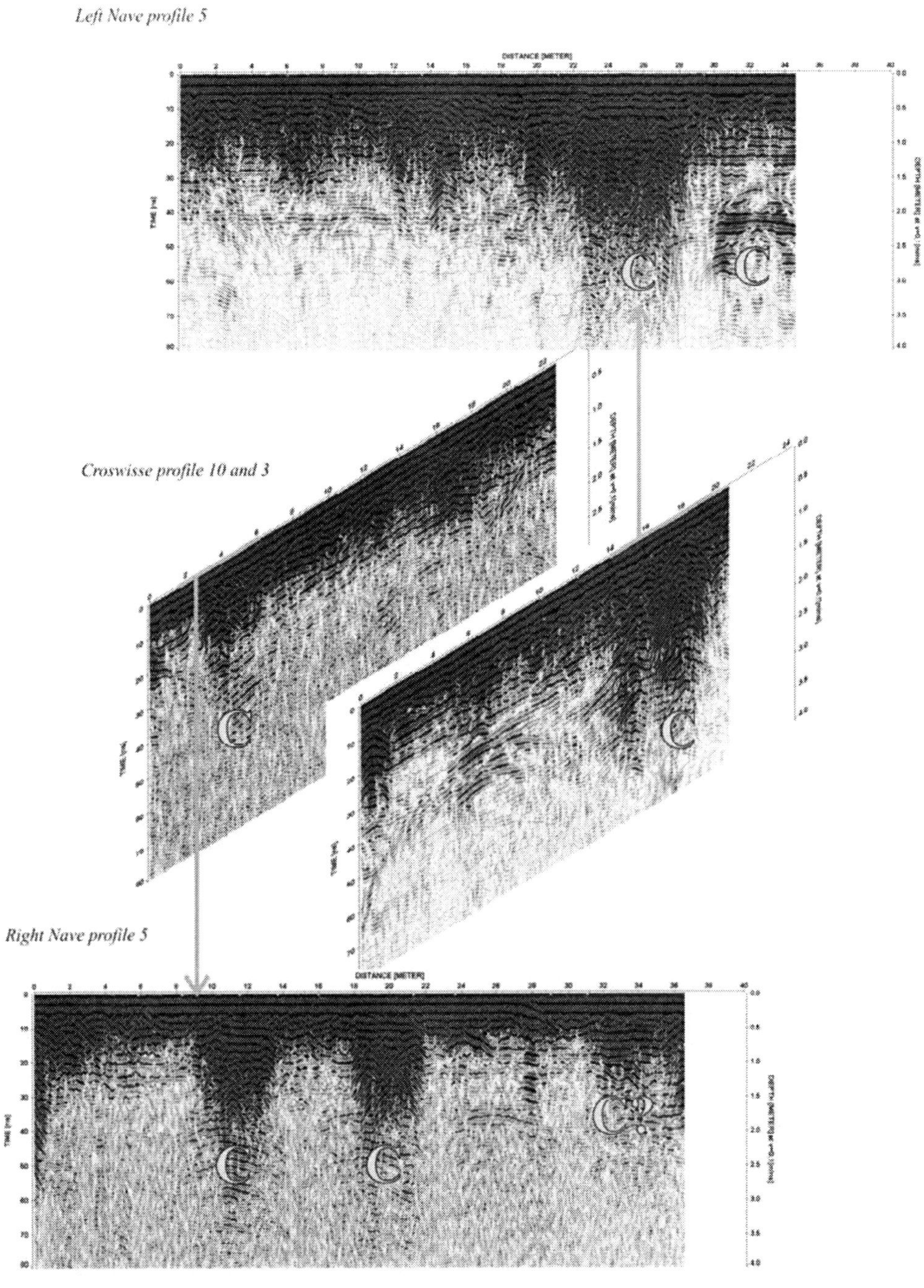

Fig. 2. Processed images from GPR profiles 5 made in the "Left Nave" and "Right Nave" subarea, and perpendicular profiles 10 and 3 from the "Crosswise" subarea. "C" indicates anomalous zone – detected crypt

The above image (Fig. 3) obtained from the profile 9 indicates an anomaly, between 0 and 2 m and at a depth of 1 m, which is very similar to the anomalies recorded above the crypts whose existence had been confirmed. It can be seen at a small depth of 0.3–0.5 m at the beginning of the profile. The perpendicular profile 13 shows strong reflections between 3 and 7 running meters, the origin of which is a crypt parallel to the profile.

https://doi.org/10.1051/e3sconf/20183503006

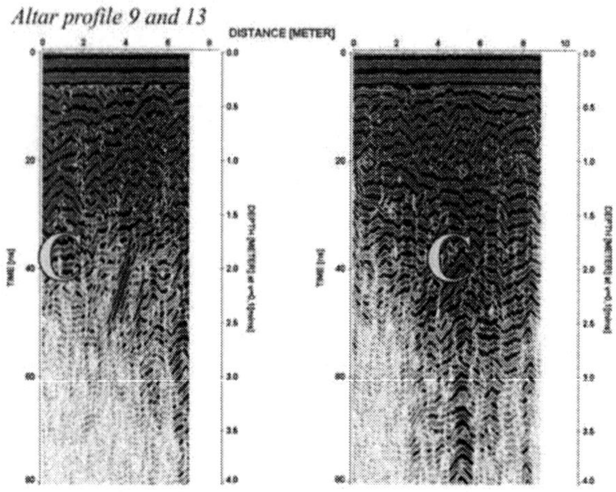

Fig. 3. Processed images from GPR profiles 9 and 13 made in the "Altar" subarea. "C" indicates anomalous zone – detected crypt

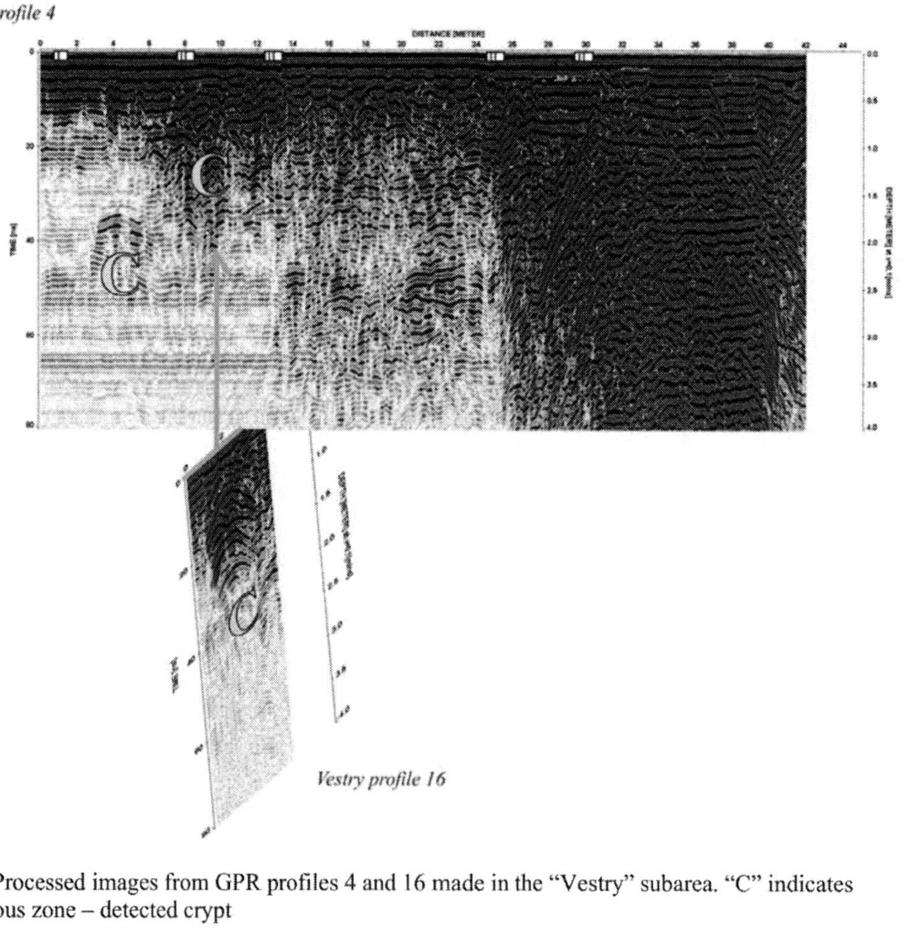

Fig. 4. Processed images from GPR profiles 4 and 16 made in the "Vestry" subarea. "C" indicates anomalous zone – detected crypt

Multiple markers on the horizontal axis indicate the passage through a door (Fig. 4). The anomalies recorded at the ends of the profiles 2 to 6 in the "Left Nave" can be observed also in the profiles 1 to 5 in the "Vestry" region. Vertical boundaries of these anomalies indicate the existence of an anthropogenic object which could be a crypt. On the profiles 1 to 4 it is located between 3 and 6 running meters. On the profile 5, which is shorter, it occurs before 3 running meters. Compared to other anomalies confirmed by the drilled boreholes, it is situated significantly deeper. On all the profiles running across the "Vestry" (profiles 4, 5, 19, 20) anomalies starting at floor beams and at a depth of approximately 2.3 m can be seen. At the end of shown profile 4 form "Vestry" subarea – between 25 and 40 running meters, high amplitude anomalous area is visible. Its source is connected with room in the basement.

Additional profiles 16 and 17 were performed in order to verify the anomalies recorded on the profiles parallel to the main axis of the church (4 and 5). These are the profiles where a clearly visible shallow anomaly can be observed, which unambiguously indicates the existence of a crypt. The ceiling of this crypt is probably at a depth of 0.5 m.

During the measurements there was also possible to identify anomalies caused by construction beams in the floor in the "Vestry" subarea. They are visible at depth 0.25 m on above image (Fig. 5) small hyperbolas (indicated by arrows).

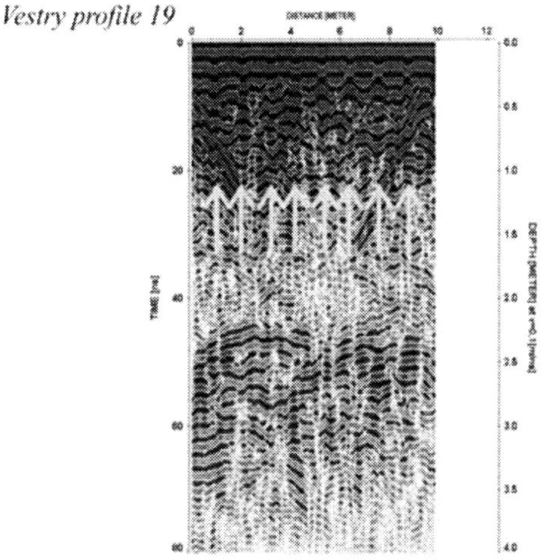

Fig. 5. Processed image from GPR profiles 19 "Vestry" subarea showing construction beams (marked by arrows)

5 Conclusion

Collected and processed data allowed interpretation of image up to 3 m in depth. The appropriately selected surveying parameters and proper methodology of these measurements was adequate to expected results. In this case study it was possible to perform instantaneous localisation and even verification of the results.

The discovered and identified anomalies which are due to the existence of crypts allow for designing a collision-free layout of a heating system or other installations in accordance with the planned concept. It was thanks to it that completely non-invasive, quick and high-resolution GPR measurements brought about high-quality results.

Considering the characteristics of the site and physical working space, the method chosen by the authors was more than adequate. In this case, the field experience of the authors [11] was crucial for the quality of obtained results. Authors previously conducted similar type of research in other churches [12]. In their opinion similar result could be obtained with use of shielded antennae with 250 and even 800 MHz frequency.

References

1. Z. Kozłowska, Rocznik Krakowski 20, 1–19 (1926)
2. U. Sowina, *Woda i ludzie w mieście późnośredniowiecznym i wczesnośredniowiecznym*, 76-104 (2009)
3. J. Muczkowski, Rocznik Krakowski 20, 40–56 (1926)
4. J. Panisova, M. Frastia, T. Wunderlich, R. Pasteka, D. Kusnirak, Archaeological Prospection 20, 163–174 (2013)
5. L.B. Conyers, Interpreting Ground-penetrating Radar for Archaeology (2012)
6. J.M. Reynolds, An Introduction to Applied and Environmental Geophysics, 2nd Edition, 289-348, 535–624 (2011)
7. A.P. Annan A.P., Ground Penetrating Radar Workshop Notes (2001)
8. Ł. Ortyl, J. Karczewski, E. Mazurkiewicz, SGEM 2017: 17th international multidisciplinary scientific geoconference: science and technologies in geology, exploration and mining: 29 June–5 July, 2017, Albena, Bulgaria: conference proceedings 17(14), 173–180 (2017)
9. M. Łyskowski, E. Mazurek, Logistyka 4, 330–336 (2013)
10. K.J. Sandmeier, ReflexW (2009)
11. J. Ziętek, K. Wawrzyniak-Guz, E. Mazurkiewicz, M. Łyskowski, *Nawarstwienia historyczne miast Europy Środkowej*, 397–405 (2016)
12. A. Strzępowicz, J. Ziętek, VIET-POL 2015: Second International Conference on Scientific Research Cooperation Between Vietnam and Poland in Earth Sciences, 117–132 (2015)

Instantaneous Attributes Applied to Full Waveform Sonic Log and Seismic Data in Integration of Elastic Properties of Shale Gas Formations in Poland

Kamila Wawrzyniak-Guz[1,*]

[1] AGH University of Science and Technology, Faculty of Geology Geophysics and Environment Protection, Department of Geophysics, A. Mickiewicza Av. 30, 30-059 Krakow, Poland

Abstract. Seismic attributes calculated from full waveform sonic log were proposed as a method that may enhance the interpretation the data acquired at log and seismic scales. Though attributes calculated in the study were the mathematical transformations of amplitude, frequency, phase or time of the acoustic full waveforms and seismic traces, they could be related to the geological factors and/or petrophysical properties of rock formations. Attributes calculated from acoustic full waveforms were combined with selected attributes obtained for seismic traces recorded in the vicinity of the borehole and with petrophysical parameters. Such relations may be helpful in elastic and reservoir properties estimation over the area covered by the seismic survey.
Key words – shale gas, Baltic Basin, full waveform sonic log, seismic attributes, elastic and reservoir properties.

1 Introduction

In shale gas deposits elastic and reservoir properties of rocks are one of the most significant parameters in hydrocarbon prospection. They can be derived from seismic surveys, well logging measurements or laboratory tests. However, integrated interpretation is still a challenge due to the scale dependence of geophysical parameters. Seismic attributes calculated from full waveform sonic log and from seismic data were proposed as a method that may enhance the interpretation the data acquired at various scales.

Silurian and Ordovician shales in Poland spread along the western margin of the East European Platform in Lublin, Podlasie and Baltic basins, reaching about 700 km in length. In this paper analysis of the Silurian (S) and Ordovician (O) shale formations from the Baltic Basin, Poland is presented. There are numerous claystones, siltstones and mudstones deposits: the Pelplin Claystone Formation (Pe Fm, S), the Pasłek Claystone Formation (Pa Fm, S) with the Jantar Bituminous Black Claystone Member (Ja Mb, S) and the Sasino Claystone Formation (Sa Fm, O). Other formations are marly, such as the Prabuty Marl and Claystone Formation (Pr Fm, O); and the other are carbonate deposits: the Kopalino Limestone Formation (Ko Fm, O).

* Corresponding author: wawrzyni@agh.edu.pl

Ja Mb and Sa Fm are considered as formations with high hydrocarbon potentials. These formations are rich in organic matter and are characterized by elevated TOC content up to 7 wt.% in investigated borehole [1] (Fig. 1).

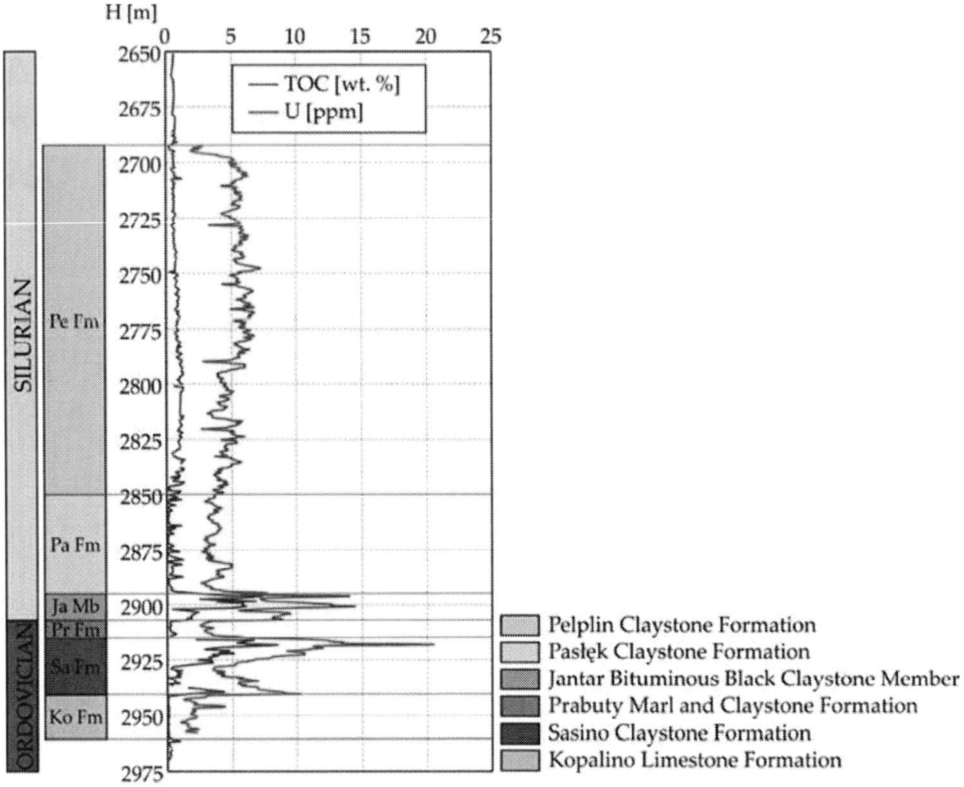

Fig. 1. TOC laboratory results and uranium concentration from spectral gamma well log in a well located in Baltic Basin, Poland (after [2])

2 Well log and seismic data

In this study the acoustic full waveforms recorded by the cross dipole sonic tool were used for instantaneous attributes calculations. Compressional (P), shear (S), and Stoneley (St) waves induced by the monopole transmitter were selected from the recordings.

10 seismic traces (5 Xlines and 5 Inlines) in the location of the investigated borehole were chosen for the analysis. The seismic traces were converted into depth domain with the use of P-wave sonic log corrected to the checkshot data.

3 Methodology

Special methodology for the calculation of attributes from acoustic full waveforms, similar to seismic attributes, was developed. The technique was previously designed for data recorded by sonic tools with monopole sources of elastic waves and successfully tested in siliciclastic and carbonate rocks of some Polish conventional hydrocarbon plays [3]. However, in this study the methodology needed to be adjusted and modified with respect to different tool design and distinct characteristic of waveforms that were recorded by Cross Dipole Wave Sonic tool (Halliburton) in shale formations from the Baltic Basin, Poland.

The idea of was to convert acoustic log into seismic data presentation (Fig. 2). In well logging orientation the vertical axis displays the depth of the borehole (i.e. "the length of the profile"), whereas the horizontal axis shows the recording time of acoustic full waveforms. Visualization of seismic data is similar, but the horizontal axis presents the profile (f. ex. CDP traces) and the vertical axis is the recording time of the seismic traces. When the axes of acoustic full waveforms were rotated into seismic survey orientation it was possible to import the acoustic full waveforms into the seismic software, such as Hampson-Russell software. Then instantaneous attributes analogous to the seismic attributes were calculated within the time windows defined separately for P, S and St waves (Fig. 3).

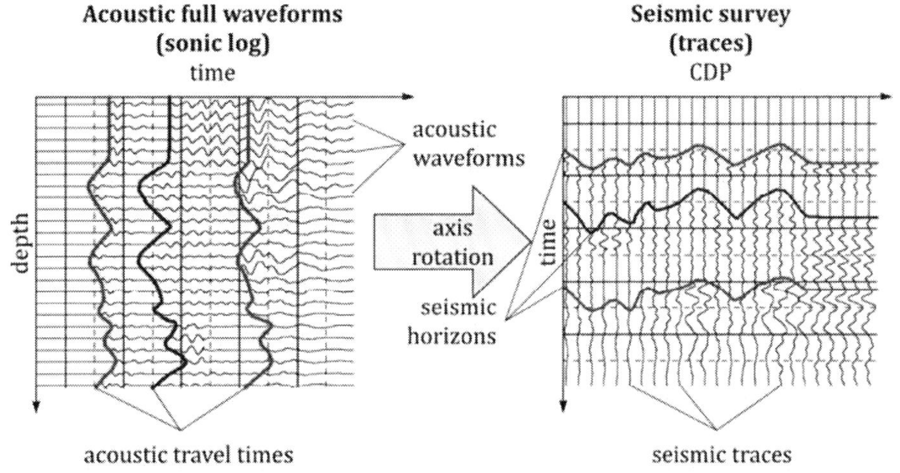

Fig. 2. Rotation acoustic well log data into seismic orientation.

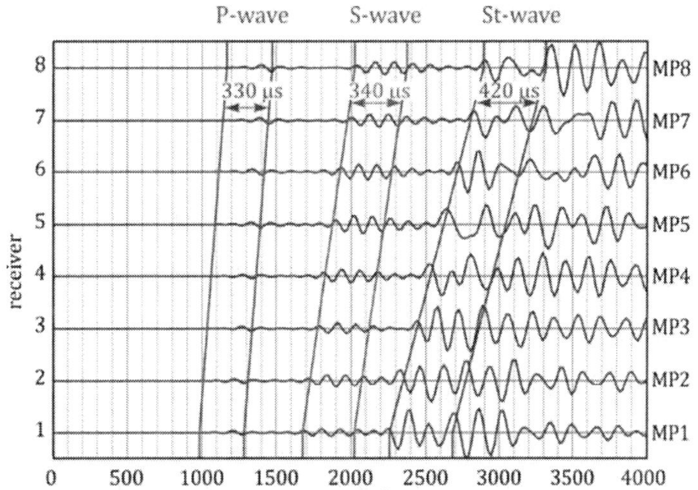

Fig. 3. Selection of P, S & Stoneley waves from the acoustic full waveforms

Attributes calculated from acoustic full waveforms were compared with the attributes obtained for seismic traces recorded in the vicinity of the boreholes. As a result, several crossplots and correlations were constructed. Such relations may be useful for lithology determination; indication of zones saturated with water and hydrocarbon and may be helpful in elastic and reservoir properties estimation over the area covered by the seismic survey.

4 Elastic and reservoir properties of shale formations investigated by attributes

4.1 Attributes applied to acoustic waves (P, S & St) and to seismic data

Instantaneous attributes can enhance subtle changes of signals that were not visible at the raw recordings. Though attributes calculated in this study were the mathematical transformations of amplitude, frequency, phase or time of the acoustic full waveforms and seismic traces, they could by related to the geological factors and/or petrophysical properties of rock formations [4–6].

For P, S and Stoneley waves (that were extracted from sonic log) the following attributes were calculated: acoustic envelope (P_AE, S_AE, St_AE), instantaneous frequency (P_InFreq, S_InFreq, St_InFreq), dominant frequency (P_DomFreq, S_DomFreq, St_DomFreq) and some combination of these attributes such as P and S acoustic envelope ratio (AE_P/S) or dominant frequency to instantaneous frequency ratio (DomFreq_P/InFreq_P and DomFreq_S/InFreq_S).

A set of attributes calculated from seismic data was more extensive and included all complex trace attributes (AE_Seis, InFreq_Seis), P-wave impedance from seismic inversion (P-Imp_Inv) and its mathematical transformations (square, derivative, integral, logarithm), and results of spectral decompositions (f. ex. SD60_20 60Hz, window 20 ms).

4.2 Analysis of instantaneous attributes for P- and S-wave from sonic log

Firstly, the distribution of instantaneous attributes over the formations was investigated. Figure 4 presents the histograms of P_AE, S_AE, P_DomFreq and S_DomFreq.

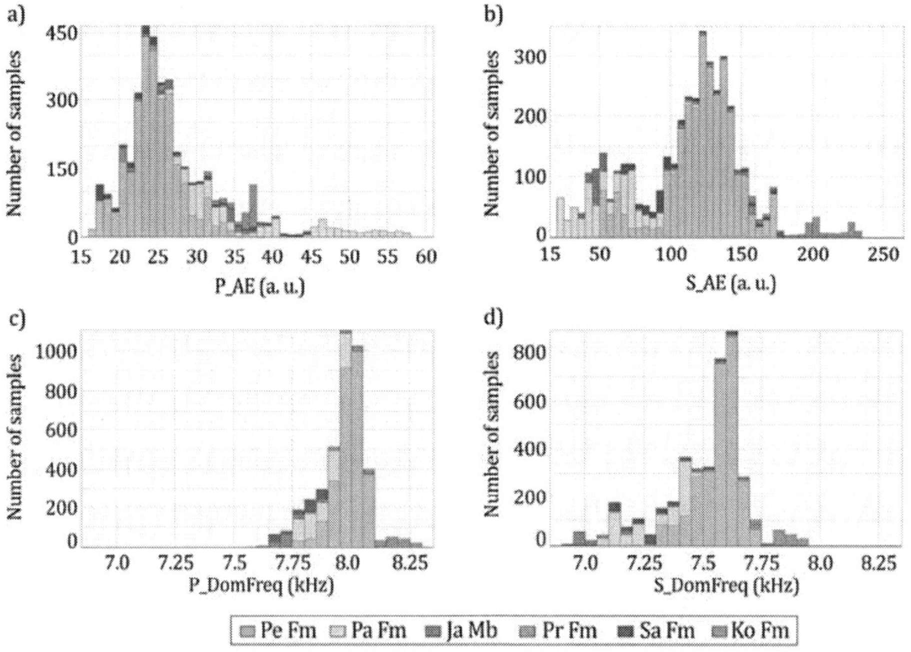

Fig. 4. Histograms of selected instantaneous attributes calculated from sonic logs: acoustic envelope of P wave (a) and S waves (b), and dominant frequency of P (c) and S wave (d)

The histograms prove diversity of the shale formations. Each formation is characterized by specific range of values of calculated attributes that is listed in Table 1.

Table 1. Basic statistics of instantaneous attributes from sonic logs over investigated formations

Statistics	Pe Fm	Pa Fm	Ja Mb	Pr Fm	Sa Fm	Ko Fm
P_AE (a. u.)						
Min-max	24.5–38.3	25.5–57.3	25.9–44.6	20.1–26.1	17.3–44.7	31.8–37.8
Arith. mean	24.5	38.5	33.2	22.6	26.5	36.2
Std. dev.	3.4	9.1	5.0	2.1	8.4	1.9
S_AE (a. u.)						
Min-max	51.2–174.4	20.2–121.2	35.9–64.8	66.6–159.7	66.3–177.7	153.3–234.5
Arith. mean	121.3	56.1	48.3	124.4	108.7	200.6
Std dev.	25.6	22.8	4.8	26.6	33.3	23.0
P_DomFreq (kHz)						
Min-max	7.88–8.10	7.58–7.99	7.64–7.85	7.86–8.05	7.67–7.89	7.97–8.23
Arith. mean	8.00	7.96	7.73	7.97	7.79	8.17
Std. dev.	0.06	0.07	0.05	0.06	0.08	0.07
S_DomFreq (kHz)						
Min-max	7.31–7.73	6.97–7.46	6.94–7.29	7.30–7.74	7.10–7.64	7.68–7.94
Arith. mean	7.57	7.29	7.05	7.57	7.34	7.83
Std. dev.	0.08	0.13	0.10	0.13	0.16	0.07
P_InstFreq (kHz)						
Min-max	7.73–7.94	7.79–7.93	7.70–7.81	7.79–7.80	7.73–7.80	7.91–7.95
Arith. mean	7.86	7.85	7.77	7.79	7.77	7.90
Std. dev.	0.03	0.02	0.04	0.01	0.02	0.04
S_InstFreq (kHz)						
Min-max	7.93–8.56	7.71–8.47	8.06–8.50	8.22–8.37	7.70–8.54	8.10–8.20
Arith. mean	8.26	8.18	8.29	8.26	8.28	8.16
Std. dev.	0.10	0.20	0.15	0.04	0.20	0.03

Acoustic envelope (AE), also known as instantaneous amplitude, is defined by the trace and its Hilbert transform. It is a magnitude of the complex trace and reflects the energy of the signal. In general, amplitudes of P waves are much smaller than S waves (Fig. 4a & 4b); however in Pa Fm considerable amplification of P wave compared to S wave is noticed. The greatest diversity of the instantaneous amplitudes of P and S waves is observed in Pe Fm where amplitudes of P and S waves can vary a lot. Ko Fm is characterized by the highest S wave energy and relatively low P wave amplitudes.

Frequency composition of acoustic signal is expressed by the dominant and instantaneous frequencies. The first one is computed as a maximum of the amplitude spectrum over a small window around the time sample. The latter is a rate of change of signal phase and is calculated from the complex trace. On the basis of the dominant frequency it can be concluded that P and S waves have visibly different frequency content (Fig. 4c & 4d). P wave packets have higher frequencies then S wave. The frequency distribution over formations is clear and consistent. The lowest frequency of acoustic waves is observed in Ja Mb whereas the highest frequencies are in Ko Fm. Pe Fm and Pa Fm are characterized by wide range of frequencies for P and S waves. Distribution of the instantaneous frequency is different (Tab. 1). P_InstFreq has similar but much narrower range of values comparing to P_DomFreq. S_InstFreq presents significantly higher values than S_DomFreq, even higher than frequency of P wave packets.

Figure 5 presents crossplots of selected attributes derived from acoustic full waveforms. Figure 5a shows dominant frequency of P and S waves. Carbonates (here Ko Fm), which are considered as rocks with high elastic parameters, have maintained high frequency. On the other hand Ja Mb and Sa Fm are characterized by lower dominant frequency. This may be related to the presence of organic matter (please compare Fig. 7c & 7d). It can be noticed that there are some points with higher values of P_ and S_ DomFreq in Sa Fm (above the grey thicker lines that indicate average DomFreq in the whole investigated interval). These points correspond to the carbonate lamina within the Sa Fm.

Figure 5b shows that siltstones and mudstones (Pa Fm, Ja Mb, Pe Fm and part of Sa Fm) have high Poisson's ratio the higher Poisson's ratio is the lower elastic properties are. Limestones of Ko Fm have high Poissons's ratio but it is related to the mineral composition: calcite has higher Vp/Vs ratio than quartz and thus increased Poisson's ratio [7]. Elastic properties are expressed here by the amplitude envelop of S wave (S_AE) the lower the amplitude envelope the lower the elastic properties.

Figure 5c confirms that weaker elastic properties (here represented by the S-wave impedance) decrease frequency of the acoustic wave. Ja Mb that is built of siltstones and mudstones with kerogen is characterized by the lowest dominant frequency and the lowest impedance. The highest frequency and impedance are observed in limestones of Ko Fm.

The last figure (Fig. 5d) reveals an interesting relation between instantaneous amplitudes of P and S waves. S_AE as a function of AE_P/S shows strong power relation. The higher elastic properties of the formation are the higher amplitudes of S wave and lower amplitudes of P-wave (i.e. lower ratio of P and S-waves AE) are.

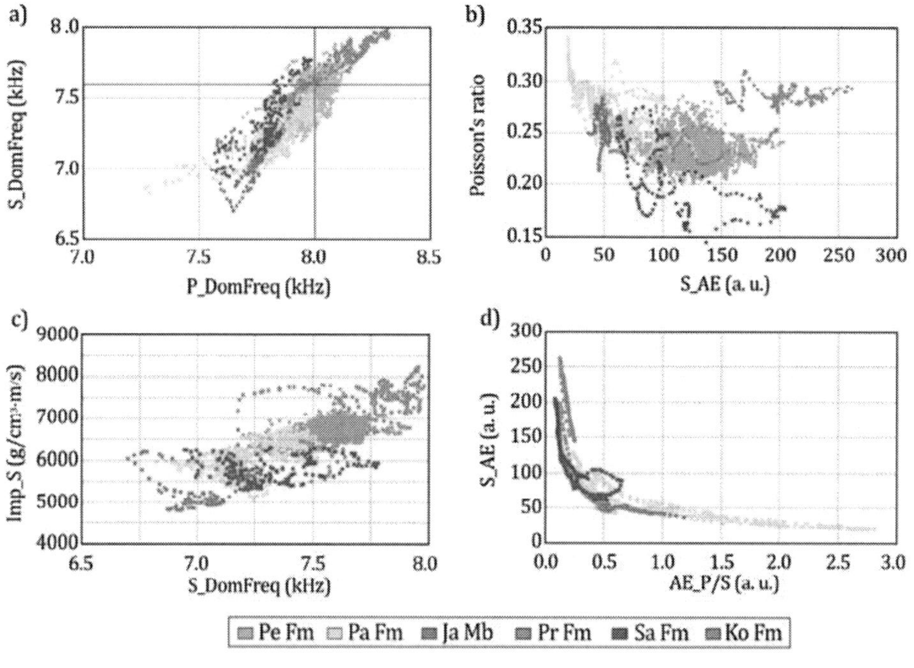

Fig. 5. Selected instantaneous attributes calculated from sonic logs for P and S waves in shale formations: relation between dominant frequency of P and S waves (a), Poissons's ratio as a function of amplitude envelope of S wave (b), correlation between impedance and dominant frequency of S-wave (c), and power function between amplitude envelopes of P and S waves (d)

4.3 Seismic attributes combined with attributes from acoustic log

Due to different vertical resolutions of seismic data and well logs it was necessary to downscale attributes from sonic log. Seismic attributes after time-to-depth conversion were sampled

 https://doi.org/10.1051/e3sconf/20183503007

with 4-m step interval. This is why the acoustic log attributes were smoothed in 4-m running window and interpolated to the seismic depth interval. Figure 6 shows resolution of seismic attributes, log data and the attributes after downscaling. It reduced significantly number of data in the crossplots (Fig. 7).

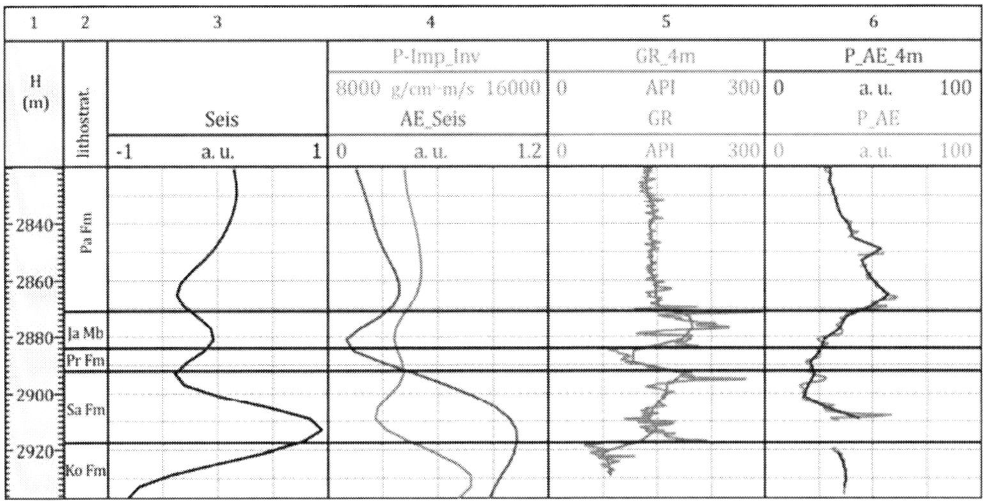

Fig. 6. Comparison of vertical resolution of seismic data (3rd & 4th tracks) with logging data (5th & 6th tracks) before and after downscaling

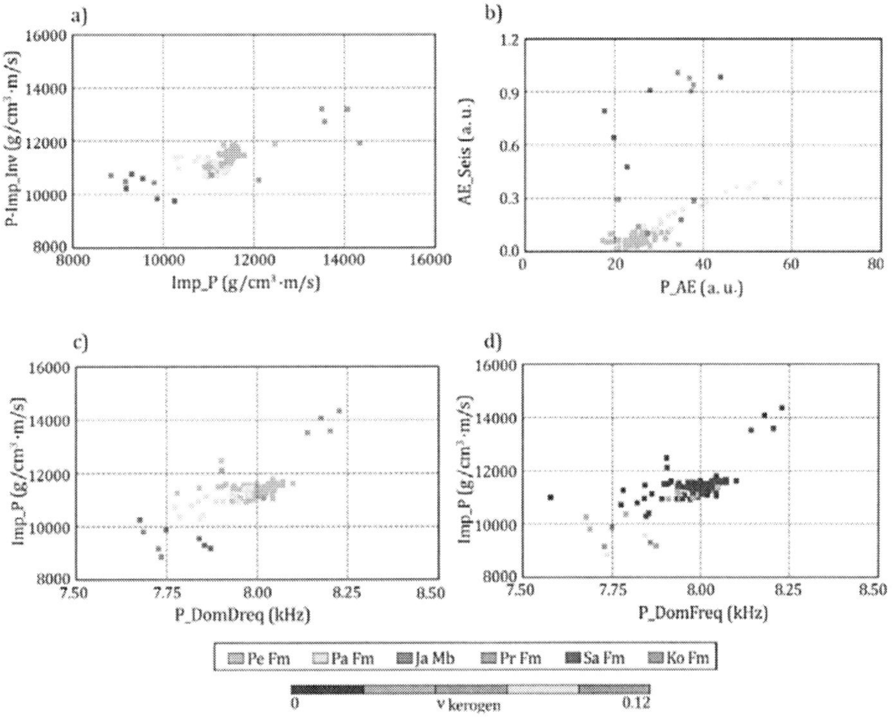

Fig. 7. Attributes from sonic log vs attributes from seismic data (a, b): comparison of impedance from log and seismic inversion (a), and amplitude envelope calculated from seismic trace and extracted P-wave packet from acoustic full waveform (b). Relation between dominant frequency of P wave and P-impedance from sonic log (c) and the same relation as a function of volume of kerogen (d)

5. Summary

Comparison of corresponding attributes from seismic and log data (Fig. 7a and b) shows that some of them represent the same information on the formation besides different scale of the measurements. For example, P-wave impedances from acoustic log and from seismic data are similar (Fig. 7a), though Imp_P has a bit higher values than P-Imp_Inv due to higher velocity of acoustic wave. The other attributes, such as AE of P-wave from sonic log and AE calculated from seismic trace (Fig. 7b) must be interpreted separately. AE of acoustic wave depends on the elastic properties of the some volume of the formation, whereas AE of seismic wave is a function of reflection coefficient at an interface between two layers with various impedances. Figure 7c & 7d shows approximately linear relation between P-impedance and dominant frequency of P-wave (both from sonic log) as a function of organic matter (volume of kerogen). The figures clearly show that presence of organic matter (i.e. elevated TOC content) decreases significantly elastic parameters of rocks it lowers density of the formation, velocity and frequency of acoustic waves.

Acknowledgements

This study was financed by The National Centre for Research and Development in the program Blue Gas, project: "Methodology to determine sweet spots based on geochemical, petrophysical and geomechanical properties in connection with correlation of laboratory test with well logs and generation model 3D" (MWSSSG) Polskie Technologie dla Gazu Łupkowego. Data for the study was allowed by Polish Oil and Gas Company, Warsaw, Poland. It was also financially supported by statutory funds No. 11.11.140.645 at AGH University of Science and Technology (AGH UST), Faculty of Geology, Geophysics and Environmental Protection (FGG&EP), Krakow, Poland.
Author thanks Kamil Cichostępski for calculations of seismic attributes.

References

1. P. Poprawa, *Shale gas potential of the Lower Palaeozoic complex in the Baltic and Lublin-Podlasie basins (Poland) (in Polish)*. Prz. Geol., **58**, 3, 226–249 (2010)
2. J.A. Jarzyna, M. Bała, P.I. Krakowska, E. Puskarczyk, A. Strzępowicz, K. Wawrzyniak-Guz, D. Więcław, J. Ziętek, *Shale Gas in Poland* ((Ed.) InTech, 2007); Available from: https://www.intechopen.com/books/advances-in-natural-gas-emerging-technologies/shale-gas-in-poland
3. K. Wawrzyniak-Guz, *Acoustic Full Waveforms as a Bridge between Seismic Data and Laboratory Results in Petrophysical Interpretation*, Acta Geophys., **64**, 6, 2356–2381 (2016)
4. A.E. Barnes, *Instantaneous spectral bandwidth and dominant frequency with applications to seismic reflection data*, Geophysics, **58**, 3, 419–428 (1993)
5. A.R. Brown, *Understanding seismic attributes*, Geophysics, **66**, 1, 47–48 (2001)
6. M.T. Taner, *Attributes revisited*; Available from: http://www.rocksolidimages.com/attributes-revisited/ (2003)
7. J.H. Schön, *Physical properties of rocks: fundamentals and principles of petrophysics* (Elsevier, Amsterdam, 2004)

Applying of the Artificial Neural Networks (ANN) to Identify and Characterize *Sweet Spots* in Shale Gas Formations

Edyta Puskarczyk[1,*]

[1] AGH University of Science and Technology, Faculty of Geology, Geophysics and Environmental Protection, A. Mickiewicza Av. 30, 30-059 Krakow, Poland

Abstract. The main goal of the study was to enhance and improve information about the Ordovician and Silurian gas-saturated shale formations. Author focused on: firstly, identification of the shale gas formations, especially the *sweet spots* horizons, secondly, classification and thirdly, the accurate characterization of divisional intervals. Data set comprised of standard well logs from the selected well. Shale formations are represented mainly by claystones, siltstones, and mudstones. The formations are also partially rich in organic matter. During the calculations, information about lithology of stratigraphy weren't taken into account. In the analysis, self-organizing neural network – Kohonen Algorithm (ANN) was used for *sweet spots* identification. Different networks and different software were tested and the best network was used for application and interpretation. As a results of Kohonen networks, groups corresponding to the gas-bearing intervals were found. The analysis showed diversification between gas-bearing formations and surrounding beds. It is also shown that internal diversification in sweet spots is present. Kohonen algorithm was also used for geological interpretation of well log data and electrofacies prediction. Reliable characteristic into groups shows that Ja Mb and Sa Fm which are usually treated as potential sweet spots only partially have good reservoir conditions. It is concluded that ANN appears to be useful and quick tool for preliminary classification of members and sweet spots identification. **Key words** – shale gas, sweet spots, Artificial Neural Networks, classification.

1 Introduction

Advanced statistical methods and artificial neural networks can provide useful information for identification of productive horizons in oil and gas reservoirs [1]. Commercial software, used by oil and gas companies usually contain modules based on them. One of the basic types of self-organizing neural networks are Kohonen's networks. This type of neural networks can be used for data classification. As a result electrofacies were created. The term electrofacies was originally defined by Serra and Abbot [2]. An electrofacies represents a unique set of log responses, which characterizes physical properties of the rocks and fluids contained in the volume investigated by the logging tools.

* Corresponding author: puskar@agh.edu.pl

For determining electrofacies, authors like Serra [3], Doveton [4], Moss [5] suggest to use cluster analysis in *n* dimensions with each wireline log as a dimension. In this paper there are shown the results of using Kohonen algorithms for identification beds with different properties recorded on the well logs. After an automatic clustering and classification, each group was described for sweet spot identification. Several neural networks were tested using two different software, one dedicated for well logs analysis and the other one dedicated for statistical analysis.

2 Materials and methods

A single well located at the Polish Baltic Basin was chosen for the ANN tests. Data set contains data from the Silurian and Ordovician gas-saturated shale formations from Pa Fm to Sl Fm (Fig. 1). The depth intervals of each formations/members range as follow: Pa Fm (3736.5–3915 m), Pe Fm (3915–3938 m), Ja Mb (3938–3951 m), Pr Fm (3951– 957.5 m), Sa Fm (3957.5–3975 m), Ko Fm (3975–3980 m) and Sl Fm (3980–3983 m). An analysed depth interval is almost 250 m with sample rate equal to 0.1 m. That means that about 2500 objects were taken into account for the ANN calculations. Shale formations are represented mainly by claystones, siltstones, and mudstones. One of the formations, Ko Fm is represented mainly by limestones. Usually, bituminous shales have significant content of bentonite and pyrite. It was also observed the high content of marly limestones. The following standard well logs were chosen as an input data for ANN:

- CALI (mm) – calliper log, used for well QC, primary lithology indicator, in shales usually shows caves,
- GR (API) – natural gamma ray log, total radioactivity of the formation, reach high value in the shale rich formation, is treated as a primary shale indicator,
- POTA (%), THOR (ppm) and URAN (ppm) – individual concentration of potassium, thorium and uranium respectively, from spectral gamma ray log. Potassium and thorium tend to be concentrated in clays while uranium and its derivatives tend to be concentrated in source rocks because of adsorption by organic matter,
- DT (μs/ft) – compressional slowness log, in shales lowers sonic velocities are observed because of presence of organic matter,
- MSFL (omm), LLS (omm) and LLD (omm) – micro, shallow and deep resistivity logs respectively,
- NPHI (%) – neutron porosity, high values in shales, caused by high hydrogen index in organic matter and presence in clays free water, adsorbed water (clay bound water) as well as lattice-water, which is a part of clay mineral structure,
- RHOB (g/cm^3) – bulk density log, thick series of shales demonstrates progressive increase in density due to the compaction, when formation is overpressure there is a break in normal shale compaction trend manifested by a drop of bulk density, organic shales are distinguished by clear low anomalies on RHOB log because organic matter has much lower density in comparison with non-source shales,
- PE (b/e) – photoelectric factor log, is a function of chemical (mineral) composition and thus is indirectly related to lithology, but is independent from porosity.

Artificial neural networks proposed by Teuvo Kohonen [6] are the most popular model of self-organizing networks. Generalized learning algorithm for Kohonen networks is the following:

- the *n*-dimensional weight vectors v_1, v_2, ...,v_i of the *j* computing units are randomly selected. Then an initial radius *r*, a learning constant *a* and a neighbourhood function α are selected,
- an input vector **ε** is selected by using the probability distribution over the input space,

- the unit k with the maximum excitation and for which the distance between v_i and $\boldsymbol{\varepsilon}$ is minimal is selected,
- using the neighbourhood function, the weight vectors are updated

$$v_i \leftarrow v_i + a\alpha(i,k)(\varepsilon - v_i) \ for \ \ i = 1,\ldots,j \tag{1}$$

- modify a and/or α and continue from the beginning or stop after N iterations. By repeating the process several times, arriving at a uniform distribution of weight vectors in input space is expected. During the learning process both the size of the neighbourhood and the value of neighbourhood function fall gradually, so that the influence of each unit upon its neighbours is reduced. From an initial distribution of random weights, and over N iterations, the networks settles into a map of stable zones. Each zone is effectively a feature classifier.

Two different and independent software were used for electrofacies identification. The first one, IPSOM module at Techlog Schlumberger software was designed for geological interpretation of well log data and facies prediction. This module provides classification solutions based on 'classical' clustering and the Kohonen networks. IPSOM provides also QC tool for the results evaluation as a probability of occurrence of predicted facies at each depth. In this study hierarchical cluster analysis was performed based on the Euclidean metrics. The second, ANN module at Statistica software developed by StatSoft is an universal statistical tool used in many different fields, e.g. economics, electronics. It has the ability to select the type of network and its parameters by the user.

3 Statistical characterization of shales

Descriptive statistics were calculated for each lithostratigraphic formation and member. The statistics were calculated for both the well logs and the results of well logs comprehensive interpretation. The average value of each petrophysical parameter and standard deviation (values are in brackets, arithmetic average and standard deviation respectively) observed and interpreted in two selected formations (which can be treated as potential *sweet spots*) are presented below:

- **Ja Mb (a part of Pa Fm)** – 130 samples; CALI (221 and 0.15 mm), GR (134 and 27 API), POTA (3 and 0.5%), THOR (10 and 1 ppm), URAN (5.5 and 1.6 ppm), DT (80 and 5 us/ft), MSFL (94 and 70 omm), LLS (52 and 33 omm), LLD (60 and 41 omm), NPHI (18 and 5%), RHOB (2.6 and 0.05 g/cm^3); data from comprehensive interpretation: PHI (4 and 1%), VCL (58 and 14%), VSAND (22 and 11%), VLIME (10 and 10%), VDOLO (3 and 3%), VMIN Fe (0.04 and 0.1%), VKEROGEN (2.2 and 0.8%) and TOC (1.06 and 0.3%wt),
- **Sa Fm** – 165 samples; CALI (217 and 4.5 mm), GR (142 and 23 API), POTA (2.9 and 0.4%), THOR (11 and 1 ppm), URAN (6.3 and 2 ppm), DT (81 and 8 us/ft), MSFL (333 and 2203 omm, but median equal 35 omm), LLS (40 and 59 omm), LLD (50 and 75 omm), NPHI (19 and 4%), RHOB (2.47 and 0.2 g/cm^3); data from comprehensive interpretation: PHI (6 and 3%), VCL (52 and 11%), VSAND (21 and 16%), VLIME (15 and 16%), VDOLO (2 and 6%), VMIN Fe (0.08 and 0.3%), VKEROGEN (3.4 and 1%) and TOC (1.7 and 0.7%wt).

Similar mean GR values were obtained for Ja Mb, Sa Fm and other shale formations. These values were much lower (average 60 API) only for Ko Fm and much higher for Sl Fm. Similar relationships were obtained for POTA, THOR and URAN logs. DT average values ranged from 69 to 81 us/ft for all formations except for Ko Fm, where they reached lower (average 55 us/ft) values. Lower densities were observed for Ja Mb and Sa Fm. On the Pe curve only Ja Mb had lowered values and Ko Fm increased, the remaining formations reached an average about 3.8 b/e. At each shale formation, the NPHI values were high, while the decrease value was recorded for Ko Fm. Resistivity curves recorded large variation within each formation, but mean values did not show differences between the formations.

The statistical analysis of the petrophysical parameters of comprehensive well logs interpretation in individual formations shows that at Ja Mb and Sa Fm the highest porosities (average 5%) is observed. In sweet spots intervals also the highest content of sandstones (about 22%) is observed. Most limestone content (average 70%) has been observed for Ko Fm, and most dolomite content (average 12%) for Pr Fm. Highest values of kerogen content and total organic carbon were observed for Sa Fm and Ja Fm. The remaining shale formations (Pa Fm, Pe Fm, Pr Fm) and Ko Fm registered the TOC content below 1%wt.

The analysis of descriptive statistics shows that individual lithostratigraphic intervals differ from others, but also often show large internal variations.

4 Classification and *sweet spots* identification

Several dozen Kohonen networks were tested to determine the electrofacies. The tests started with a 10 × 10 grid. Finally, the best network was chosen, with a 2 × 3 grid. The networks created with exactly the same parameters and input in the two tested software gave very similar results. This article presents the results of two networks. The results were compared with the results of classical cluster analysis (Fig. 1). Network called IPSOM (Techlog) was created based on 12 input logs (described in the paragraph 2). Network called ANN 14_6 (Statistica) as an input included additionally TOC and PHI logs. We have decided to add an extra logs because when exactly the same input logs were applied for exactly the same type of networks, the results were almost the same. We have decided add two logs, total porosity (PHI) and total organic carbon content (TOC), obtained from comprehensive well log interpretation as a parameters which have high influence on the possible gas presents. In both cases, there were obtained 6 groups, but obtained division was no identical. By analysing the last two tracks shown in the Figure 1, it can be noted that the IPSOM network has allocated thinner layers within the individual lithostratigraphic units (shaded alternately on the Fig. 1). However, within the potential *sweet spots*, both networks showed a very similar distribution. By comparing the results of Kohonen networks to the results of 'classical' clustering analysis it can be concluded that: on the basis of cluster analysis, the division is approximately proportional to the lithostratigraphic division, while the Kohonen networks have shown the variation within these units.

Figures 2 and 3 show the distribution of mean values (square inside the box) and the range of variability (box frame sets minimum and maximum) of the standardized petrophysical parameters of comprehensive interpretation in created groups based on the IPSOM Techlog (Fig. 2) and ANN 14-6 Statistica (Fig. 3) networks.

Results of analysis:

- *sweet spots* can be identified by using ANN. For the IPSOM network within the Ja Mb and Sa Fm, groups 3 and 5 were identified. For the ANN 14_6 network within Ja Mb groups 1 and 5 were identified and within Sa Fm 1 and 4 were identified. Group 3 on IPSOM correspond to group 1 on ANN 14_6. In this groups high values of porosity and TOC are observed. Group 5 in IPSOM correspond to group 5 in ANN 14_6. Average values of each log are similar.
- it is therefore possible to conclude that in both *sweet spots* intervals there are levels with great potential for hydrocarbon presence in the top of the member/formation and in both *sweet spots* the lower part of the intervals has worse properties,

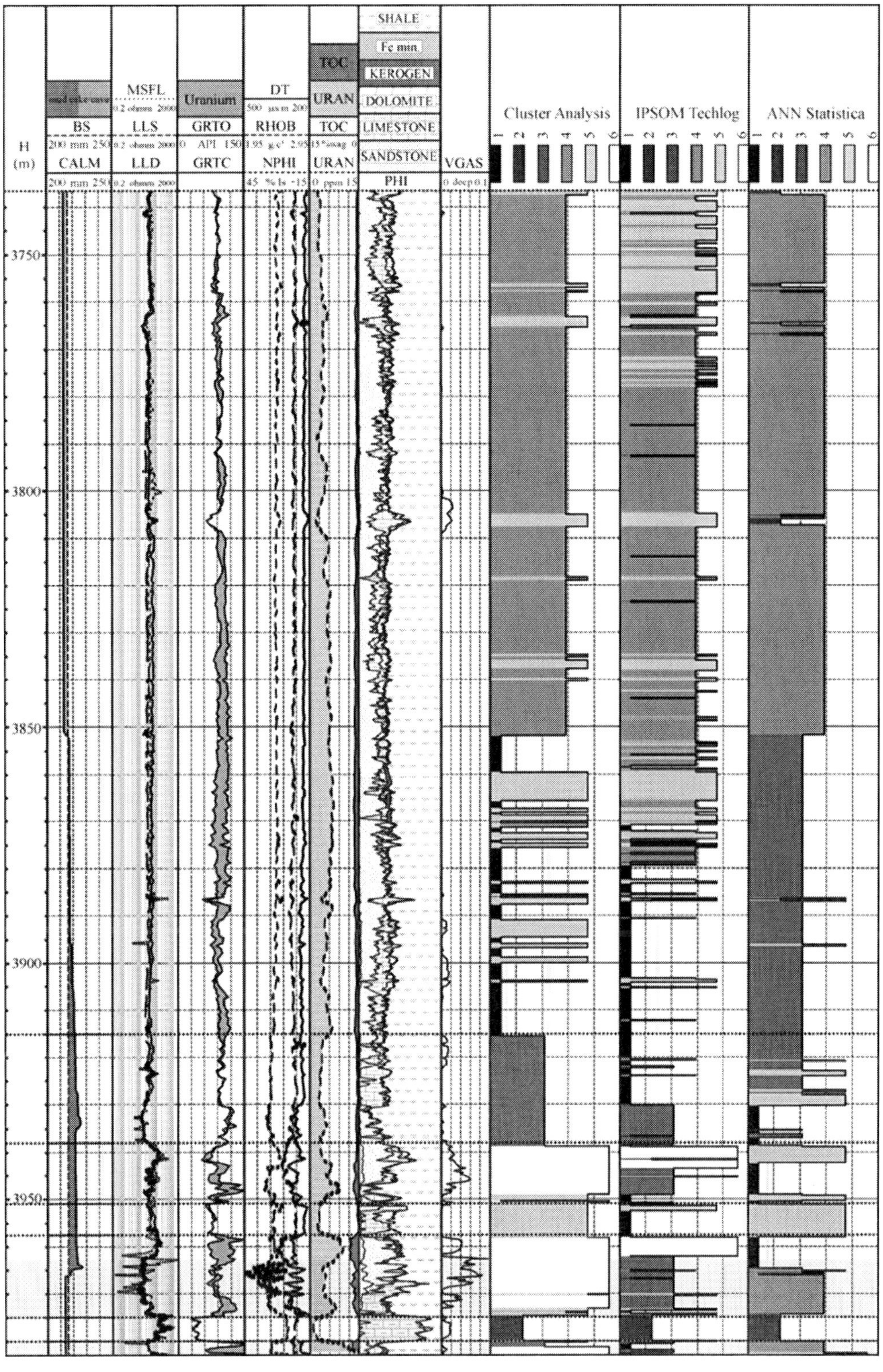

Fig. 1. Well logging data from measurements (first 5 tracks), interpretation (lithology track, TOC and gas content) and results of 'classical' cluster analysis and artificial neural networks application

https://doi.org/10.1051/e3sconf/20183503008

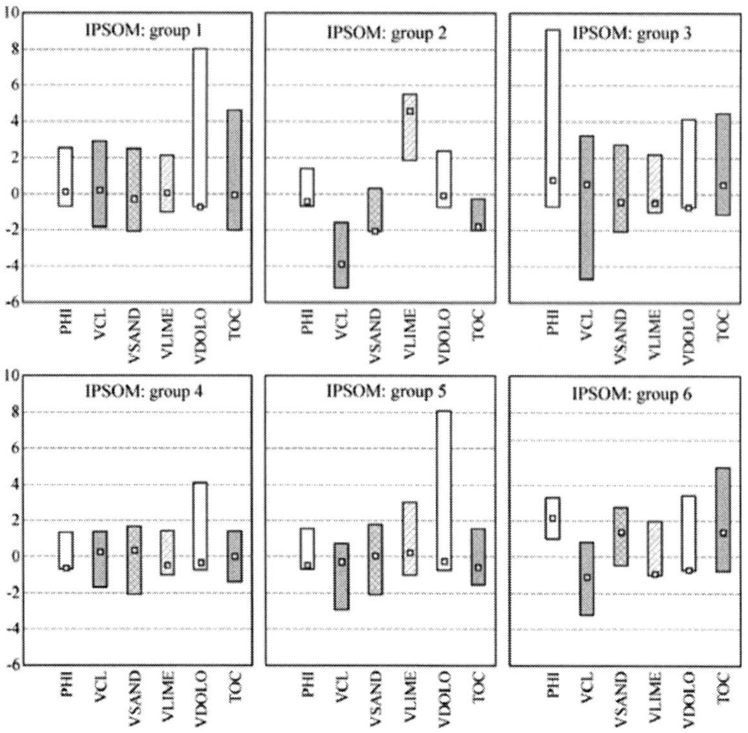

Fig. 2. Average values (square inside the box) and minimum – maximum (box), calculated for standardized results of comprehensive interpretation in each group based on IPSOM module

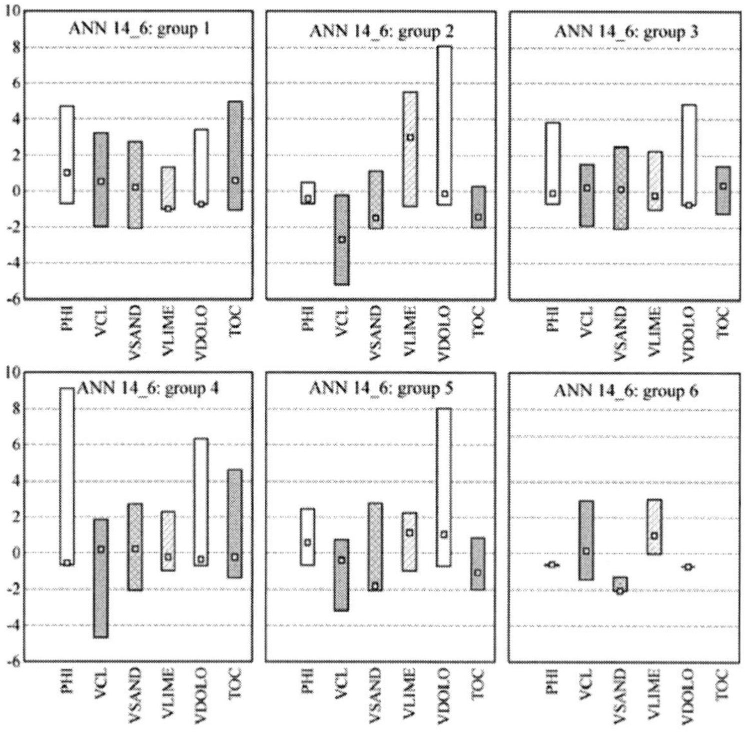

Fig. 3. Average values (square inside the box) and minimum – maximum (box), calculated for standardized results of comprehensive interpretation in each group based on ANN Statistica module

- it is worth to check the interval at which IPSOM shows electrofacies 1, because it has a high TOC value. This electrofacies was found at Pa Fm, Pe Fm and Pr Fm,
- in both networks electrofacies number 2 correspond to the Ko Fm interval. In this group limestone content is extremely high compared to other electrofacies,
- a comparison of 'classical' cluster analysis results to neural networks shows a similarity in the general distribution of clusters, but also differences in the detailed analysis of individual small layers,
- using two different software for Kohonen networks didn't show significant difference in terms of using identical inputs,
- adding two additional variables (PHI and TOC) to the ANN14_6 network resulted in large variations in the grouping compared to IPSOM. The selection of variables for analysis is one of the key elements of correct analysis.

4 Conclusions

The ANN were used for the classification and grouping of data accordingly to natural petrophysical features of rocks. Three were found groups corresponding to the potentially gas-saturated zones. The Kohonen networks confirm and refine the cluster analysis results.
- Complex analysis showed diversification between shale gas formations and adjacent beds.
- *Sweet spots* intervals were separated from the shale formation. It shows that internal diversification in each gas formation is present.
- Electrofacies determined on the basis of standard logs turned out to be useful in discrimination of reservoir and sealing parts of formation.
- All statistical methods that were applied to the rock classification gave satisfactory results.

It is concluded that artificial neural networks and especially self-organizing maps appear to be a useful and quick tool for preliminary classification of members and hydrocarbon-saturated zones identification. Statistical methods offer petrophysicists an additional tool for improving standard well log analysis. This offers possibility to define zones of interests based on the criteria petrophysical parameters defined from well logs interpretation.

Acknowledgements

Data was allowed by POGC Warsaw co, Poland for the MWSSSG Polskie Technologie dla Gazu Łupkowego project (2013-2017). Techlog 13 software was used under the AGH UST grant from Schlumberger. Statistica 12 software was used under the AGH UST grant from StatSoft.

References

1. E. Puskarczyk, J. Jarzyna, Sz. J. Porębski Geol. Q **59**, 1 (2015)
2. O. Serra and H. Abbot, SPE. 55th Annual Fall Technical Conference and Exhibition Dallas, Texas, paper No. 9270 (1980)
3. O. Serra, Edu. Serv., **272**, Buenos Aires (1987)
4. J. Doveton, AAPG Computer Applications in Geology, No. 2, 169, Tulsa (1994)
5. B. Moss, Geological Society Special Publication **122** (1997)
6. T. Kohonen, Biol. Cybern., **43**, 1 (1982)

Comprehensive Interpretation of the Laboratory Experiments Results to Construct Model of the Polish Shale Gas Rocks

Jadwiga A. Jarzyna[1,*], *Paulina I.* Krakowska[1], *Edyta* Puskarczyk[1], *Kamila* Wawrzyniak-Guz[1], and *Marcin* Zych[1]

[1] AGH University of Science and Technology, Faculty of Geology Geophysics and Environmental Protection, A. Mickiewicza Av. 30, 30-059 Krakow, Poland

Abstract. More than 70 rock samples from so-called *sweet spots*, i.e. the Ordovician Sa Formation and Silurian Ja Member of Pa Formation from the Baltic Basin (North Poland) were examined in the laboratory to determine bulk and grain density, total and effective/dynamic porosity, absolute permeability, pore diameters size, total surface area, and natural radioactivity. Results of the pyrolysis, i.e., TOC (Total Organic Carbon) together with S1 and S2 – parameters used to determine the hydrocarbon generation potential of rocks, were also considered. Elemental composition from chemical analyses and mineral composition from XRD measurements were also included. SCAL analysis, NMR experiments, Pressure Decay Permeability measurements together with water immersion porosimetry and adsorption/desorption of nitrogen vapors method were carried out along with the comprehensive interpretation of the outcomes. Simple and multiple linear statistical regressions were used to recognize mutual relationships between parameters. Observed correlations and in some cases big dispersion of data and discrepancies in the property values obtained from different methods were the basis for building shale gas rock model for well logging interpretation. The model was verified by the result of the Monte Carlo modelling of spectral neutron-gamma log response in comparison with GEM log results. **Key words** – shale gas rock, porosity, density, TOC, mineral composition.

1 Introduction

Shale gas deposits belong to the unconventional hydrocarbon (HC) resources. They played important role in the energy balance all over the world since the prices of oil and gas from conventional deposits started enormously increase. They will be still important because conventional HC resources are now deeper and deeper, more difficult and complicated in the aspect of porosity/permeability and saturation determination. Therefore, world companies elaborate the technology of effective shale gas deposits exploitation. Poland belongs to the countries rich in shale gas formations. The most prospective are the Silurian and Ordovician shaly formations located in three Polish sedimentary basins: Baltic, Podlasie, and Lublin in

* Corresponding author: jarzyna@agh.edu.pl

the north-east and central-east parts of Poland [1, 2]. Polish shale gas formations differ from the most known in literature American Barnett, or other shale plays because of deeper depth of present deposition, different mineral composition, and changeable TOC values [3–5]. Results presented in the paper are typical for the Silurian and Ordovician *sweet spots* in the Baltic Basin.

2 Materials and methods

Laboratory experiments were done on 72 rock plugs cut from the Silurian and Ordovician cores drilled in three on-shore boreholes in the Baltic Basin. Depth intervals, stratigraphic and lithological characteristics of the geological material are presented in Table 1. Depth intervals were similar in BH-2 and BH-3 wells. Samples from BH-1 well were sited deeper. Sequence of the Palaeozoic formations drilled in three boreholes in the study was as follows (going down): Silurian (Llandovery), claystone/mudstone Pa Formation with Ja Member built of bituminous claystone at the bottom, next – Ordovician (Ashgillian) marl, limestone and claystone/mudstone Pr Formation and at the end of the studied intervals – Ordovician (Caradocian/Llanvirnian) bituminous claystone and claystone/mudstone Sa Formation.

Table 1. Rock material for laboratory measurements

Well (Total no. of samples)	Depth interval, [m]	Silurian (Llandovery)		Ordovician (Carad./ Llanvirn.)	
		No. of samp.	Lithology	No. of samp.	Lithology
BH-1 (27)	3185.89-3234.42	5	claystone/ mudstone	16	clay/mudst.
				1	limestone
				1	sandstone
				4	tuff
BH-2 (27)	2896.38-2938.96	6	claystone/ mudstone	17	clay/mudst.
				1	limestone
				3	tuff
BH-3 (18)	2870.36-2908.76	7	claystone/ mudstone	7	clay/mudst.
				1	limestone
				3	tuff

2.1 Rock characteristics

Majority of the investigated rock samples were claystones/mudstones from two *sweet spots*: Silurian Ja Member (15 samples) and Ordovician Sa Formation (45 samples). Basic statistics of the selected petrophysical properties of these two shale gas rocks are presented in Table 2. Symbols used in Table 2 are explained in paragraph 2.2.

2.2 Laboratory investigations of rock samples

Mercury injection porosimetry (MIP) was done using AutoPore IV 9500 Micromeritics Co and following parameters were determined: bulk density (at 36.17 psi), ρb [g/mL], apparent (skeletal) density, ρs [g/mL], effective porosity, Φ_{eff}, total pore area, TPA [m^2/g], average pore diameter, APD [μm], and permeability, K [mD]. Density was also determined with the use of He pycnometer AccuPyc 1330, ρHe [g/ccm].

Table 2. Basic statistics of the selected petrophysical properties: the Ja Member and Sa Formation; symbols: Min – minimum, Max – maximum, Avg – average, Me – median

Type of measure.	Parameter	Min	Max	Avg	Me	Min	Max	Avg	Me
	Stratigraphy	Silurian Ja Member				Ordovician Sa Formation			
MIP	ρb [g/ccm]	2.28	2.54	2.42	2.37	2.23	2.97	2.43	2.51
MIP	ρs [g/ccm]	2.35	2.57	2.36	2.44	2.29	2.98	2.49	2.55
He pycn.	ρHe [g/ccm]	2.55	2.72	2.63	2.66	2.56	3.07	2.68	2.73
WIP	GD_{WIP} [g/ccm]	2.50	2.75	2.62	2.60	2.46	2.79	2.66	2.73
MIP	Φ_{eff} [%]	0.93	3.08	1.70	2.77	0.29	7.51	2.32	1.67
NMR	Φ_{NMR_t} [%]	0.61	1.90	1.10	1.06	0.08	5.96	2.00	0.83
NMR	Φ_{NMR_eff} [%]	0.10	0.38	0.20	0.24	0.01	0.73	0.29	0.18
NMR	$S_{w,irr}$ [%]	0.66	0.90	0.81	0.75	0.16	1.00	0.79	0.80
NMR	CBW [%]	0.49	3.31	1.48	1.60	0.15	3.51	1.62	2.72
WIP	Φ_{WIP} [%]	2.93	11.94	6.11	6.05	0.83	10.09	5.50	7.04
DLP	Φ_{DLP} [%]	2.31	9.65	4.98	3.87	0.76	7.69	4.58	5.63
XRD	OM [wt%]	0.40	7.90	3.87	3.65	0.00	4.90	1.44	0.85
NaI(Tl)	U [ppm]	3.24	18.64	8.17	8.65	2.03	9.80	5.13	3.89
	Th [ppm]	8.90	12.64	11.87	12.08	4.72	26.20	11.70	10.51
Rock-Eval	TOC [wt%]	0.36	7.15	3.42	3.22	0.01	4.31	1.26	0.71
	S1 [mgHC/g r-k]	0.25	2.73	1.56	1.62	0.02	3.63	0.81	0.33
	S2 [mgHC/g r-k]	0.30	10.19	4.68	3.60	0.01	0.34	0.09	0.08
MIP	TPA [m^2/g]	0.41	6.52	3.22	6.02	0.11	15.25	4.83	3.44
N2 A/D	S_{BET} [m^2/g]	2.90	10.30	6.79	6.55	0.70	40.10	13.94	13.20
N2 A/D	TPV [ccm/g]	0.016	0.034	0.023	0.024	0.01	0.05	0.03	0.03
N2 A/D	Vmicro [ccm/g]	0.001	0.002	0.001	0.001	0.001	0.008	0.003	0.002
N2 A/D	Vmezzo [ccm/g]	0.006	0.018	0.011	0.011	0.002	0.035	0.017	0.016

Nuclear Magnetic Resonance (NMR) experiments were done using the Maran Ultra 23 MHz spectrometer of Resonance Instruments, Ltd and provided total and effective porosity, Φ_{NMR_t} and Φ_{NMR_eff}, respectively; irreducible water volume, $S_{w,irr}$, and clay bound water, CBW. Specific method - Dual-Liquid Porosimetry (DLP)/Water Immersion Porosimetry (WIP) [6, 7] was used to determine total porosity, Φ_{WIP} [%] and dual liquid porosity, Φ_{DLP} [%], corrected for swelling of water filled samples, together with grain density GD_{WIP} [g/ccm]. Quantachrome Autosorb-1C automatic gas adsorption apparatus was applied to measure physical adsorption/desorption of nitrogen (N_2 A/D) at 77 K. Nitrogen adsorption/desorption isotherms were used to characterize the specific surface area, S_{BET} [m^2/g] and pore size (diameters < 350 nm) distribution [8, 9]. The Brunauer – Emmet – Teller (BET), Barrett – Joyner – Halenda (BJH) approach, as well as Quenched Solid Density Functional Theory (QSDFT) were adopted for interpretation of adsorption data. From the nitrogen adsorption/ desorption measurements total pore volume (pore diameter less than 350 nm), TPV [ccm/g] was calculated and QSDF theory provided with volume of micropores of diameters less than 2 nm, Vmicro [ccm/g] and mezzo pores of diameters 2–50 nm, Vmezzo [ccm/g]. Rock-Eval 6.0 pyrolysis as the organic geochemical analysis oriented to hydrocarbon deposits recognition allowed determination among others of total organic carbon, TOC, free hydrocarbons, S1 and hydrocarbons generated through thermal cracking (residual hydrocarbons, S2) [10]. These parameters are good indicators of rocks rich in organic matter (OM). TOC is well correlated with individual S1 and S2 parameters and also with the sum of them thus, each parameter may be used interchangeably for characterization of source rocks. Uranium and thorium contents, U [ppm] and Th [ppm] measured using apparatus with crystal NaI(Tl) SKW IUO4 (TESLA) inform on the level of organic matter OM [wt %] and shale volume VCL [wt %], respectively, on the basis of natural radioactivity. Values presented in Table 2 and relationships illustrated in chapter 3 present the typical characteristics of the *sweet spots* of considered shale gas formations, which are an input data to construct rock model.

3 Relationships between parameters

Correlations in Figures 1 and 2 illustrate selected examples of mutual relationships between properties comprising various types of porosity, density, and volume of organic matter. They are presented separately for Ja Member and Sa Formation to underline the differences in the bituminous claystone/mudstones building *sweet spots*. The goal of the presentation of these simple correlations is to show the relationships as an introduction to multidimensional regression and revealed by mutual connections that gas is placed in the bituminous claystone/ mudstone formation in the pore space and also is connected to organic matter. Correlation between the total volume of pores with a diameter less than 350 nm, TPV, and effective porosity from NMR experiment, Φ_{NMR_ef}, shows that the smallest pores are connected among themselves and gas inside can be removed. The relationship between bulk density ρb from MIP and organic matter content calculated together with mineral composition (Fig. 1b) documents distinct density decreasing with organic matter increase and confirms the high importance of density determination. Dispersion observed in plots is caused by heterogeneity of bituminous claystone/mudstones [11, 12].

A good correlation is observed between organic matter content calculated together with the mineral composition on the basis of elements recognition by chemical analysis followed by mineralogical interpretation and TOC from Rock-Eval geochemical measurement (Fig. 1c) despite of different methods of determination. Also, the significant relation is visible in Figure 1d between grain density, determined from water immersed porosity method [7], and TOC. The relationship in Figure 1d confirmed the necessity of precise density measurements and understanding the difference between bulk, skeletal (from MIP) and grain density from WIP method in shale gas rocks. Similar relationships are shown for Sa Formation (Fig. 2a, b). The main feature of these plots in comparison to previous ones is larger number of data but also higher dispersion of points due to geological features of rocks [11, 12]. Two data sets are distinctly visible in Figure 2a. The first one is more dispersed, marked by oval, and the second one, more consistent. shows TPV porosity increase with the OM decrease. These two sets of data mean that organic matter and small diameter porosity exist alongside in the bituminous claystone/mudstone.

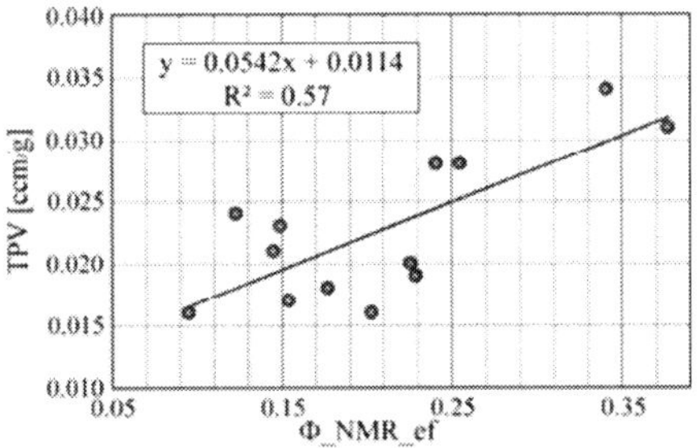

Fig. 1a. Relationship between the total volume of pores less than 350 nm vs. effective porosity from NMR laboratory experiment, Ja Member

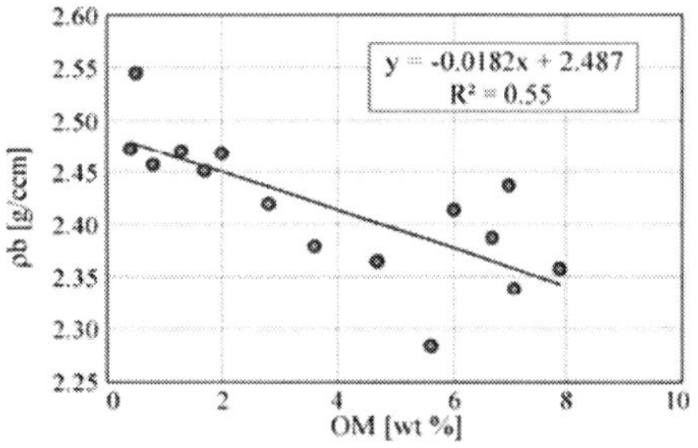

Fig. 1b. Relationship between bulk density from mercury injection porosimetry (MIP) vs. organic matter content, Ja Member

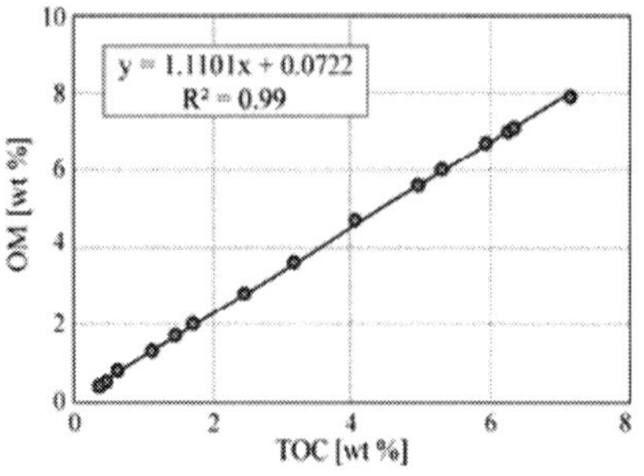

Fig. 1c. Relationship between organic matter content vs. Total Organic Carbon, Ja Member

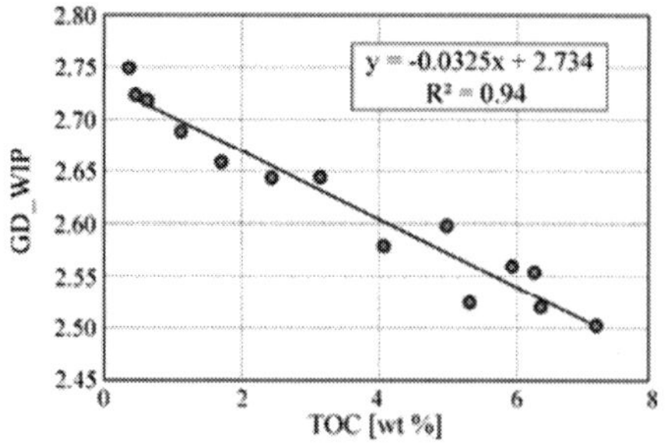

Fig. 1d. Relationship between grain density from WIP method vs. Total Organic Carbon, Ja Member

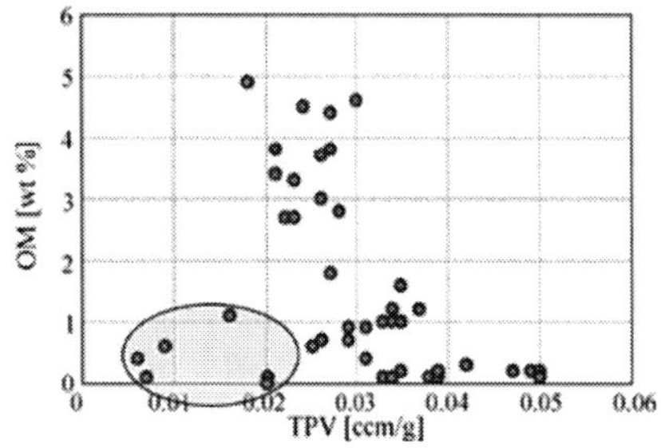

Fig. 2a. Relationship between organic matter content and total porosity including pores < 350 nm, Sa Formation

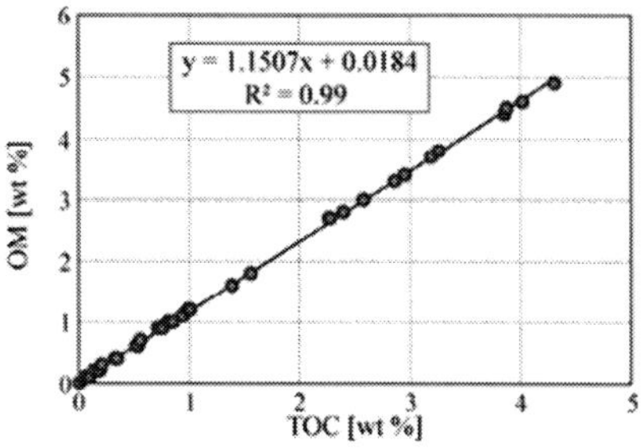

Fig. 2b. Relationship between organic matter content and Total Organic Carbon, Sa Formation

Simple linear relations were checked between all parameters in the study (Table 1) to check how big part of regression they explain. All of them were included in multiple regression to construct shale gas rocks model based on porosity and density as values determined continuously along the borehole axis in well logging. Parameters used as inputs into multiple regression were selected by interpreters on the basis of simple regression because of a limited number of cases in which there were complete data set with all quantities.

3.1 Multidimensional regression

Multidimensional regression was tested between density, porosity, TOC, and other available parameters. An exemplary result of bulk density (MIP) prediction on the basis of selected parameters (Table 3) for Ja Member is presented as equation (1):

$$\rho b = 2.7898 - 0.19\Phi_{DLP} - 0.05CBW + 0.26\Phi_{NMR\_t} + \\ - 0.17S_{w,irr} - 0.83TOC - 0.05V_{i+s} - 0.07V_{chl} \quad (1)$$

Symbols in equation (1) and Table 3 are the same as in Table 2 and subchapter 2.2. CBWmeans clay bound water [%], V_{i+s},V_{chl} – volumes of illite+smectite and chlorite [%] from XRD mineral analysis. Correlation coefficients between independent and dependent variables are included in Table 3. They illustrate the strength of relationship and contribution in evaluation which quantities are important in prediction. The correlation coefficient for multidimensional regression (eq. 1) is equal to R = 0.91, determination coefficient R^2 is equal to 0.83 and corrected R^2 = 0.59.

Similar equation (2) was obtained for effective porosity (MIP) predicted on the basis of known parameters in Sa formation:

$$\Phi eff = 27.99 - 0.42\Phi_{DLP} + 1.02GD_{WIP} + 0.49CBW + 0.50S_{w,irr} - 9.32\rho b + 9.50\rho s + \\ - 6.60TOC - 2.16S1 + 0.08S2 - 1.75\rho He + 0.01U - 0.43Th + 0.18VCL + 6.69OM \quad (2)$$

In that case correlation coefficient in multidimensional regression is equal to R = 0.96, determination coefficient R^2 is equal to 0.93 and corrected $R^2 = 0.63$.

In Figure 3a, b there are presented regressions for predicted bulk density (ρb_pred) and effective porosity (Φeff_pred) vs. their measured values. High determination coefficients for bulk density and effective porosity correlations are equal to 0.99 and 0.93, respectively. They confirmed the ability of petrophysical properties prediction on the basis of the determination of various quantities, but Table 3 showed that contribution of selected parameters to variability explanation is different.

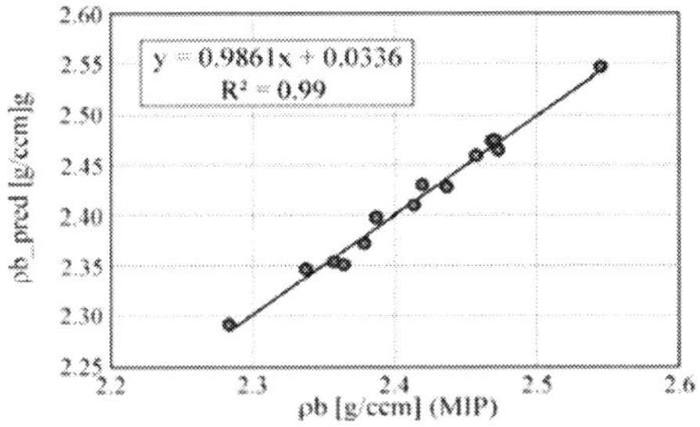

Fig. 3a. Relationship between predicted (eq. 1) and measured bulk density (MIP), Ja Member

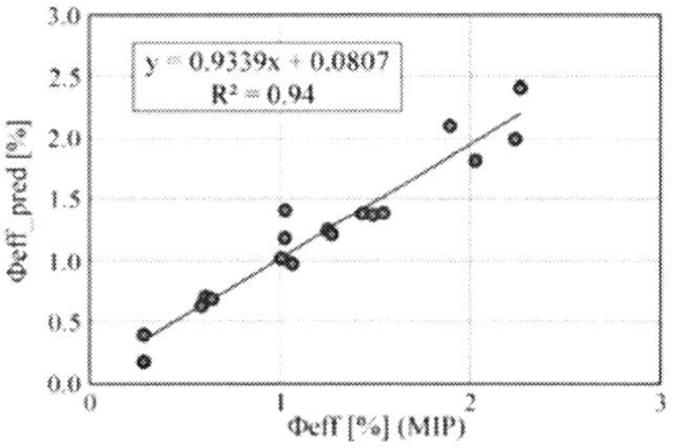

Fig. 3b. Relationship between predicted (eq. 2) and measured effective porosity (MIP), Sa Formation

Table 3. Correlations between independent variable, ρb and dependent variables, Ja Member

	Φ_{WIP}	Φ_{DLP}	$S_{w,irr}$	CBW	ρs	TOC	S2	ρHe	V_{i+s}	V_{chl}	OM	ρb
Φ_{WIP}	1,00	0,96	–0,23	0,61	0,49	–0,71	–0,68	0,69	0,27	0,39	–0,72	0,40
Φ_{DLP}	0,96	1,00	–0,19	0,87	0,30	–0,69	–0,71	0,69	0,22	–0,08	–0,97	0,69
$S_{w,irr}$	–0,23	–0,19	1,00	–0,23	–0,34	0,41	0,53	–0,11	0,33	–0,17	0,42	–0,26
CBW	0,61	0,87	–0,23	1,00	0,77	–0,83	–0,74	0,65	0,07	0,19	–0,82	0,04
ρs	0,49	0,30	–0,34	0,77	1,00	–0,81	–0,69	0,43	0,14	0,20	–0,81	0,98
TOC	–0,71	–0,69	0,41	–0,83	–0,81	1,00	0,96	–0,79	–0,06	–0,46	1,00	–0,73
S2	–0,68	–0,71	0,53	–0,74	–0,69	0,96	1,00	–0,82	0,13	–0,48	0,96	–0,60
ρHe	0,69	0,69	–0,11	0,65	0,43	–0,79	–0,82	1,00	0,00	0,38	–0,78	0,35
V_{i+s}	0,27	0,22	0,33	0,07	0,14	–0,06	0,13	0,00	1,00	–0,38	–0,05	0,13
V_{chl}	0,39	–0,08	–0,17	0,19	0,20	–0,46	–0,48	0,38	–0,38	1,00	–0,46	0,13
OM	–0,72	–0,97	0,42	–0,82	–0,81	1,00	0,96	–0,78	–0,05	–0,46	1,00	–0,74
ρb	0,40	0,69	–0,26	0,04	0,98	–0,73	–0,60	0,35	0,13	0,13	–0,74	1,00

In bulk density case the most important role played skeleton density, organic matter content, TOC, grain density, and CEC (Tab. 3). Effective porosity calculated from the modified equation (2), without S1, S2 and OM, gave very similar result but parameters of correlation were as follows: R = 0.95, $R^2 = 0.89$ and corrected $R^2 = 0.70$.

As a final result, the multidimensional regression was calculated to obtain relationships between bulk density and porosity from well logging and analyzed petrophysical parameters to build a statistical model of shale gas formation. Mutual connections between parameters measured in the laboratory and used as independent variables in the multidimensional regression, and quantities measured in well logging, showed the influence of selected variables on the output result. The regression equation for RHOB in the gas formation of Ja Member predicted on the basis of considered parameters is presented as equation (3):

$$RHOB = 0.08 - 1.19CBW + 1.57GD_{WIP} + 1.08\rho b - 0.78\rho s - 0.32CEC - 0.10S_{wirr} \quad (3)$$

R equals to 0.95 $R^2 = 0.90$ and corrected $R^2 = 0.82$. The most informative quantities, in that case, were clay bound water CBW and grain density from water immersed method GD_{WIP}.

3.2 Final confirmation of the built model

Final confirmation of the constructed model quality was the comparison of the cumulated mass concentration of selected elements (H, Si, Ca, Fe and Al) calculated using Monte Carlo method for the spectral neutron-gamma device response with the results of GEM device measurements. Rock model was composed on the basis of highly correlated among themselves parameters which were checked in the simple and multidimensional regressions. In Figure 4 results of modelling of spectral neutron-gamma response are presented as mass concentration of selected elements as a function of the detector in SN-G device position along the geological profile.

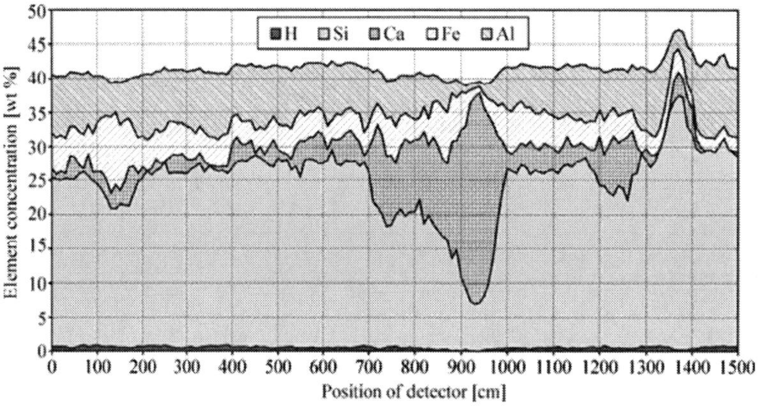

Fig. 4. The cumulated mass concentration of hydrogen, silicon, calcium, iron, and aluminum in the rock model calculated using Monte Carlo method for the spectral neutron-gamma device response

4 Conclusions

Variety of measured parameters allowed differentiation of the claystone/mudstone *sweet spots* among other Silurian and Ordovician formations of similar lithological characteristics. Simple statistics of measured petrophysical (reservoir and geochemical) properties and mineral composition provide good characteristics of *sweet spots* and can be treated as the primary indicators of the potential shale gas plays. Analysis of the results of simple linear regression showed mutual relations between properties and revealed important features of shale gas rocks. Cross-plots analysis enabled selection of the most informative parameters to model construction. Similar petrophysical parameters worked as the input data in the multidimensional regression for Ja Member and Sa Formation despite the difference in stratigraphy and heterogeneity of lithology. The most effective parameters providing characteristics of typical *sweet spots* in Polish shale gas plays consists of: porosity from different methods, bulk and skeletal density from MIP, grain density from WIP method, He pycnometer density, $S_{w,irr}$ from NMR experiment and mineral components: volume of illite, smectite and organic matter. Monte Carlo modeling worked effectively as confirmation of the modeled rock formation.

Acknowledgements

Data was allowed by POGC Warsaw co, Poland for the MWSSSG Polskie Technologie dla Gazu Łupkowego project (2013-2017). Authors thank prof. dr. Leszek Czepirski and his team for N vapors adsorption/desorption (77 K) measurements, dr. R. Semyrka for mercury injection porosimetry, prof. dr. A. Derkowski and T. Topór, M.Sc. for WIP porosimetry, D. Dworak, M.Sc. for Monte Carlo modeling and dr. T. Zorski for cooperation in defining problems.

References

1. H. Kiersnowski, in: Nawrocki J. (ed.), *Shale gas as seen by Polish Geological Survey Polish Geological Institute* – National Research Institute 26–31 (2013)
2. P. Poprawa, Przegląd Geologiczny (in Polish, Abstract in English), **58**, 3226–249 (2010)
3. D. Więcław, M.J. Kotarba, P. Kosakowski, A. Kowalski, I. Grotek, Geological Quarterly, **54**, 2, 159–182 (2010)
4. T. Topór, A. Derkowski, P. Ziemiański, L. Marynowski, D.K. McCarty, Int. Journal of Coal Geology, **179**, 45–59, (2017) https://doi.org/10.1016/j.coal.2017.05.001
5. T. Topór, A. Derkowski, P. Ziemiański, J. Szczurowski, D.K. McCarty, Int. Journal of Coal Geology, **180**, 46–56, (2017) https://doi.org/10.1016/j.coal.2017.07.005
6. U. Kuila, D.K. McCarty, A. Derkowski, T.B. Fischer, M. Prasad, Fuel, **117**, 1115–1129 (2014)
7. T. Topór, A. Derkowski, K. Utpalendu, T.B. Fischer, D.K. McCarty, Fuel, **183**, 537–549 (2016)
8. M. Thommes, K. Kaneko, A.V. Neimark, J.P. Olivier, F. Rodriguez-Reinoso, J. Rouquerol, K.S.W. Sing, Pure and Applied Chemistry, **87**, 9–10, 1052–1069 (2015)
9. F. Rouquerol, J. Rouquerol, K.S. W. Sing, P. Llewellyn, G. Maurin, *Adsorption by Powders and Porous Solids: Principles*, Methodology and Applications (Academic Press, 2014)
10. M.J. Kotarba, M. Lewan, D. Wieclaw, *Shale gas and oil potential of Lower Palaeozoic strata in the Polish Baltic Basin by Hydrous Pyrolysis*. Fourth EAGE Shale Workshop. Shales: What do they have in common? 6–9 April, Porto, Portugal, (2014)
11. K. Wawrzyniak-Guz, J.A. Jarzyna, M. Zych, M. Bała, P.I. Krakowska, E. Puskarczyk,*Analysis of the heterogeneity of the Polish shale gas formations by Factor Analysis on the basis of well logs*. In: Extended abstracts of the 78th EAGE Conference and Exhibition, 30 May – 2 JuneVienna, 2016, Tu SBT3 07, (2016)
12. J.A. Jarzyna, M. Bała, P.I. Krakowska, E. Puskarczyk, A. Strzępowicz, K. Wawrzyniak-Guz, D. Więcław, J. Ziętek M. *Shale Gas in Poland*, in: Advances in Natural Gas Emerging Technologies, DOI: 10.5772/67301 (2017) http://www.intechopen.com/articles/show/title/shale-gas-in-poland

Application of Laser Scanning for Creating Geological Documentation

Michał Buczek[1,*], *Martyna* Paszek[1], and *Anna* Szafarczyk[1]

[1] AGH University of Science and Technology, Faculty of Mining Surveying and Environmental Engineering, A. Mickiewicza Av. 30, 30-059 Krakow, Poland

Abstract. A geological documentation is based on the analyses obtained from boreholes, geological exposures, and geophysical methods. It consists of text and graphic documents, containing drilling sections, vertical cross-sections through the deposit and various types of maps. The surveying methods (such as LIDAR) can be applied in measurements of exposed rock layers, presented in appendices to the geological documentation. The laser scanning allows obtaining a complete profile of exposed surfaces in a short time and with a millimeter accuracy. The possibility of verifying the existing geological cross-section with laser scanning was tested on the example of the AGH experimental mine. The test field is built of different lithological rocks. Scans were taken from a single station, under favorable measuring conditions. The analysis of the signal intensity allowed to divide point cloud into separate geological layers. The results were compared with the geological profiles of the measured object. The same approach was applied to the data from the Vietnamese hard coal open pit mine Coc Sau. The thickness of exposed coal bed deposits and gangue layers were determined from the obtained data (point cloud) in combination with the photographs. The results were compared with the geological cross-section. **Key words** – intensity, terrestrial laser scanning, geological cross-section, geological profile.

1 Introduction

The graphical presentation of the current state of the geological structures is usually required for the creation of a geological documentation, its supplements, and carrying out a register and a resources balance. The geological mapping of the excavation at the present moment includes making maps, cross-sections, profiles and other diagrams.

The geological cross-section method is one of the basic methods to estimate the resources. Nowadays, there is plenty of software facilitating geologist's work. The meticulous fieldwork is still the base for the whole procedure of creating cross-sections and maps which may be used for subsequent calculations.

The field-based geological mapping methods of exposed rock formations are still mainly the conventional techniques using a measuring tape, a piece of paper and a pencil. No less important is geologist's experience and reliability. Recently, researchers postulate application of

* Corresponding author: mibuczek@agh.edu.pl

new remote sensing techniques like point cloud from a stereoscopic pair of photographs [1], or a model from standard color slides [2].

The cross-section accuracy depends on many factors such as type and formation of the deposit, documentation purpose etc. The results accuracy can reach the centimeter level, but lower accuracy is sufficient for the business and industry conditions [3].

Recently, in surveying measurements, the Light Detection and Ranging (LIDAR) is becoming more and more popular, commonly called 3D laser scanning. It is a surveying method to measure a terrain surface. It measures the distance to a target from a device (laser scanner) mounted inside an airship, a land vehicle, or on a station (eg. tripod). The device sends pulsed laser light and measures the reflected pulses with a sensor, simultaneously computing the measured point's coordinates (X, Y, Z) and the direction of the laser beam [4]. The accuracy and the speed of the terrestrial LiDAR allow to apply it in many different tasks: structures' health monitoring [5], rivers [6] and forest measurements [7], and creating geological documentation [8].

As a result, the point cloud data is obtained. Despite the geometrical information (coordinates) it may contain extra information such as intensity of the reflected beam (intensity), RGB information about a color of the scanned point, or normal vectors of each point etc. The laser signal intensity depends mostly on the scanned surface material [9]. Except that, range, incidence angle, meteorological and atmospheric conditions, albedo and many others influence the intensity measurement [10, 11].

The paper presents an approach to create the cross-section of the deposit with a terrestrial laser scanner (TLS). The approach to determine the type of rocks based on the laser beam intensity is presented. The results are compared with the results from conventional methods. The application of LIDAR technology allows one to enhance the quantitative elements of the geological documentation.

2 Test site measurements

The LIDAR technology has been already applied successfully in many geological tasks like digital and schematic models, digital elevation models (DEM), volume computation, and many others [8, 12]. The research focused on the creation of a geological cross-section of a mine wall from a point cloud obtained from LIDAR measurements. The assumption was made that higher accuracy of LIDAR data can improve the resource evaluation from the model. Acquiring the raw point cloud is faster than conventional methods. In some conditions, like landslides, the remote sensing technology is also more safe, because one does not have to enter the endangered area [13].

The study was based on the two test sites: AGH experimental mine scanned with Leica C10, and Coc Sau open pit mine scanned with GeoMax ZOOM300. Depending on the laser scanner's construction and software, the intensity is recorded in a different format. The parameters of the laser scanners are presented in Table 1.

Table 1. The parameters of the used terrestrial laser scanners

	Leica C10	**GeoMax Zoom300**
Type	pulse-based	Time-of-Flight
Accuracy of length measurement	4 mm (up to 50 m)	6 mm/50 m
Range	up to 350 m	300 m
Carrier wavelength	532 nm	905 nm
Beam divergence	0.24 mrad	0.37 mrad
Intensity (range of values)	0–1	0–255

2.1 The Used data

The first point cloud depicts a part of the longwall in AGH experimental mine (Fig. 1a). The scan was taken from one station under favorable measuring conditions: uniform lighting, close distance to the object, clear and dry air, and constant temperature. The point cloud containing around 2.3 million points was smoothed with the Gaussian filter to compute the mean value of intensity.

The chosen part of the wall contains 3 surface types (called Rock A, Rock B, and Rock C) of different color and lithology. For each material 5 samples from the point cloud were chosen (Fig. 1b). Figure 1c depicts the field-based geological drawing of the measured test site.

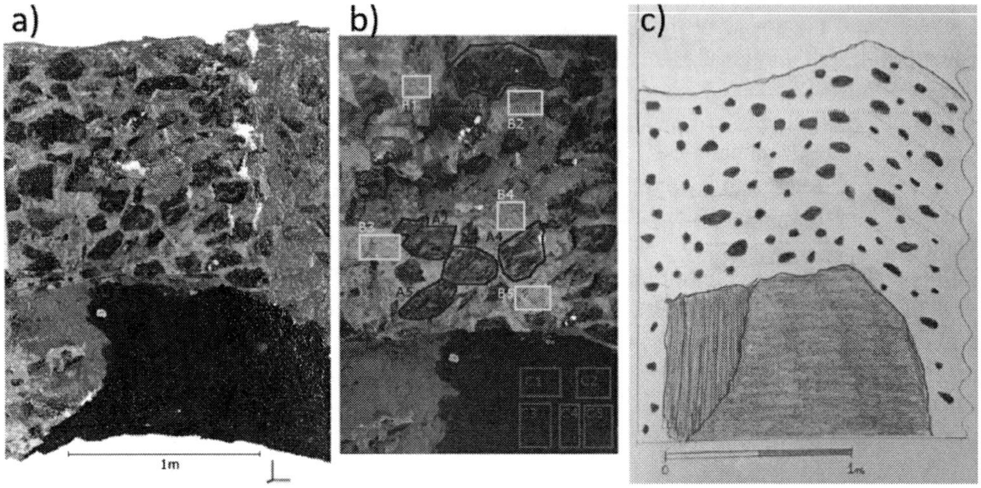

Fig. 1. AGH test site: a) point cloud (presentation type: intensity), b) photograph with marked rock samples (red – Rock A, yellow – Rock B, green – Rock C), c) field geological profile

2.2 Methodology

The intensity value histograms with 64 intervals were made for each selected rock sample. The obtained datasets were characterized by the normal distribution. The normal distribution was assumed on the base of the central limit theorem of Lindenberg-Levy and Lapunow, due to a large number of observation in the samples. The basic attribute of the normal distribution is that the 68% of the observations are within one standard deviation (1σ), and 95% of the observations are within two standard deviations of the mean (2σ). Based on that attribute the ranges (1σ and 2σ) were computed.

The averaged ranges (obtained from 5 samples) were computed for each rock type. The final values were used to divide the point cloud into separate groups, depending on the intensity of each point. The Gauss filter, applied previously, reduces the noise and number of outliers. The generated groups contain points with intensities assigned to different materials.

The ranges may intersect with each other, depending on the characteristics of the materials. For those ranges, the rock type is disambiguated. The level of certainty depends on lithology of measured materials, each rock may have different characteristics. One cannot simply assume apriori the level of certainty used to separate materials from the point cloud. For different pairs of rocks, the boundary may be more or less vague. Thus, the new point clouds for each material were compared with the photographs and geological drawing to check its correctness.

2.3 Results

Leica C10 laser scanner stores intensity values within the range 0 to 1. The intensity ranges of each type of lithology and theirs averaged values were presented in Table 2.

Table 2. The sample characteristics and the intensity ranges

Rock type	Number of points in sample	Intensity histogram	Intensity range (1σ)		Average intensity range (1σ)		Intensity range (2σ)		Average intensity range (2σ)	
A1	4333	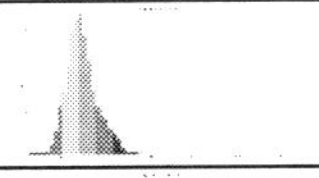	0.141	0.171			0.124	0.188		
A2	2893	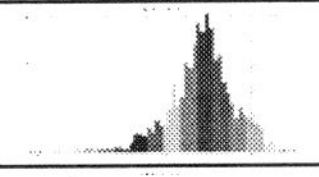	0.144	0.174			0.129	0.189		
A3	2665	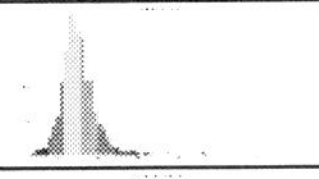	0.149	0.171	0.147	0.171	0.140	0.186	0.134	0.186
A4	2405	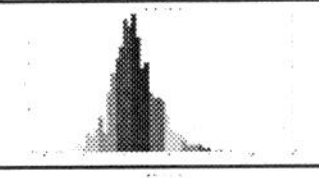	0.155	0.173			0.140	0.187		
A5	2405	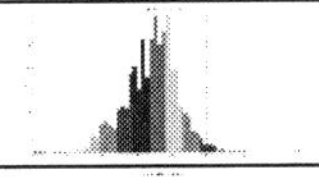	0.148	0.168			0.135	0.180		
B1	584	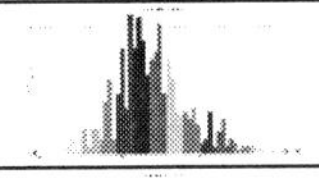	0.187	0.201			0.186	0.217		
B2	1096	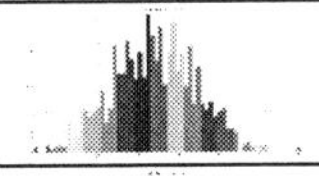	0.182	0.203			0.173	0.212		
B3	1845	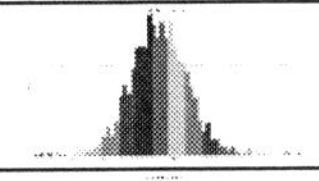	0.194	0.212	0.190	0.207	0.186	0.220	0.183	0.217
B4	629	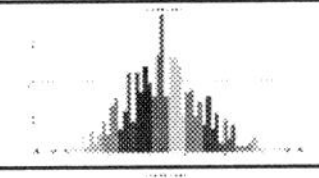	0.188	0.201			0.183	0.206		
B5	886	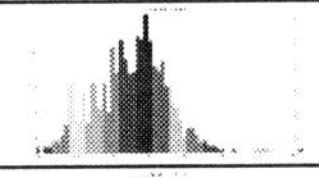	0.198	0.220			0.188	0.230		
C1	1343	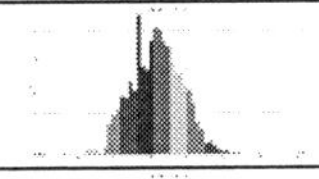	0.128	0.145			0.121	0.152		
C2	1685	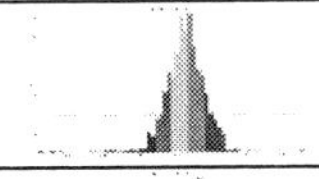	0.133	0.149			0.125	0.158		
C3	3161	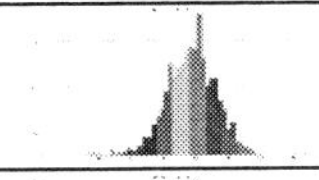	0.133	0.149	0.134	0.150	0.125	0.157	0.126	0.158
C4	2381	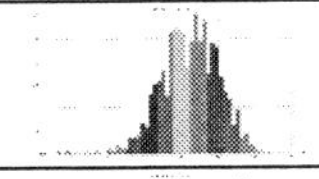	0.137	0.151			0.129	0.160		
C5	2484	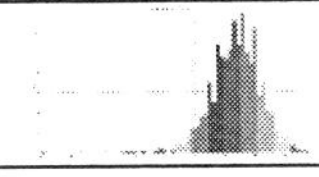	0.138	0.155			0.130	0.161		

For the standard deviation 2σ, there were intersections for pairs Rock A – Rock B (0.183–0.186) and Rock A – Rock C (0.134–0.158). For the standard deviation 1σ, there was only one, narrow intersection for the pair Rock A – Rock C (0.147-0.150). Therefore the standard deviation 1σ was used in analyses. For each point cloud, the mesh network was generated. It may be used as a 3D model for further analyses, the basis for the creation of intersection (Fig. 2).

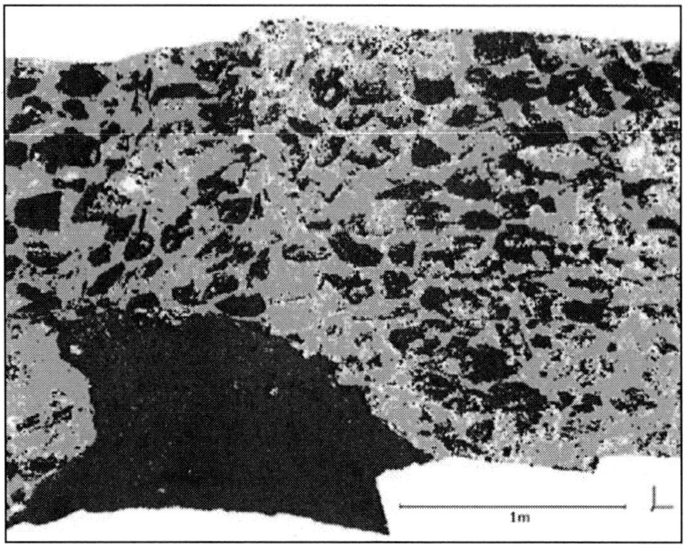

Fig. 2. The 1σ point cloud profile of the AGH experimental mine (blue – Rock A, green – Rock B, red – Rock C)

The 3D model may be used for quantitative analyses like volume computation. The mesh networks were created from the original point cloud and from point clouds for each rock type. The surface area of the whole dataset was 6.27 m^2. The area of Rock A, Rock B and Rock C was 1.92 m^2 (30.6%), 2.08 m^2 (33.1%) and 1.32 m^2 (21.1%), respectively. The areas' sum cannot be equal to the whole area due to the unclassified points. To compare the results, the percentage of the area was calculated with the manual method from the calibrated photograph. The area of Rock A equals to 25.5%, and Rock C to 19.5%. The rest of the photograph may be assumed as the Rock B. Differences in values between two counted methods may be caused by the accuracy of picking areas to computation. In the conventional method, small patches of rock may be hard to distinguish. Also important is the accuracy of drawing boundary of the patches. The point cloud method is fully automated and uninfluenced by the geologist skills and precision.

3 Case study

3.1 The Used data

The second point cloud was generated from the data obtained in Coc Sau open pit mine (Fig. 3). The Coc Sau open pit mine is placed next to Cam Pha city, in the Quang Ninh province of Vietnam. Presently, the mine's depth is up to 200 m above the sea level. The measures were conducted with laser scanner GeoMax Zoom300 storing intensity values within the range from 0 to 255. The pointset depicts the slope outside the excavation zone. Additionally, the data was modified and transformed to local coordinate system. The point cloud was noisy and its density was low.

Fig. 3. The point cloud of the highwall slope in Coc Sau open pit mine

3.2 Results

The corresponding method to the method presented in chapter 2 was applied. The analysis of the photographs and colors of the point cloud (Fig. 3) allowed to identify three different type of rocks (materials). The names of the materials were respectively: Rock D (yellow color), Rock E (dark gray color), and Rock F (light gray color). Due to the low quality of the input data, the intensity ranges were computed only for standard deviation 1σ. The average intensity range for 1σ is between 198.75 and 222.37 for the Rock D, 160.125 to 180.25 for the Rock E, and 174.69 to 196.36 for the Rock F. There were only the ranges for the Rock E and the Rock F intersected each other (174.69 to 180.25) and for these separate range was created.

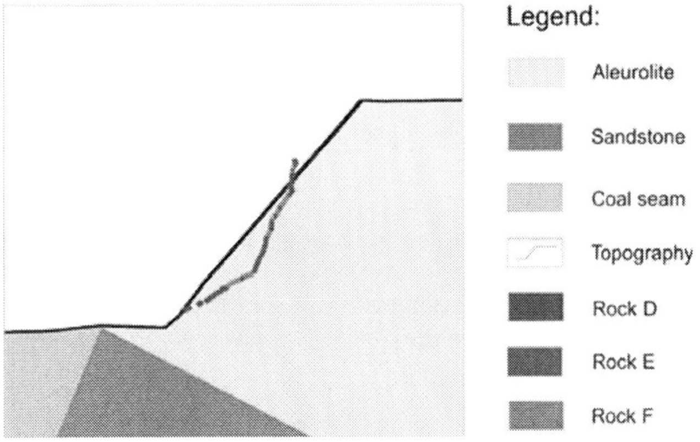

Fig. 4. Geological profile (source: geological department of Coc Sau) with profile generated from point cloud

The geological profile was generated, based on the computed intensity ranges. The result was compared with the geological intersection profile for the same place (Fig. 4) in the same scale. The LIDAR measures points only on the visible surfaces. Therefore point clouds may contain gaps and the products from post-processing may be incomplete. The red curve is separated in the bottom part due to the lack of data in this place.

The point cloud depicts the real shape of the surface. Thus, this method is more accurate than any conventional method in geology to create cross-sections. Also, matching points with types of rocks based on the intensity allows identifying lithological layers with centimeter accuracy. The geological profile presents only one type of rock for the measured place.

4 Conclusions

The hand-drawn geological profile depicts the lithological changes visible with the bare eye, but it isn't cartometric. Photographs of the observed object may be used as a source for the geological documentation. To do so, the photogrammetric and geometric conditions have to be preserved like camera calibration, lack of picture deformations, station in front of the photographed object etc.

The application of modern remote sensing techniques like LIDAR in earth sciences is labor-saving and faster than the conventional approaches. The collected data are cartometric and preserves the geometrical features of the observed object regardless the localization of the station. The point cloud data allows creating object's digital model. Thus, the TLS results can improve the quality of the results from the conventional methods.

The data processing procedure was tested on the AGH test site with the laser scanner Leica C-10. The intensity's ranges for each material with the 1 and 2 sigma probability level were computed. The ambiguous ranges may be assumed as the transition zone between two different rocks.

The intensity values depend on the type of the material, laser scanner, and the weather conditions during the measurements. Therefore, to get extract information about rock types, the previous calibration is necessary. The calibration field should contain samples taken from the investigated object. To fulfill the demand of the similar conditions, the calibration should be made in-the-field, during the object's measurements.

The obtained values allowed to create the profiles of the measured objects and conducting some quantitative analyses such as area computation. The comparison of the obtained results with the conventional methods reveals that the accuracy of the second one depends on the geologist's skills and precision, the complexity of the presented object and the accessibility of it. The application of remote sensing methods, like LIDAR, provides higher accuracy, uninfluenced by the human.

The possibilities of processing the point cloud into a drawing of the object's visible surface and object's intersections were presented. The results are always cartometric, thus they may be used in the geological documentation. Other documents, like intersections and photographs, may help to interpret the point cloud.

Authors propose the application of the TLS technology to create intersections and profiles due to the full cartometric, higher accuracy, and field works time-saving. Additionally, remote sensing technologies allow depicting inaccessible places. The digital models allow one for conducting the quantitative analyses. One of the benefits is the ease of creation of the 3D model, which can be used to compute the excavation progress.

Acknowledgment

Authors would like to thank HUMG's Mine Surveying Department for helping in contacting with Coc Sau mine and providing data. Authors also would like to thank AGH for providing access to the AGH experimental mine.

References

1. S. Martín, H. Uzkeda, J. Poblet, M. Bulnes, R. Rubio, COMPUT GEOSCI, 51, 90–100 (2013)
2. K.S. Dueholm, A.A. Garde, A.K. Pedersen, J STRUCT GEOL, 15, 933–937 (1993)
3. D.M. Ragan, *Structural geology: an introduction to geometrical techniques* (Cambridge University Press, 2009)
4. Rozporządzenie Ministra Spraw Wewnętrznych i Administracji z dnia 9 listopada 2011 r. *w sprawie standardów technicznych wykonywania geodezyjnych pomiarów sytuacyjnych i wysokościowych oraz opracowywania i przekazywania wyników tych pomiarów do państwowego zasobu geodezyjnego i kartograficznego*, (MSWiA, 2011)
5. H. Park, H. Lee, H. Adeli, I. Lee, COMPUT-AIDED CIV INF, 22, 19–30 (2007)
6. R. Williams, J. Brasington, D. Vericat, D. Hicks, EARTH SURF PROC LAND, 39, 167–183 (2014)
7. M. Simonse, T. Aschoff, H. Spiecker, M. Thies, Proceedings of the scandlaser scientific workshop on airborne laser scanning of forests, 2003, 252–258 (2003)
8. S.J. Buckley, J. Howell, H. Enge, T. Kurz, J GEOL SOC LONDON, 165, 625–638 (2008)
9. D. Costantino, M. G. Angelini, INT ARCH PHOTOGRAMM, 1, 207–212 (2013)
10. J. Zaczek-Peplinska, K. Osińska-Skotak, K. Gergont, *Inżynieryjne zastosowania geodezji* (Wydawnictwo Politechniki Poznańskiej, Poznań, 2012)
11. K. Tan, X. Cheng, REMOTE SENS-BASEL, 8, 251 (2016)
12. I. Piech, B. Kwoczyńska, D. Słowik, INF EKO TER WIEJ, 323–334 (2016)
13. J. Maciaszek, R. Gawałkiewicz, A. Szafarczyk, *Geodezyjne metody badania osuwisk* (Wydawnictwo AGH, Kraków, 2015)

The Impact of the Rock Mass Deformation on Geometric Changes of a Historical Chimney in the Salt Mine of Bochnia

Anna Szafarczyk[1,*], and *Rafał* Gawałkiewicz[1]

[1] AGH University Science and Technology, Faculty of Mining Surveying and Environmental Engineering, A. Mickiewicza Av. 30, 30-059 Krakow, Poland

Abstract. There are many ways of the geometry measurement of slim objects, with the application of geodetic and photogrammetric methods. A modern solution in the diagnostics of slim objects is the application of laser scanning, with the use of a scanner of a scanning total station. The point cloud, obtained from the surface of the scanned object gives the possibility of generating not only information on structural surface deformations, but also facilitates obtaining the data on the geometry of the axis of the building, as a basic indicator of the characteristics of its deformation. The cause of the change in the geometry of slim objects is the impact of many external and internal factors. These objects are located in the areas of working or closed underground mines. They can be impacted by the ground and they can face the results of the convergence of cavities. A specific structure of the salt rock mass causes subsequent convergence of the post-exploitation cavities, which has the influence on the behaviour of the terrain surface and the related objects. The authors analysed the impact of the changes in the rock mass and the surface on the changes of the industrial chimney in the Bochnia Salt Mine.
Key words – rock mass deformations, chimney, salt mine, historical objects.

1 Introduction

Underground mining exploitation is always accompanied by the formation of a subsidence trough, which causes changes in the inclination of the surface and all the objects situated within its range (especially directly over the exploitation). To determine forces affecting buildings, in general, a rigid junction between the constructions on the surface with the ground should be taken into account. This means that the inclination of the construction is, in the value and direction, compliant with the inclination of the formed trough. In fact, the vector of the object's inclination is not compliant with the resultant direction of changes in the inclination of the ground in the vicinity of engineering constructions.

Industrial chimneys are particularly sensitive to external factors. They are affected by static and dynamic burden and this way they are getting continuously deformed. The inclination of a chimney in mining areas is initiated by exploitation, which generates diminishing the compactness of the ground and facilitates the influence of other factors (wind, sun, ground

* Corresponding author: szafarcz@agh.edu.pl

heterogeneity). The complexity of the phenomenon and its variability is proved by the examples of the recording of geometric states of many chimneys in Poland, localized in the zones of mining influence. In many cases, the direction of the inclination is determined by the depth of the fundaments of the construction and its localization in the external part of the developing trough. In case of chimneys with deep fundaments and situated on the slope of the developing trough, their axes usually incline in opposite direction than the surface of their foundation. This is caused by the specifics of the development of post-exploitation troughs, where the under-surface layers are subdued to stronger deformations and dislocations than the surface layer making the overburden.

Modern measurement methods of the geometry of slim objects provide the users not only discrete data on the state of the construction (axis geometry), but also multi-point sets defining the state of its surface (structural deformations). Their proper mathematical analysis at the stage of post-processing allows generating discrete or continuous image of the central axis, as the main parameter of the characteristic of the chimney's state. One of such slim constructions, having the status of historical monument, is an industrial chimney in the Salt Mine „Bochnia". In the article the authors presented the results of control measurements of the chimney's axis with the application of the scanning total station VX Spatial Station Trimble'a and the analysis of the impact of the changes in ground of the rock mass in the vicinity of the construction's localization.

2 The measurements of the strains of industrial chimneys

Geodetic measurements of industrial chimneys (slim objects) are made to define their geometric characteristic (the value of strains). The range of these activities includes [1]:

- measurement of tilts (fundament of the construction or its base) – requires making observations with the method of precise levelling of the network of stabilized benchmarks in the stem or fundament of the chimney;
- measurement of verticality (deviation from verticality) – making the resultant of the tilt and deflection of the chimney's stem can be every time carried out from newly established measurement base in the vicinity of the observed object.

The analysed industrial object is an element of the liquidated boiler station of the old saltworks [2]. Now it is one of the elements of the historic complex of the Salt Mine "Bochnia – Campi street 15. It was registered to the list of historical monuments of the Malopolska Voivodeship as A-238 of 11th December 1981 [3]. The object is located on the slope of subsidence trough (in the borders of the mining area). So far the points of the control network (benchmarks) on the stem or the fundament have not been distributed. The lack of height control line points on the surface of the construction limits the set of information in the area of the full diagnostics of the preservation of the construction and makes impossible to determine the value of the tilt, its direction and optional vertical translation (depending on the calculation methods). Observations of existing, characteristic architectonic elements (ledges, plinths), do not guarantee necessary accuracy of the obtained measurement results. The only element allowing us to define the influence of changes in the ground inclination on the state of the body of the chimney (deviation from the vertical), is the analysis of the changes in the height of the points localized in the vicinity of the chimney points (on the ground and on the wall) and the points of mine's height network (on the exploitation level) in the range of the impact of the excavations on the ground surface and the rock mass. Thus the article analysed the change in the deviation of the chimney referring to the height data of the existing points of the height control line.

The authors focussed on the analysis of the changes in the deviation of the chimney's axis from the vertical (compared to the measurement data of 2007), which did not require the reference of the observations to the same points of the spatial control line (XYH). In case of destroying the points of the geodetic benchmarks, there is a possibility of restoring the points in the terrain or constructing a new, similar geometric construction and marking the coordinates of these points (to give spatial orientation), with the support to other surveying techniques, e.g. GNSS. In this situation the identical system of coordinates is obtained is obtained, which creates the possibility of precise comparison of the results of subsequent measurement series. In such a case the information on the change of the value of the vector of the deviation of the chimney's axis from the vertical change in the azimuth of this vector in time is obtained, which for this brick chimney is connected with the changes in the ground and the plasticity of the halite rock mass (convergence of the cavities).

In the publication the results of the analysing of the characteristics of the geometry of the chimney based on the data of two surveying methods e.g., bisector methods surrounding the tangential and laser scanning with the application of the scanning total station VX Spatial Station by Trimble.

3 The characteristic of the historical chimney

The analysed historical industrial chimney is the construction of full ceramic brick on the cement–lime mortar, located in the area of the liquidated Salt Mine "Bochnia". With the historical building of the old panel saltworks makes a unique complex of post-industrial architecture, despite the fact that these objects were not technologically linked.

It makes the remnants after the engine room of the liquidated of the mining enterprise. Its height is $H = 45.4$ m (height defined between the upper verge of the crown of the chimney and the base of the cleaning hole). In case of the analyzed body of the chimney one should divide it into three parts [5, 6]:

- lower – a flue, of the geometry of a cuboid of base 6.55 × 6.55 m and height 3.8 m;
- central – base of changeable cross-section in the shape a regular octagon made with a proper security of external areas of spherical triangles;
- upper – the covering of changeable circular cross-section (from 2.668 do 1.602 m), enforced by steel rings of the cross-section of 8 × 80 mm consisting of three segments joint with screw locks (Fig. 1), allowing the regulation of the tension power, which, during the inclination measurements with geodetic methods facilitate the identification of the purpose (observation level).

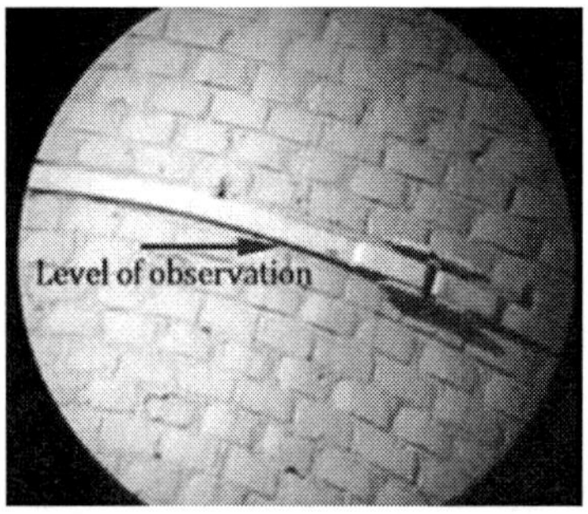

Fig. 1. Rings equipped with screw locks making the marked observation levels 4 Analysis of the deformation of the terrain surface in the region of the chimney

The historical chimney, although localized on the areas under the influence of mining exploitation (completed) was not equipped with the points of the observation network (control points). Thus it is impossible to obtain precise information on the tilt of the fundament, which influences the behaviour of the body of the chimney. To show the characteristic of the ground change in the vicinity of the chimney the height data of precise levelling network of wall benchmarks (control points) stabilized on the buildings (localized in meridian system) of new hoisting machine and steam machine, were used (Fig. 2). The analysis of the change on the surface of the terrain in the analysed region of placing the industrial chimney was carried out based on the data coming from the Surveying Division in Salt Mine „Bochnia".

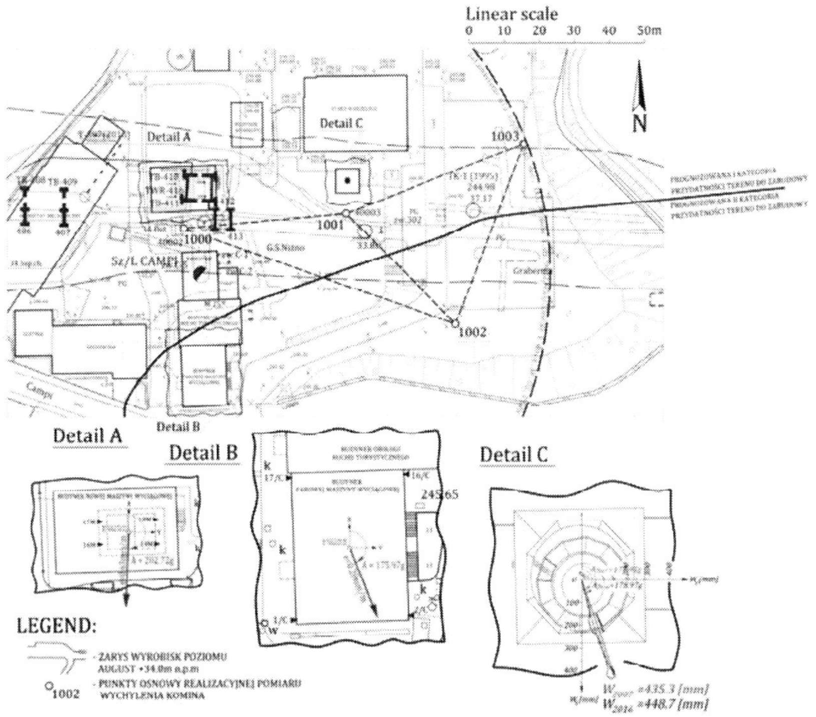

Fig. 2. Localization of the measurement stands on the background of situation details on the surface of mining excavations on Level August (the shallowest level connected with Shaft Campi)

The analysis of the change in height points in the analysed period of time 2007–2016, allowed the measurement the phenomena of inclinations and the subsidence of fundaments in these buildings, which, due to a small distance (44–67 m) from the chimney can be regarded as representative values of the changes in the region of the chimney.

The analysis of the inclinations and subsidence in selected buildings was based on the algorithm described by [7]. To geometrically interpret vertical dislocations, the equation of plane was fit into the measured points (4 points for each object according to Fig. 2), defined by the formula:

$$u_i = x_i \cdot e_x + y_i \cdot e_y + z_0 \tag{1}$$

where:

- u_i – subsidence of the control point,
- x_i – coordinate x of the reference point to the system of the object with the beginning in the gravity centre of the object,
- y_i – coordinate y of the reference point to the system of the object with the beginning in the gravity centre of the object,
- e_x – inclination of the plane alongside axis *x*,
- e_y – inclination of the plane alongside axis *y*,
- z_0 – mean subsidence of the body (vertical dislocation of the starting point of the system).

The solution of the system of equations are the unknowns of the inclination components of the plain inclination e_x and e_y and the value of mean subsidence of body z_0, calculated with the smallest square method, according to the relationships:

$$X = \left(A^T A\right)^{-1} \cdot A^T L \qquad (2)$$

where:

A – matrix of the coordinates of benchmarks in the local system,

L – matrix of the difference in the height of benchmarks in the discussed period of June 2007 – November 2016.

The applied algorithm of calculations, allowed the determination of basic values defining the characteristic of spatial changes in both buildings. The error of fitting the plane was preceded by the determination of adjustment corrections, according to the formula:

$$v = A \cdot X - L \qquad (3)$$

The values were put in Table 1.

Table 1. The basic values of spatial changes of the buildings of: the new hoisting machine and steam hoisting machine in the period of June 2007 – November 2016

Parameter		The building of the new hoisting machine	The building of the steam hoisting machine
		Value	
The value mean subsidence Uz_0 in [mm]		–28.5 ±0.5	–51.5 ±0.5
The components of the inclination of the building	e_x in [mm/m]	0.632 ±0.292	0.258 ±0.063
	e_y in [mm/m]	0.027 ±0.118	–0.102 ±0.078
The value of the biggest inclination e in [mm/m]		0.632	0.277
The azimuth of the biggest inclination in [g]		202.7181	175.9678

5 Analysis of the deformation in the body of the chimney in 2007–2016

The protection of construction objects in the mining areas is ordered by the regulations of the Geological and Mining Law [8]. This protection means periodical diagnostic of the construction in the form of (inter alia) control measurements my by surveying services. This means the determination of geometric parameters of constructions crucial for the safety of the building and its users. Moreover, in the framework of interdisciplinary expertises, the border states of the bearing capacity (excessive strains in the construction or ground, excessive scratches on the construction or vibrations) and usability were defined [9]. Today chimneys are built of reinforced concrete or steel, while brick chimneys belong to the group of historical objects. Thus the safety of brick construction will depend on the ability to safe interception of the influence of:

- curvatures and horizontal strains of the surface;
- inclinations of the surface;
- mining-originating tremors.

In case of the halite medium, the decisive influence on the safety of the construction is inflicted by curvatures and strains of the ground surface layers and the change of the ground slope in the place of the construction fundament. These parameters mainly decide on the direction and the values of geometric strains of slim constructions.

The article presents the results of the measurements of the deformations of the stem of the brick chimney with the application of commonly applied classical measurement method, i.e. bisector of the tangential directions, carried out according to [4] and modern laser scanning making alternative for traditional methods. Measurement data obtained with the use of the scanning total station VX Trimble'a, were post-processed, according to algorithms described in [5].

In case of the application of the laser scanning methods to define the geometry of the stem chimney from the points of the established control line (Fig. 2), the algorithm of the approximation of circles directly from the selected point cloud (band) was applied on selected levels (in case of the circular cross-section of the construction) described in [5]. Laser scanning actually generates dense point data, which, if properly interpreted and processed allow us making 2D or 3D documentation in the modelling process.

To analyse deformations of the chimney's the method of the approximation of circles into the bands of the points projected on the surface of the chimney with the smallest square method. This means that the geometric analysis of the construction was based on the determination of physical parameters of approximated circles on the defined study levels, i.e.:

- coordinates of the centres of circles Xi, Yi, Hi (Hi – height of ith level);
- radiuses of the circles of subsequent study levels.

Analysis of the coordinates of the centres of circles of subsequent study levels directly allows the determination of the tilt vectors W_i and their directions in the accepted local or geographic system of coordinates. Multi-point cover of the chimney surface (Fig. 3) provides surplus observations, which increase the accuracy of the results on the stage of the analysis and minimize the influence of human factor (including the aiming error), on the final results of the analysis. One should emphasize that in the above example the mean of the points lying on a certain level in the whole circuit of the chimney's shield is 53 points (we accepted the band of height-attributing of points to cross-section on level $h = \pm 5$ mm). Basic parameter defining the present geometric state of the slim construction is the resultant vector of the inclination of its stem W or the inclination of generalized axis Wz (Table 2). According to the formula contained in [10], the border value of the inclination of the axis of the studied chimney (regarded dangerous) is 2.13 m.

Table 1. The most important parameters of the chimney's deformation

The date of the measurement:	**June 2007**		**November 2016**		**November 2016**	
Measurement method: / Parameter	Laser scanning method		Method of adhering tangents		Laser scanning method	
Components of the inclination of the generalized axis of the chimney in [mm/m]	e_x	e_y	e_x	e_y	e_x	e_y
	–10.9	3.9	–11.1	3.81	–11.3	3.79
Total inclination of chimney's axis e in [mm/m]	11.1		11.7		11.9	
Vector of the inclination of chimney's stem W in [mm]	435.3		448.7		443.7	
Inclination of the generalized chimney's axis Wz in [mm]	503.9		531.2		540.3	
The angle of the direction of the inclination of the generalized chimney's axis Aw in [g]	177.93		178.97		179.40	

Fig. 3. View of the historic chimney – point cloud – the image of the power of the reflected signal (the measurement of the geometry of the chimney with the use of the scanning total station VX Trimble)

Conclusion

The application of laser scanning technology in the inventory processes and control measurements of tall and slim constructions simplifies and accelerates the implementation of surveying tasks both indoors and outdoors. This allows comprehensive geodetic assessment of geometric deformations of the object with the preservation of high accuracy. The presented in the article example of the application of a modern scanning total station equipped with the integrated digital camera and the obtained results of the inclination, on selected observation levels, referring to classical methods prove great usefulness of this class of instruments and high accuracy. This is confirmed by the difference of the values of the vectors of the inclination of chimney's stem of the applied measurement methods on the level of 5 mm.

As a difference from the bisect method, which requires a definite measurement space to preserve high accuracy, such modern scanning total stations facilitate the measurement. Due to the precision of the measurement and multi-point cover of the measured area (surplus observations) the image of structural deformations and the control of the data is obtained, although the process of post-processing is more complicated. In practice, the high number of data, indirectly facilitates precise determination of elements allowing the users mutual adjustment of the sets of points coming from various check points. Precision of angular–linear measurements guarantees high accuracy of determination of vectors of the inclinations of various levels in the relation to the base, physical parameters of the horizontal cross-sections of the shield of the chimney, i.e. radiuses and coordinates of the centres of circles, but also the registration of local structural deformations of its coating. Moreover, laser scanning, due to the possibility of increasing the density of observations, full control over the results of measurements is possible and points charged with gross errors are eliminates from the data sets. Their cause can be the reflection of the impulse from more than one surface. In the relation with this, in case of the surfaces possible to be defined mathematically, the neighbouring points make the control of the point model.

The presented in the article example of the chimney located in the mining area and subdued to the influence of the carried out exploitation, showed a real character of the deformations of the stem of the construction. The resultant vector of inclination is compliant with the direction of the localization of excavation of subsequent exploitation levels and the direction of their collapse (structure of the deposit). Results of measurements of 2007 and 2016 indicate slight growth of the vector of the inclination of stem chimney by the value of 8.4mm and the change of azimuth by 1.47 g, in the direction of the slope of the subsidence trough (localization of the deepest mining excavations).

Acknowledgements

Article was financed from the grant Badania Statutowe 11.11.150.195.

References

1. R. Gawałkiewicz, J. Maciaszek, *The application of laser scanning in the diagnostics of objects influenced by mining exploitation.* Zeszyty Naukowe „Geodezja" – półrocznik AGH, vol. 12, issue 2/1, UWND Kraków (2006)
2. Enclosure to the resolution no. XXXIII/289/17 of the City Council of Bochnia of 27th April 2017. Municipal Programme or the revitalization of the City of Bochnia for 2017–2026
3. The list of objects in the Register of Historical Monuments (Buildings) of the Małopolska Voivodeship (state for June 2017)
4. J. Gocał, Engineering Surveying, vol. II – issue 2 amended, Polskie Przedsiębiorstwo Wydawnictw Kartograficznych. Kraków – Wrocław (1994)
5. R. Gawałkiewicz, *The example of the application of laser scanning in the inventory and monitoring of tall industrial objects.* Geomatics and Environmental Engineering, vol. 1, no. 4. pp. 111 – 126, AGH Kraków (2007)
6. A. Żaboklicki, M. Gębski, M. Żmijewski, *Technical expertise referring to the possibilities and conditions of the application of industrial chimney in Salt Mine "Bochnia" as a tourist look-out tower.* EMGIE projekt Sp. z o.o. Kielce (2006)
7. W. Janusz, *The method of the calculation of vertical dislocations and deformations of the fundament of the building with the identification of the symptoms of its discontinuity,* Prace Instytutu Geodezji i Kartografii, vol. LI, issue 109 (2005)
8. Geological and Mining Law of 9th June 2011
9. J. Kwiatek, *The protection of buildings in mining areas.* Warsztaty 2000 nt. Zagrożenia naturalne w górnictwie, pp. 45–56 (2000)
10. O. Niemczyk, Bergschadenkunde. Essen (1949)

Vertical Displacement of the Surface Area over the Leakage to the Transverse salt Mine in 1992–2012

Tomasz Lipecki[1,*]

[1] AGH University Science and Technology, Faculty of Mining Surveying and Environmental Engineering, A. Mickiewicza Av. 30, 30-059 Krakow, Poland

Abstract. The leakage of water in the salt mine caused considerable deformation of the surface. This article shows the vertical displacement in the area of leakage to the mine excavation, measured by precision leveling, carried out from the first days of leakage in 1992 until 2012. The geological and hydrogeological conditions of the mine, as well as the associated water hazards were described, which in conjunction with the inconvenient location of the excavation site in the northern frontage of the Carpathians and also inadequately conducted mining operations, contributed to the risk of flooding mine. The analysis of the vertical movements of the surface – subsidence and uplift – were present as well as the process of formation of the depression trough in the form of maps and graphs. The analyzes were based on 49 measurement series, starting from the first days of the disaster within the next 20 years. The course of development of the depression trough and the condition of the surface after stopping the water from the rock mass has been shown, which caused the surface to uplift. **Key words** – salt mining, surveying, subsidence monitoring.

1 Introduction

Salt mining has been operating in Poland for more than seven centuries, starting with excavations near the surface of the earth and now reaching below 1000 meters below the surface level. Due to the mining activity, which particularly violate the surface of the rock, it presents a great deal of danger. The catastrophic leakage of water along with rock material into the mining pit of one of the mines is the proof of that. Minor spills have been occurring in this area for more than 100 years, therefore the geological-mining service was aware of the existing threat, that was associated with the breach of the protective plaster mine by mining activities prior to the age. In spite of this, no one predicted such a catastrophic leak in 1992, which caused massive damage beneath the surface, flooding and destroying some of the mines in the historic mine, and also in areas where rapid deformation have emerged. In 2012 the 20th anniversary of this event took place, which allowed to do some analyzes covering this period of time and a summary of revealed process of the surface deformation. This article focuses on the vertical surface movements which are one of the basic indicators of deformation. The area where the depression trough has developed has been monitored

* Corresponding author: lipecki@agh.edu.pl

and its growth dynamics has been observed in almost 50 series of elevation observations of precision levelling on a network of controlled points, which were established for this purpose, covering the area in question. Based on them, the course of the deepening of the depression trough and the present state of the surface of the area is presented, the same as in 2012, in which practically no vertical movement of the surface was observed.

2 Characteristics of the mine area – a historical outline of the transverse area

In the course of 700 years of operation, the salt mine carried out mining operations at 9 levels reaching 327 m below the earth's surface. 26 shafts were drilled, around 2050 chambers and over 300 km of excavations. The volume of selected voids is approximately 7.5 million m^3. Some of the excavations from level I to I were considered historic and were protected by creating a historical zone. Depth of the deposit is from 30 to 330 m below the surface and the thickness is up to 400 m. The salt layer can be divided into two contrasting parts: a solid layer in the higher parts and the layer below a deck. The first structure (at a depth of 60–140 m below the surface) are solids of salt composed of various shapes and sizes, distributed irregularly in the rock. And this is where the oldest chambers and historical part of the mine are located. The deck layer is strongly corrugated, with layers of salt of different lithium with ganglionic rocks. Figure 1 shows a geological section through the layer.

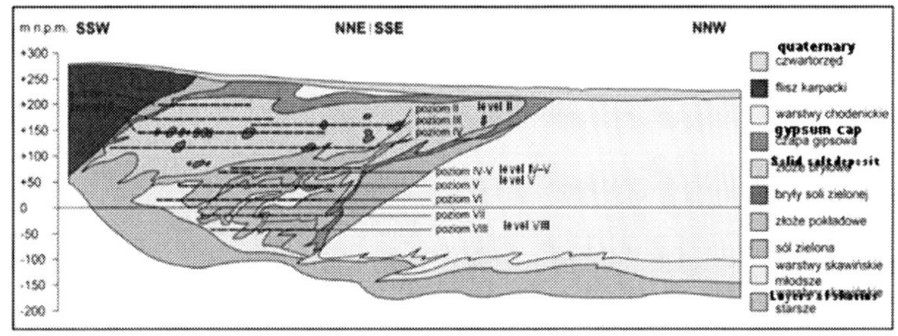

Fig. 1. Geological section through the central part of the deposit [1]

The most important rule in salt mining is to prevent the formation of hydraulic contacts between the mining excavations and the water stored in the rocks. The natural cover in this area are loamy anhydrite chalk, which together with so-called the igneous rock around the reservoir are a natural dam to protect the layer from water from the tertiary waters of the area, which are consisting of the stratocycles, consisting mainly of sandstones filled with deep water, migrating between the upper layers of the salt deposit [2]. Then there are also tributaries of the surface and subsurface layers of the Quaternary. They also come through the shafts and the slits in the rock. Such unfavorable hydrogeological conditions, together with some mining works that broke the insulation decades ago, have allowed uncontrolled water inflows to the mine [3]. This threat is connected with the phenomenon of suffication, consisting in leaching and flushing of rock particles from water permeable rocks, mainly dusty, resulting in the formation of underground channels and caverns, which weaken the stability of the rock surrounding the excavation.

Discussed crosswise was a excavation already hollowed in the nineteenth century at a depth of 150 m below the surface and was about 400 m long, forming a research pavement and then transformed to the transporting one. Unfortunately, the final section of the crosswise, at a length of 15 meters, went beyond the boundaries of the layer, thus piercing the gypsum plaster cast, which secured the layer from aquifer tertiary deposits. Over the excavation are chambers operated in the nineteenth and twentieth centuries, and few in the upper levels.

3 Disaster in the transverse

The damage of the protective layer in the transverse has opened the hydraulic connection with the waters in the sandstones of chodenite layers already in 1910. Leakage was not eliminated, due to the small inflow, but constant monitoring of this phenomenon was implement. By 1992, about 226 thousand m^3 of water were imported to the mine. Despite the decline of water supply, the water still penetrated lower levels, washed out cavities and cracks, and threatened uncontrolled leakage. This contributed to the decision to carry out security work in the form of liquidation of the leak. Since 1990 covering the leakage has started, but was performed with some interruptions, related to the elimination of voids and croaks around the transom. In 1992 there was a sudden and uncontrolled leakage of variable efficiency, reaching up to 330 l/min. The outflow appeared in various places and was a mixture of fresh water with silt, sand and clay content (Fig. 2).

Fig. 2. Place of transverse leakage

In spite of the rescue operation, it was not possible to fully protect the tributary and one month later another flow occured and reached 5000 l/min and on 30 September it reached even 20 000 l/min. It was only after several months that the inflow decreased and stabilized at the level of 1.5 l/min, which made it possible to build a dam about 7 m from the transverse face. Unfortunately, this caused the waters to pile up behind the dam and generated further water repellents until 1994 when the leak was successfully managed.

Fig. 3 Deformation of the track and destruction of the wall and cracking of the soil in the monastery's garden [4]

Water supply to the transom was the largest and catastrophic leakage of water into the mine, and could have destroyed it. It caused many losses and damage, and the multi-month struggle for survival of the mine was enormous. Water rapidly penetrated the lower excavations, practically destroying level IX, which was eventually liquidated by flooding. Water flooding the mine also threatened the city located above the mine. As a result of the disaster, significant deformations occurred on the northern front of the mine, and after a spill in 1992, a subsidence of a 0.3 m drop was observed, which deepened at an extremely fast rate from day to day. The trough covered by its impact railways and threatened the buildings of the nearby monastery, eventually causing the wall to collapse. (Fig. 3). These spots are located north of the transverse front, which indicates the size of the area covered by the water inflow. Figure 4 shows the surface deformation in relation to the transverse run that is visible at the bottom of the picture.

Immediately, after the inflow was revealed, protective work was started, which was based not only on the execution of the dam but also on the drilling of wells [3, 5]. Observation wells and subsequent ones forming the so-anti-filter and injection barrier [4]. Due to the unsatisfactory effectiveness of these activities in 1993, the drainage holes of the dam were made and the rock injections were started by drilling around the transverse in the so-called. "Range". In total, 102 injection holes were drilled, including almost 2000 m^3 of mortar ore. By the year 2000, the outflow stabilized at around 120 l/min, similarly saturated NaCl to about 3 g/l and insoluble (from 240 g/l to 10 g/l).

4 Surveying monitoring

During the catastrophic effects of leakage, a number of studies based on geophysical measurements, such as gravimetric and electrofusion, were conducted. However the most frequently performed tests were surface deformation measurements in the form of precision leveling on an established control network [6] covering the extent of the fallen depression from September 1992. For the control of successive states also used InSAR [7] and laser scanning [8]. Initially, the network consisted of 55 ground points distributed regularly every 50 m and two wall-mounted repairs on the walls of the monastery. After further flows of water, the net was enlarged in the north and east direction and was thicken to a mesh of 25 m as well as a number of wall markers were stabilized in the surrounding buildings (Fig. 4).

Fig. 4. Location of observation points of fallen trench along the course of transverse

These points were also stocked with precision centering marks for performing situational measurements. Due to the renovation of the railway network, the redevelopment of the deformed area in the following years was destroyed but some of the points survived, allowing the analysis of surface changes, for over 20 consecutive years after the catastrophe in the transverse, which is presented in this article.

Initially, the height of the points were determined at several intervals (June – October 1992), then monthly, later quarterly (November 1992 – September 1994) and after calming the impact, as the surface movements ceased, observations are made once per year since 1995. Average errors in point height (± 0.5–2 mm) were obtained by referencing the network to a fixed repertoire, which was set 3.5 km away from the deformation region and later referring to GPS points, which also served as reference for angular linear measurements. In this way, the spatial coordinates for some parts of the points, with the accuracy of ± 3 mm, were obtained. In total, between 1992 and 2012, 49 series of researches and several series of situational measurements were carried out, which allowed to perform a number of analyzes of the phenomenon of deformation at the discussed area throughout the 20 years.

5 Analysis of the results of conducted measurements

Initial measurements from June to September 1992 showed no significant changes in the surface area. Maximum reductions reached 22 mm in the two points of the observation network during that time. A sudden vertical displacement occurred at the beginning of September as a consequence of another rapid flooding of the water to the transverse, along with a large amount of rock material. The largest subsidence within 4 days reached almost 40 cm, and in the following week increased by another 21 cm. The situation of the depression formed at that time is shown in Figure 5. The reduction covered almost 3 ha and its volume amounted to over 3,800 m^3, which gives an idea of the amount of material displaced in the transverse region.

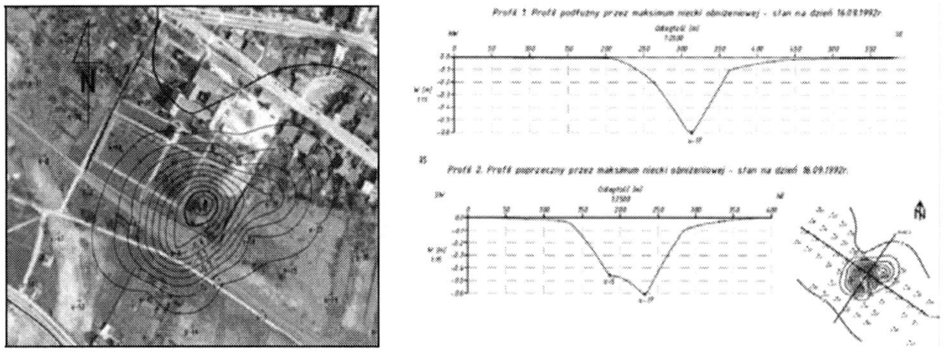

Fig. 5. Subsidence basin –September 1992

Over the next month, the surface moved twice, resulting in further deepening of the subsidence. At its lowest point, the decrease was registered at 1.5 m. Till April 1993, no surface movements had been registered and the recorded monthly cycles of deepening of the area had a value of several millimeters.

At the end of April 1993 there was another sudden spill, causing the trough to deepen to 2 m. After this incident the area was settling at an uneven speed until the end of 1994, when it was mastered and the leak was closed. The map below (Fig. 6) shows the state of deformation in 1995, after mastering the leakage. The trough covers an area of almost 8 hectares and its volume reaches 30000 m^3. The largest decreases occur all the time at the same point of the observation network, reaching the value of 2.3 m. The comparison of the spatial shapes of the basin from 1992 and 1995, when recorded its greatest development, is presented in Figure 7.

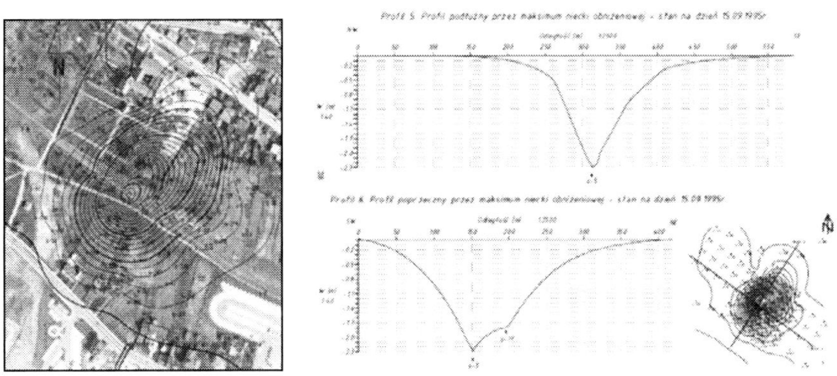

Fig. 6. Subsidence basin – state for 1995, after mastering the leak

Since the leak has been mastered, geodetic measurements have been started in annual cycles. For the next 15 years, the area did not show any significant vertical movements, and the subsidence returned to pre-disaster levels of several mm per year.

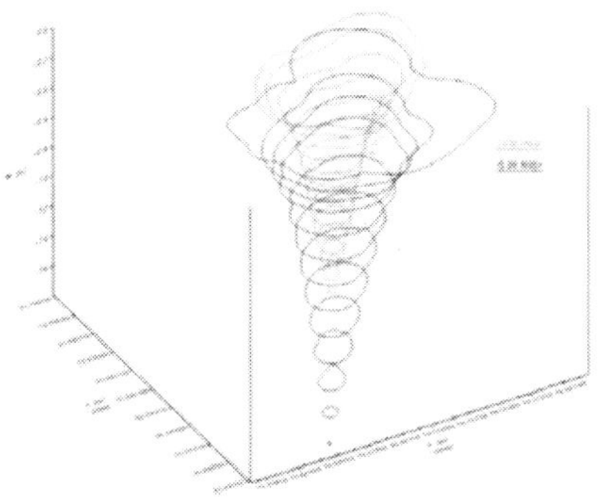

Fig. 7. The development of the subsidence basin in 1992–1995

Figure 8 shows the state of the subsidence basins of the 2012 subsidence. Long-term and slow summation of the reductions has led to the expansion of the basin to more than 12.5 hectares and its volume has increased to almost 37,000 m^3. The center of the basin has decreased by an estimate of about 2.4 m (unfortunately the point of the network has been damaged).

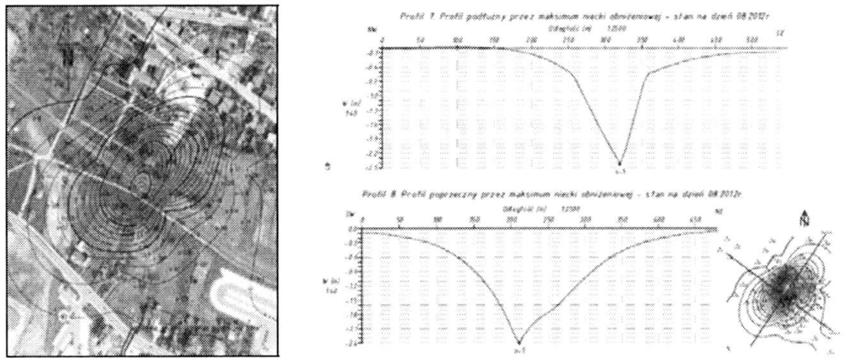

Fig. 8. Subsidence basin – August 2012

The 2012 series did not show any real reductions compared to 2011, but began to observe, since 2007, a few millimeters uplifts, significant from the point of the accuracy of the observations. It has been reported even in the years 2007–2008 that the uprising reached up to 20 mm. This is probably due to the rebuilding of the water column in the ground, after the water has completely stopped draining from the front of the transverse dam [9]. The following years of observations confirmed the revealed tendency to rise at an average of several mm per year in the discussed area. On the map shown in Figure 9, the summary uplifts from 2007 to 2012 are shown, reaching up to 50–70 mm, which suggests that the average annual value of the uplifts oscillates within 10 mm/year. The centers of reductions and uplifts initially overlapped, but later the maximum uplift shifted north.

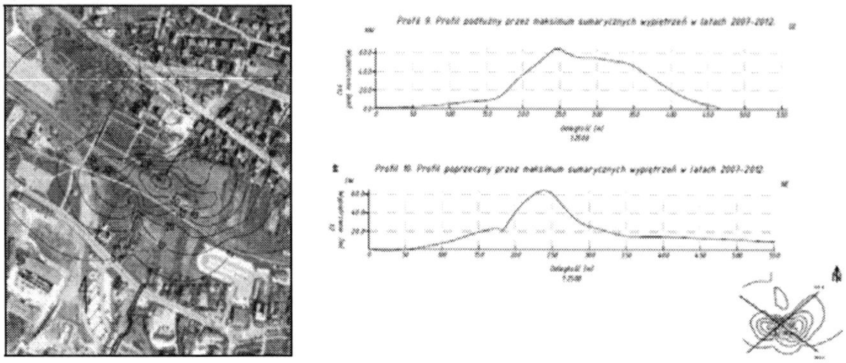

Fig. 9. Uplifts recorded in the period 2007–2012

The nature of elevation changes can be shown not only on the basis of surface analysis. Very good dynamics of changes are reflected in the graphs, that show the behavior of individual points of the research network. The first period, when catastrophic changes occurred (1992), presents a graph of changes in the height of one of the points shown in Figure 10a, and 10b presents the reductions over the period of 20 years (1992–2012). On the first one, there is a visible, instantaneous process of revealed deformation, however after this period, the reductions take a character similar to the curve striving for, almost asymptotically, zero altitude changes. The end of this curve is again increasing and as mentioned before, it presents the process of upliftment, which as you can see, is not a random phenomenon.

a)

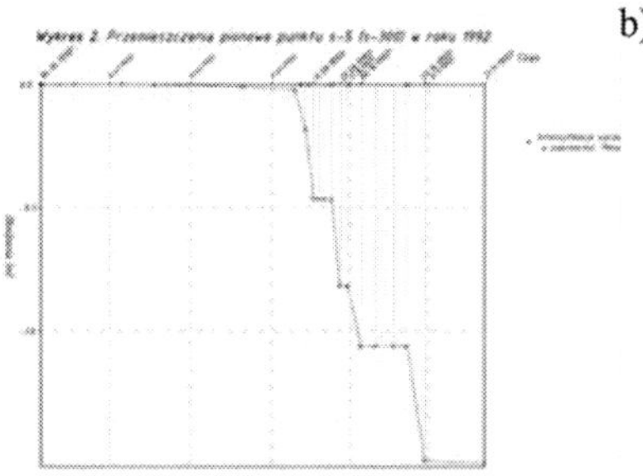

b)

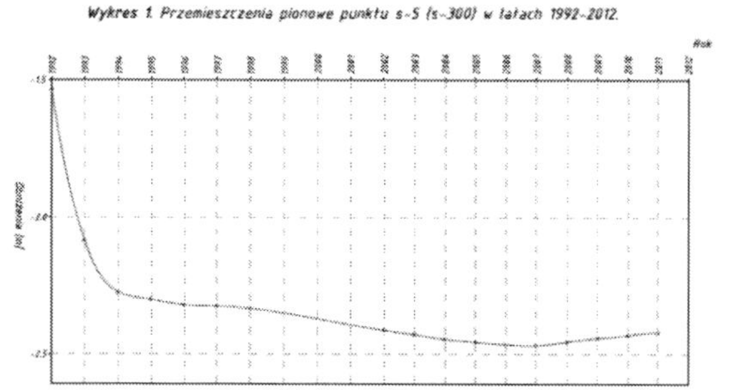

Fig. 10. The dynamics of reduction points in 1992 (a) and during the period 1992 to 2012 (b)

Conclusions

The leakage in the salt mine was, so far, the biggest disaster the mining service had to face, threatening the mine`s existence (primarily historic excavations listed at the UNESCO) and the city because of the deformations, that threatened nearby monastery and destroyed the railroad track. All vertical movements observed on the surface were preceded by intense leaks into the transverse, thus the largest land degradations were recorded between 1992 and 1994.

The character of that process was irregular with a settling velocity of several to several hundred mm per day. In the following years the orogen calmed down and the subsidence returned to the typical values of salt orogen. Since 2007, a regular uplift of the surface has been observed, covering almost the whole area of the uplift, which is probably related with the restoration of the natural hydrogeological conditions, after stopping dewatering the transverse. A similar phenomenon is also reported in the paper [10], in which is characterized the uplift of the rock with high permeability and low mechanical stiffness, which activates after a few months of injection. Based on the recent measurement cycles, it can be stated that the ground movements are small now and do not pose a threat to the infrastructure rebuilt on it, but the reductions and uplifts that still appear, indicate that the orogen is constantly subjected to processes which are associated with the reaction to the violation of its structure.

Acknowledgements

The research was financed by statutory fund WGGiIŚ: 11.11.150.195, AGH University.

References

1. A. Garlicki, A. Szybist: *Ogólne założenia dla zabezpieczenia kopalni Soli Wieliczka oraz nowy obraz geologiczny złoża wielickiego.* Likwidacja zagrożenia wodnego dla zabezpieczenia kopalni Soli Wieliczka, III spotkania PSGS, Wieliczka, (1995)
2. K. D'Obyrn, K. Brudnik: *Ograniczenie zagrożenia wodnego przez stabilizację górotworu w centralnym rejonie kopalni „Wieliczka"*, Górnictwo i Geologia, t. 7, z. 4, (2012)
3. W. Stachoń: *Likwidacja największego w historii kopalni soli Wieliczka zagrożenia wodnego*, Reinforcement and sealing of rock and construction at the beginning of 21st century, VSB, conference matterials, Ostrava 15–16.02. (2001)
4. K. Brudnik, J. Przybyło: *Historia katastrofy, prowadzonych prac ratunkowych i inżynieryjno-technicznych zmierzających do likwidacji zagrożenia kopalni i powierzchni*, WEB presentation: http://slideplayer.pl/slide/2795853/, (2008)
5. A. Gonet, K. Brudnik, S. Stryczek: *Zabezpieczenie Kopalni Soli „Wieliczka przed zagrożeniem wodnym w otoczeniu poprzeczni „Mina"*, Bezpieczeństwo Pracy i Ochrona Środowiska w Górnictwie, Miesięcznik WUG, nr 12 (40), (1997)
6. G. Kortas (red.): *Ruch górotworu i powierzchni w otoczeniu zabytkowych kopalń soli*, Wydawnictwo Inst. Gosp. Surowcami Mineralnymi i Energią PAN. Kraków, (2004)
7. Z. Perski, R. Hanssen, A. Wojcik, T. Wojciechowski: *InSAR analyses of terrain deformation near the Wieliczka Salt Mine, Poland*, Engineering Geology 106 pp. 58–67, www.elsevier.com/locate/enggeo, (2009)
8. T. Lipecki, W. Jaśkowski: *Application of laser scanners to determine the shape of mine excavations for safety assessment, using the example of the cross-cut Mina in the salt mine Wieliczka*, Reports on Geodesy No. 2(87), pp. 239–250, Warszawa, (2009)
9. A. Maj: *Wypiętrzanie terenu pod wpływem zmian warunków hydrogeologicznych na przykładzie skutków zamknięcia wycieku w poprzeczni Mina w kopalni soli Wieliczka*, Prace Instytutu Mechaniki Górotworu PAN, t. 14, nr 1–4, s. 103–111, Instytut Mechaniki Górotworu PAN, (2012)
10. A. P. Rinaldi, J. Rutqvist: *Modeling of deep fracture zone opening and transient ground surface uplift at KB-502 CO_2 injection well, In Salah, Algeria*, International Journal of Greenhouse Gas Control, vol. 12, pp. 155–167, January (2013)

Classical Measurement Methods and Laser Scanning Usage in Shaft Hoist Assembly Inventory

Wojciech Jaśkowski[1,*], *Tomasz* Lipecki[1], *Wojciech* Matwij[1], *Mateusz* Jabłoński[1]

[1] AGH University of Science and Technology, Faculty of Mine Surveying and Environmental Engineering, A. Mickiewicza Av. 30, 30-059 Krakow, Poland

Abstract. The shaft hoist assembly is the base of underground mining plant. Its efficiency and correct operation is subject to restrictive legal regulations and is controlled on a daily visual assessment by shaft crew and energomechanics. In addition, in the regular interval, the shaft hoist assembly is subject to a thorough inventory, which includes the determination of the geometrical relationships between the hoisting machine, the headframe and the shaft with its housing. Inventory measurements for shaft and headframe are used for years of conventional geodetic methods including mechanical or laser plumbing and tachymetric surveys. Additional precision levelling is also used for measuring shafts of hoisting machines and rope pulleys. Continuous modernization of measuring technology makes it possible to implement the further methods to the above mentioned purposes. The comparison of the accuracy and the economics of performing measurements based on many years of experience with comprehensive inventory of shaft hoist assembly using various research techniques was made and detailed in the article. **Key words** – headframe, shaft, laser scanning, survey economics.

1 Introduction

The article presents an analysis of the inventory measurements of symmetric double strut headframe construction and two hoisting machines (Fig. 1). The headframe is located above a 160-meter shaft and equipped with two rope pulleys platforms, located at a height of 37 m and 43.7 m above the framework of the shaft. At the bottom level there are three overturned rope pulleys, including two on one side (A) and one on the other (B). Above the single one there is located another one on upper platform (B).

The spatial distribution between elements of the headframe, shaft and hoisting machine elements is variable over time. Changes in the location are caused by mining works, mining tremors, weather conditions (sunlight, wind) and impact of the lifting equipment on the headframe construction [1]. The detailed rules (shown below) related to the Geological-Mining Law as parameters of safe transport of the shaft indicate: the verticality of the headframe's stem, the inclination of the rope pulleys and the hoisting machine drum, actual values of friction angles and values of clearance in the mining shaft.

* Corresponding author: jaskow@agh.edu.pl

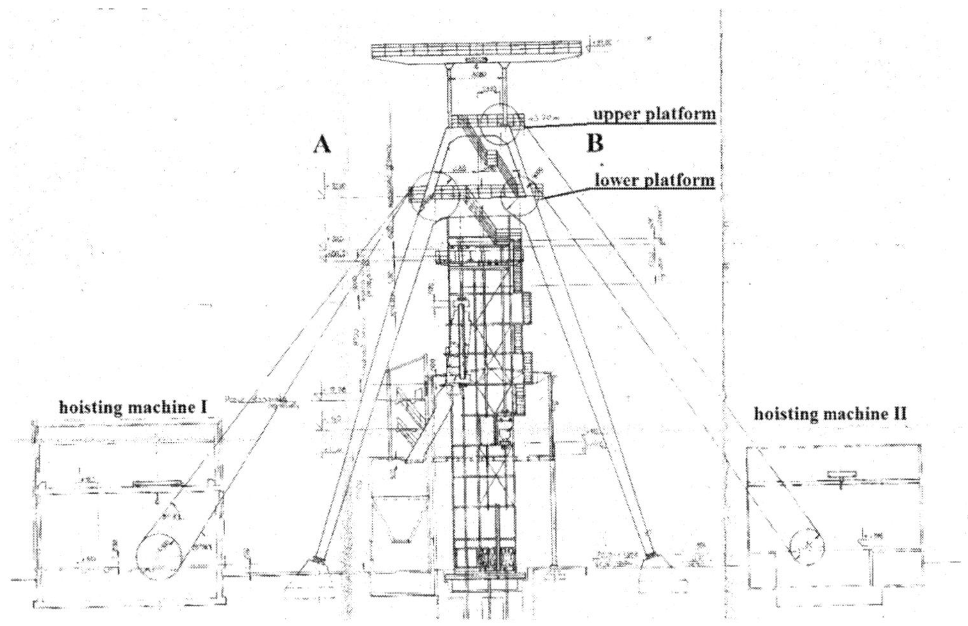

Fig. 1. Construction scheme of headframe

2 Legal basis in Poland

Principles of performing control measurements in Poland are regulated by following legal acts:

- Regulation of Minister of Energy from 23 November 2016 on detailed requiremenets for the operation of underground mining plants (Journal of Laws No. 1118, 2017, item 1),
- Polish Standard – Shafts, headframes and hoisting equipment (PN-G-09051) – p.1.2.9–1.2.14, p. 2.2.

In accordance with §43 section II of the Regulation, the mine and geological service of the mining should carry out inventory and control measurements of mining excavations and equipment, and objects of the mining plant during traffic and decommissioning, as well as observations of deformation of the land and buildings subject to the influence of mining exploitation. According to §29 the group of buildings include shafts, headframes with buildings, hoisting machines and pithead buildings. The most important rule in the Regulation (§542.2) refers to the headframe's verticality, which cannot exceed 1/500 of the headframe's height, measured from the axis of the top rope pulley to the framework. In practice, measurements determine inclination of the headframe at a certain section of this height – usually from framework or the lowest available level of the headframe, up to the level of the pulley platform and extrapolates to the height of the wheels axle. Further provisions also specify, among other things, permissible deviation of shafts of hoisting machines and rope pulleys, friction angles of ropes on the above-mentioned elements and permissible deviations from guide clearance [2, 3].

Almost all of the above regulations require very precise measurement at several millimetres. For example, for the analysed object, the height of the headframe from the beams to the axle of the upper rope pulleys is 44.8 m. According to the regulations, its inclination should not exceed 89.6 mm, which means that the measurement error on a 10 mm level can significantly affect the result of headframe's inclination.

3 Methodology of inventory measurement and analysis

In order to determine the deviation vector of the main axis of the headframe from the verticality, the following geodetic methods can be used: optical and mechanical or laser plumbing, space resection, projection to the lower levels or checkpoints and characteristic points tachymetric measurement. These are discrete methods that define the state of the axis of the structure indirectly for one observation cycle [4, 5]. In addition to the above methods, it is necessary to mention the need for precision levelling of the rope pulleys and hoisting machines shafts (to determine inclination of its axles).

In order to make a complete inventory of the mining shaft, a number of complementary measurement techniques should be used, which significantly increases the length of the survey. A collection of classic geodetic methods can be partially replaced by laser scanning, which allows surveyor to get as much information as possible in the shortest possible time. The scope of these data is incomparably greater than that obtained by traditional methods. In addition, the use of scanning can be important from an economic point of view as well as for the safety of the workers performing the measurements.

Modern surveying technologies in the form of 3D laser scanning surveys allow to perform analysis with so many measured points that the space mapped in discrete form looks real. Scanning measurements take a very short time (a few minutes at the station) and the amount of information thus obtained is impossible to get by other measurement methods. This enables analysis and digital modelling to be performed on any of a set of points that define the object components [6].

Inventory measurement was performed twice: in the first series in 2011, the precision tachymeter was used to measure the geometry of the headframe and in the second series in 2016 the terrestrial laser scanner (TLS) was used to measure the entire shaft hoist assembly system.

Due to the speed and accuracy of the measurements, as well as the lack of need to pause for long time operation of shaft equipment, the first series in 2011 was scheduled to perform measurements of angular-linear precision laser total station. Basic angular-linear measurements were focused on the characteristic points that were marked in accordance with the Regulation at various levels of the headrfame and the footings on each leg struts. These points should enable the capability of measuring the distance between them directly and should be elevated points, whose height should be determined together with the height of the benchmarks located at the struts base. The collection of points was transformed into a simplified headframe model and a deflection analysis was performed on this basis (Fig. 2).

a) b) c)

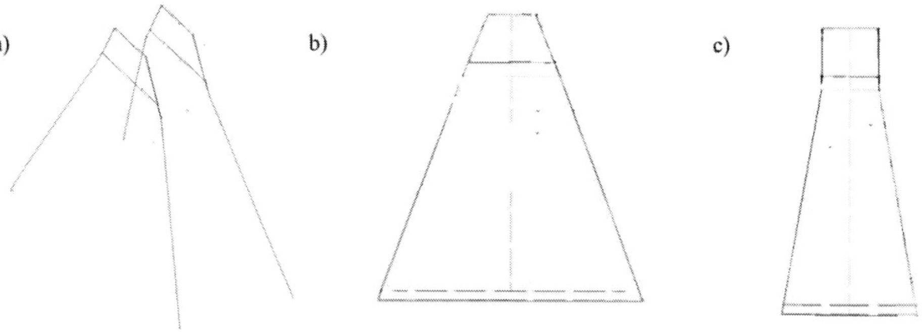

Fig. 2. (a) spatial view, (b) front view, (c) side view of the main axis of the headframe based on the control points, observations made in 2011

After the statutory five years, in 2016, the second series of measurements was carried out using a laser scanner Leica ScanStation C10 that allows obtaining information about the measured object with an accuracy of each point not worse than ± 6 mm for a single measurement scanner position [7]. For the measurement of the control-network, the connecting points of the scans (checkers) and for the control of the characteristic points, a precise tachymeter was used.

The second measurement series made in 2016 enabled cross-sections implication [8, 9] and more accurate indication of inclination changes in the vertical axis of the headframe. It also enabled creating a graph of changes in the headframe's deviation on individual sections. Test procedure is based on measurements of the headframe and its stem.

Based on the laser scanner's measurements, digital models of structures were created allowing for further numerical analyses. It consisted of determining the characteristic points located on the axes of the hoisting machines, determining the transverse profiles at the level of the girders at intervals of 4 m, and determining the coordinates of the points representing the established levels and respective planes of the rigid guides and angular guides. The obtained data was used to develop the spatial arrangement of the drum axis of the machines and the rope pulleys, the straightness of the guides and the angular guides, the determination of the guide clearance and the position relative to the rigid guides.

4 Results of analysis

The control-network was adjusted using the exact method, which allowed to designate junction points for scans with a mean position error of ± 2 mm. As a result, a single point error was less than 10 mm. Conversion to the global system was allowed by static GNSS measurement performed on control network positions [9]. The measurement product of the second series was a cloud of points representing the structure of headframe and hoisting machine (Fig. 3). The verticality is determined by analysing digital spatial model of the object. Measurement done with the above accuracy is quite sufficient to determine the position of characteristic points, later used to determine the symmetry centres of appropriate levels of the object or its fragments.

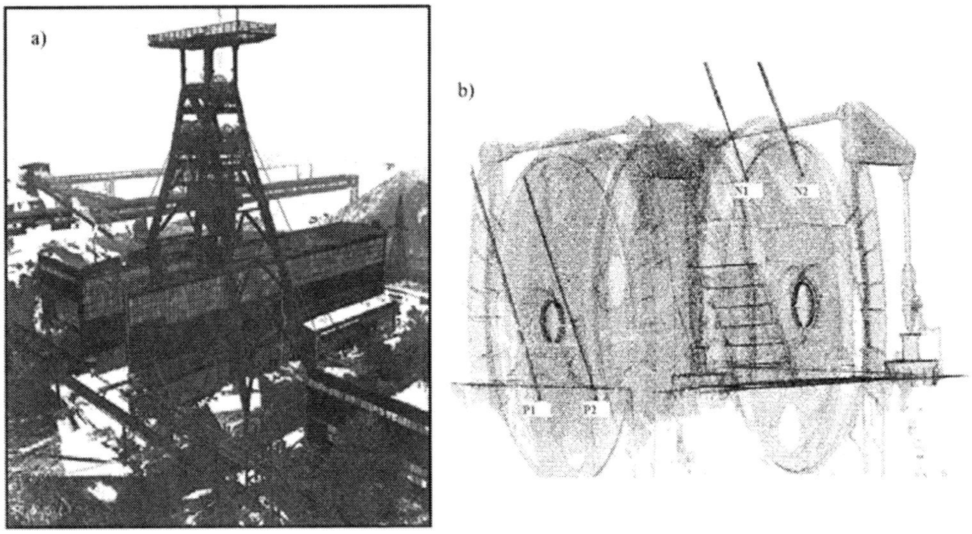

Fig. 3. Point cloud representing the (a) headframe and (b) hoisting machine. P1, P2, N1, N2 – points where line slides off from the hoisting machine's drum

Based on point cloud cross-sections of headframe's stem were made (Fig. 4a, b). It allowed to calculate their centres and what enabled to create a graph of changes in the headframe's deviation on individual sections. The goal is to designate the vector length direction and the horizontal pivot axis of the headframe (Fig. 4c).

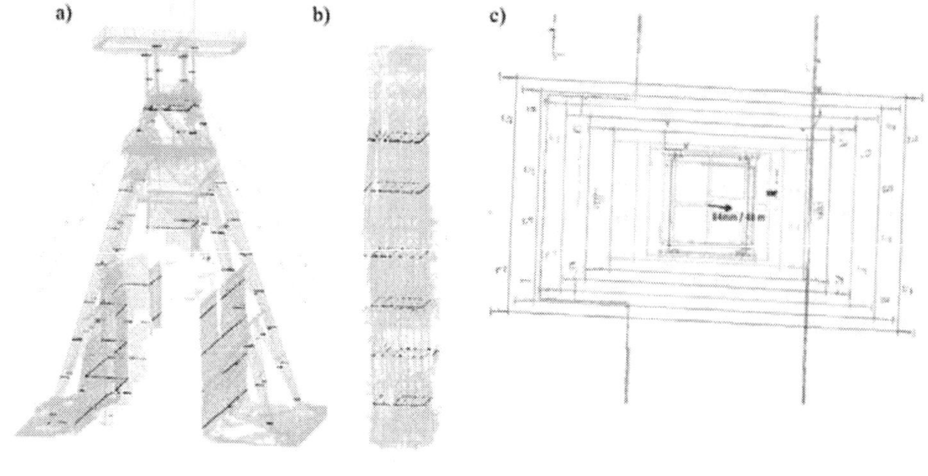

Fig. 4. Point cloud of the (a) headframe and (b) its stem (cross-sections marked with black); (c) top view of cross-sections with marked vector of inclination (status for 2016)

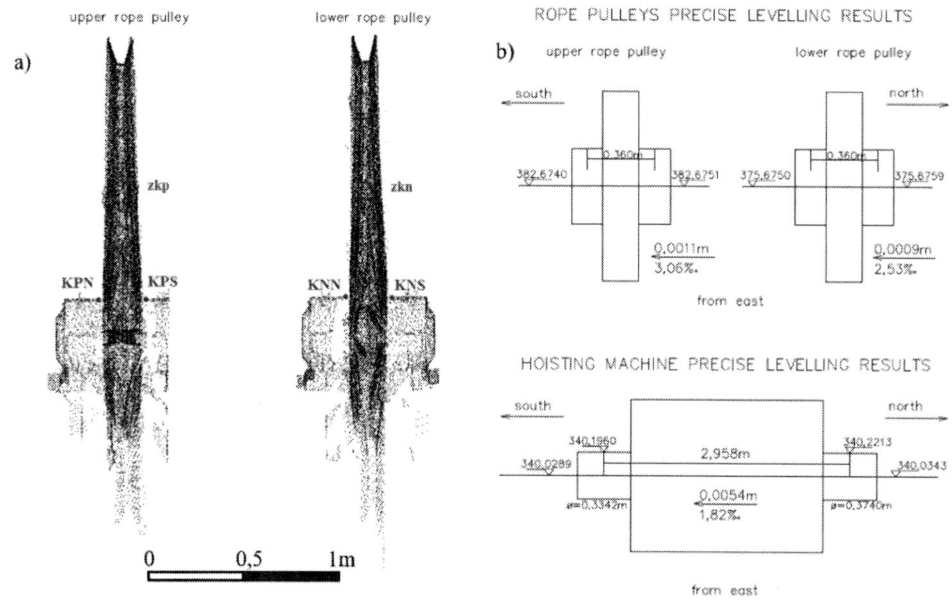

Fig. 5. View of the point clouds (a) of the rope pulleys together with the marking of the levelling points (KPN, KPS, KNN, KNS) and points where rope slides from the pulleys (zkp, zkn); (b) an example view of the results obtained (2016)

Scanning data was also used to determine the spatial relationship between the rope pulleys and hoisting machines and the verticality of the rope coming down into the shaft (Fig. 5). The points where the rope slides from the pulleys and the hoisting machines drums allowed to calculate the actual friction angles. The hoisting machine's drum width was determined by two methods: classically by tachymeter and by a laser scanner. The analysis showed a difference of several millimetres, which has no real effect on the results obtained, and falls within the accuracy of the measurement with both methods. Data from the precise levelling of the shafts of the wheels and hoisting machines were obtained with the help of a precision levelling. This data was also compared with the results from point cloud analysis and also differences in data accuracy have been reported. The cloud of points has information about the height up to few millimetres, precise levelling is much more accurate. Due to the fact that width of the pulleys is between 0,3–0,4 m, precision of measurement is becoming more important (Fig. 5).

In the second series scanning measurement was extended of shaft. Measurement were performed there from the roof of shaft cage. To register scans spheres were used as target points, allowing for measurement without time-consuming mechanical plumbing, saving several hours of shutting down the shaft for the measurement period (Fig. 6).

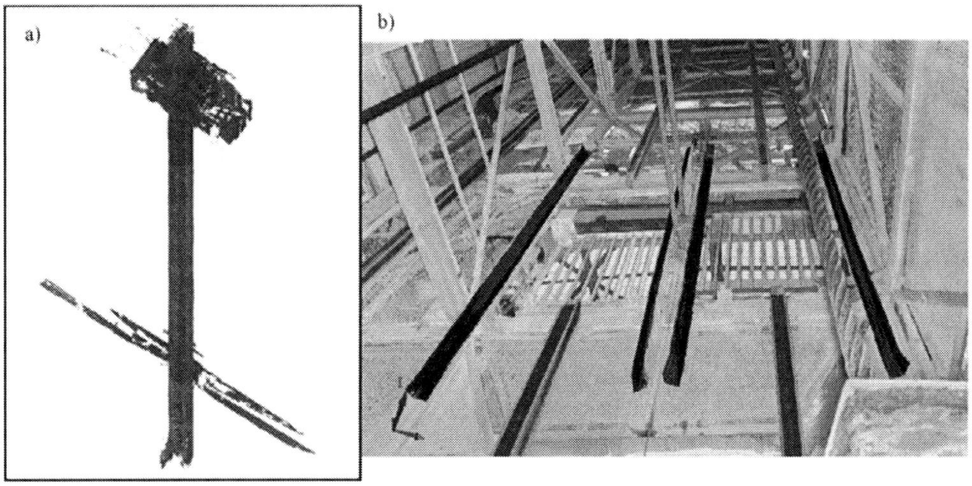

Fig. 6. Point clouds representing (a) the entire shaft with (b) its equipment

This enabled an analysis of guiding clearance and angular guiding in the shaft (Fig. 7), and the determination of the characteristic points needed to determine the verticality of the rope coming down the shaft, and the apparent and actual friction angles. Based on point cloud also the deviation of shaft verticality graph could be done.

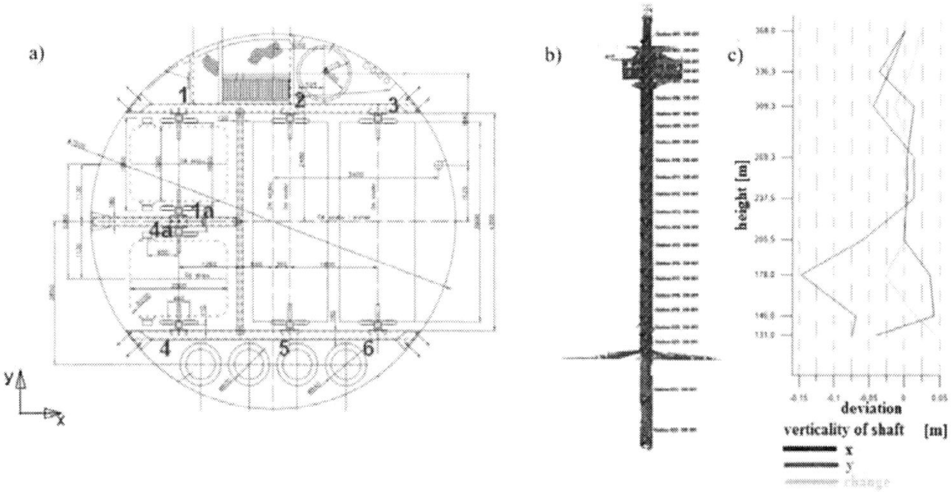

Fig. 7. (a) schematic diagram of the shaft with marked guides (in red), top view; (b) positions of the laser scanner in the shaft; (c) graph showing the verticality of the axis of the shaft

5 Conclusions

The accuracy of a single point in the point cloud is estimated to be less than 10 mm. This is sufficient in the case of the implementation of the profiles and the calculation of the course of the main axes and guide planes. For the shaft hoist assembly (divided into the west-skip and east machine) the azimuth graphs of ropes, rope pulleys and hoisting machine shafts axes, angles of friction, inclination of shafts and verticality of descent rope to the mining shaft from the level of the rope pulleys were created. During measurements nominal spacing for individual shaft compartments (east and west) and the width and length of vessels along with sliding were also designated.

Measurement of the entire hoist assembly: shaft, headframe with its stem and hoisting machine, four people team completed in about 12 hours, including control measures allowing to check the usefulness of scanning data to specific issues. Thanks to the use of laser scanning, the team was able to skip the most time-consuming tasks like mechanical plumbing in the shaft and the headframe. The tachymetric measurements was also partially eliminated. It is therefore possible to assume that the time taken to measure and thus to shut down the shaft from traffic has fallen by half. Working time, its nature and the number of people involved in the measurement also indicate an increase in safety of the measurement team.

Laser scanning becomes a very serious substitute for many measurement methods and complements the entire study. It shows how modern technologies get into the consciousness of surveying departments. However one should be aware of restrictions in the use of laser scanning in shaft conditions (for example high humidity or dust in the exhaust shafts). In addition, the availability of measuring equipment and time consuming data processing and performance of available computers can still be a problem for surveying departments. Data obtained from several dozen laser scanner stations require efficient graphics cards and large memory of computers, together with specialized knowledge of data processing.

Acknowledgment

This article was developed within the statutory work of AGH US 11.11.150.195.

References

1. Z. Szczerbowski, M. Jóźwik, *Examples of natural and mining conditions of deformation of mining shaft* (XXV Winter School of Highland Mechanics 'Geotechnics and Special Construction', Zakopane, (2002)
2. Regulation of Minister of Energy from 23 November 2016 on detailed requiremenets for the operation of underground mining plants (Journal of Laws No. 1118, 2017, item 1)
3. T. Lipecki: *A comprehensive assessment of geometric objects and shaft devices using laser scanning* (Akademia Górniczo-Hutnicza in. Stanisława Staszica w Krakowie, ISBN: 987-83-7464-602-4
4. J. Pielok: *Inventory of Pomerania mine shafts* (Monthly WUG, No. 3, (2004)
5. M. Jabłoński, W. Jaśkowski, *The usage of telemetry measurement methods in order to determine shaft tube deformations* (SGEM 2017 conference proceedings, 17, 697–704, (2017)
6. B. Collin, P. Carreaud, H. Lancon, *High efficiency techniques for the assessment of railways infrastractures and buildings* (Transportation Research Procedia 14, pp. 1865–1874, (2016)
7. Leica Geosystems AG, Leica ScanStation C10 specification, (www.leica-geosystems.com/hds, Heerbrugg, Switzerland)

8. M. Pejić, *Design and optimisation of laser scanning for tunnels geometry inspection* (Tunnelling and Underground Space Technology 37, pp. 199–206, (2013)
9. W. Wang, W. Zhao, L. Huang, V. Vimarlund, Z. Wang, *Application of terrestrial laser scanning for tunnels: a review* (Journal of Traffic and Transportation Engineering, 1(5), pp. 325–337, (2014)

Modeling and Analysis of Integrated Bathymetric and Geodetic Data for Inventory Surveys of Mining Water Reservoirs

Agnieszka Ochałek[1,*], *Tomasz* Lipecki[1], *Wojciech* Jaśkowski[1], and *Mateusz* Jabłoński[1]

[1] AGH University of Science and Technology, Faculty of Mining Surveying and Environmental Engineering, A. Mickiewicza Av. 30, 30-059 Krakow, Poland

Abstract. The significant part of the hydrography is bathymetry, which is the empirical part of it. Bathymetry is the study of underwater depth of waterways and reservoirs, and graphic presentation of measured data in form of bathymetric maps, cross-sections and three-dimensional bottom models. The bathymetric measurements are based on using Global Positioning System and devices for hydrographic measurements – an echo sounder and a side sonar scanner. In this research authors focused on introducing the case of obtaining and processing the bathymetrical data, building numerical bottom models of two post-mining reclaimed water reservoirs: Dwudniaki Lake in Wierzchosławice and flooded quarry in Zabierzów. The report includes also analysing data from still operating mining water reservoirs located in Poland to depict how bathymetry can be used in mining industry. The significant issue is an integration of bathymetrical data and geodetic data from tachymetry, terrestrial laser scanning measurements. **Key words** – bathymetry, inventory, reclamation.

1 Introduction

Exploitation of natural deposits and subsequent processing of acquired resources lead to noticeable changes not only in relief but also in water balances of mining terrains. Mining activity is often causes arising of water reservoirs, which should be inventoried both during the exploitation and after finishing the mining output.

Inventory surveys of operating mining water reservoirs, which formation is related to mining exploitation, is necessary at every stage of the mining output process. At the stage of preparation for the exploitation, measurements are based on recognizing the bottom of the reservoir and its underwater natural resources. There are documentation such as: plans for the development of the deposit, detailed technical projects of the stages of exploitation and land development plans prepared, as well as initial plans of land development after the completion of mining output [1]. During the exploitation, the bathymetrical measurements are used to make whole the geological and mining documentation. All the records are obligatory due to Polish geological and mining law and Polish norms. Except for it, hydroacoustic surveys also provide compilations of the quantity of extracted and left underwater deposits.

* Corresponding author: ochalek@agh.edu.pl

According to Polish Geological and Mining Law, in case of in case of liquidation of a mine, an entrepreneur is obliged to take the necessary actions to protect the environment and to reclaim land after mining activity. The water direction of land reclamation gives new opportunities of enriching natural environment by rising the landscape values or increasing the biodiversity [2, 3]. In this case the main purpose of researches might be determining a better direction of reclamation of post-mining reservoir and more advantageous way of its usage, for example: water and recreation direction of reclamation – post-mining reservoir as a bathing place or a fishery for anglers [4]. There are many various parameters such as: inclination of underwater coasts, structure and slitting or vegetation caver which should be taken into account during the inventory surveys of water reservoirs due to land reclamation.

The basis of the safe usage of water reservoirs influenced by mining activity is their accurate morphometric identification. Obtaining information about depth or bottom shapes of water reservoirs or identification and verification of underwater objects are complex processes which require comprehensive elaboration of collected data [5].

2 Technology of bathymetric measurements

In practice inventory of inland water reservoirs is connected with bathymetric measurements on the research area. Nowadays, possibilities of carrying out the hydroacustic surveys are constantly enhancing and becoming more and more advanced. Studies of lakes morphometry require an accurate representation of their bottom geometry in the form of a bathymetrical map or more often a three-dimensional bottom model [6].

Contemporary hydrographic works involves a lot of modern and innovative electronic equipment. The rudimentary hydroacoustic device which obtains information about the depth of inventoried water reservoir is a single beam echo sounder (SBES) equipped with a converter that produces and sends one single beam with a vertical acoustic axis (Fig. 1).

The sent impulse bounces off the bottom or obstacles and returns after some time to the device. This time depends on the depth and the speed of sound propagation in water. At the same time, the velocity is connected with density of water, so its temperature and salinity. The sound propagation in the freshwater tank is approximately 1500 m/s. Knowing the velocity and the return time of acoustic signal, it is easy to determine the depth of the reservoir.

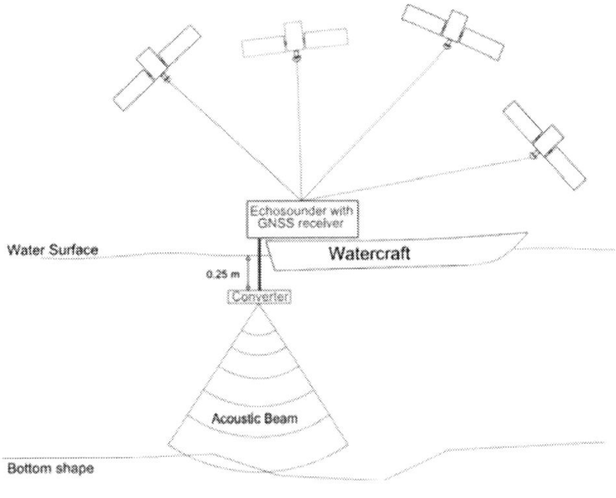

Fig. 1. Single Beam Echo Sounder – operation rules

Echo sounders used on small watercrafts used a beam width of more than 50°. Such beam with provides a relatively large space of research and gives more information about underwater environment. However, it offers less angular resolution and surveys at lower depths. Using narrower beams (for example 10°) allows a deeper penetration, but converters with narrow cone angle need stabilization and are usually installed in research echo sounders. The cone wide angle converters work better in freshwater reservoirs, while the narrow cone version is suitable for both types – freshwater and salt water reservoirs.

Technology development contributed to produce multi beam echo sounder (MBES) [7]. This is an innovative solution that combines the advantages of an echo sounder and a sonar. This device, during measurement, sends multiple independent acoustic beams at different angles, but at the same frequency. The same amount of independent impulses is received and give the information about depth and the shape of the bottom surface. In comparison to SBES, MBES has a lot of advantages [8]. For sure, it is a device which can be easily used at medium and large depths or to measure reservoirs with large surfaces. All the obtained data might be used not only to illustrate the sonar bottom configuration, but also to identify underwater objects [9].

3 Research area and methodology of measurements

There were two measurement cycles on two water reservoirs – Dwudniaki Lake in Wierzchosławice and "Mały Zakrzówek" in Zabierzów carried out in 2016 and 2017 in order to collect the adequate amount of bathymetric and geodetic data for further analysis. In this report the data were also acquired from the measurements at still operating open-pit mine Kunice IV, where deposits are extracted from underwater. All the research areas are in southern Poland and are shown in red colour at the Figure 2.

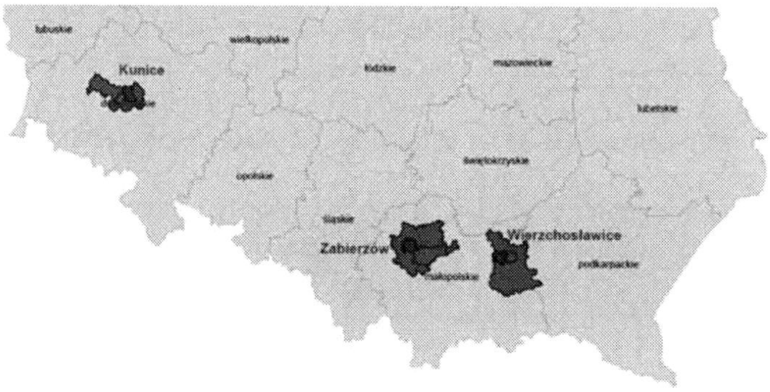

Fig. 2. Research area – localization of inventoried water reservoirs

The integrated bathymetric system based on the watercraft, multi beam echo sounder – Lowrence Elite-7 CHIRP and DGPS receiver built-in it the device was used to hydroacoustic measuremens at two above mantioned water reservoirs. The echosounder – Lowrence Elite-7 CHIRP combines two technologies: the classic echo sounder and CHIRP sonar (Compressed High Intensity Radiated Pulse) [7].

Asopposed to simple 2D sonar that sends single frequency, devices with CHIRP transmit range beams in different frequencies. Every returned frequency is individually interpreted. Sweeping range of frequencies guarantees a much wider information and the CHIRP sonar provides a higher resolution and legibility of recieived data [10].

There were not only bathymetric measurements carried out during the inventory surveys of the flooded quarry in Zabierzów. Additionally, there were some geodetic works made such as surveys to establish control points on the research area and terrestrial laser scanning in order to visualize the walls of the former excavation. There were Leica ScanStation C10 laser scanner used to make the inventory. Static measurements and tachymetry were performed with integrated Topcon HiPer Pro receivers and Topcon OS-103 tachymeters.The geodetic works allowed to register scans in global coordinate system (PL-2000, EPSG 2178). Moreover, for the determining the accuracy of the echosounder, an additional direct measurements of deptha at three profiles were carried out by using line with a centimeter scale and a weight at the end.

It is worth mentioning that there were the route profile project in *.gpx file format created before the bathymetric surveys. Next it was imported to echo sounder and used during the measurements. It has enabled to track the route and an efficient navigation of watercraft. However, it sould be emphasized that the projected survey profiles do not always fulfil their purpose, especially on operating open-pit mines where floating conveyors and mining equipments are in constant move. Measurements conducted in that conditions should be monitored on a regular basis, avoiding obstacles.

4 Processing of the data

In the geodetic aspect the bathymetric measurements are aimed to create maps, cross-sections, and a DTM of the bottom, including the positioning of this model in a global reference system. The data acquired from hydroacoustic surveys has to allow for the accurate reproduction of invetoried reservoir's details. Therefore, the information about the spatial coordinates of watercraft's localization (or more specified the echo sounder's converter) and the depth information are needed. Innovative devices also provide information about the frequency udef for survey [kHz], water temperature [°C] and velocity of the watercraft [kn]. All the data is displayed in the echosounder's screen and then collected in the database, which should be properly sorted and veryfied afther the survey. The significant aspect of database management is the validation of measured points which would be taken into account during the analysis of the bottom shape of water reservoirs. The next issue is the corecction of definited depth due to the immersion of echo sounder's converter (Fig. 1).

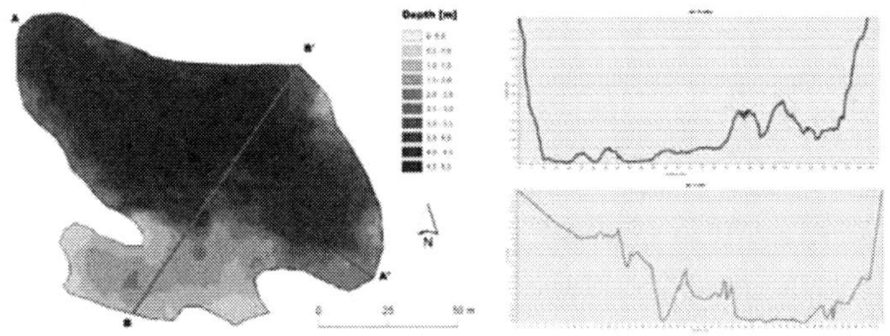

Fig. 3. The morphometric characteristic of flooded quarry „Mały Zakrzówek"

Table 1. Morphometric characteristics of inventoried reservoirs

Morphometric characteristic	Reservoir's name	
	Jezioro Dwudniaki	**„Mały Zakrzówek"**
Area F_0, [m^2]	200 800	9200
Lenght L, [m]	676	148
Mean width $B_{śr}$, [m]	297	62
Max width B_{maks}, [m]	425	70
Mean depth $h_{śr}$, [m]	1,20	2,8
Max depth h_{max}, [m]	3,65	5,5
Volume V_0, [m^3]	243 500	25 650
Number of depth points	5336	1821

Such structured and comprehensive database can be used in GIS software, which can give a lot of possibilities of presenting the result of researches. An example of detailed vizualization in the form of a bathymetric plan and a cross section through the longest profile of the reservoir which describe the shape of the bottom of the inventoried lake was preseted (Fig. 3). Commonly used determining of reservoirs' morphometric characteristic has been replaced by simple operations in GIS environment, assuming that a the inventoried reservoir is represented by the digital data model consisting of a polygonal contour and a layer connected with depth (the raster model). Obtained morphometric characteristics of the invetoried lakes are shown in Table 1.

Bathymetric data is often presented as traditional (paper or vector) two-dimensional maps. However, nowadays this way of presenting to many clients is no longer satisfactory, geovizualization and three-dimensional models of terrain and surface are essential. One of the effective methods of representing and analyzing surfaces is the Digital Terrain Models (DTM). DTM can be represented for example by Triangulated Irregular Network (TIN), which is an irregular grid of triangles used to present complicated and varied surfaces. The GPS positioning and echo-sounding measurements allows the use of a third dimension in the processing of the obtained bathymetric data. The process of creating DTM and 3D visualization of two inventoryed water reservoirs were carried out in the ESRI ArcMap 10.2 program (Fig. 4).

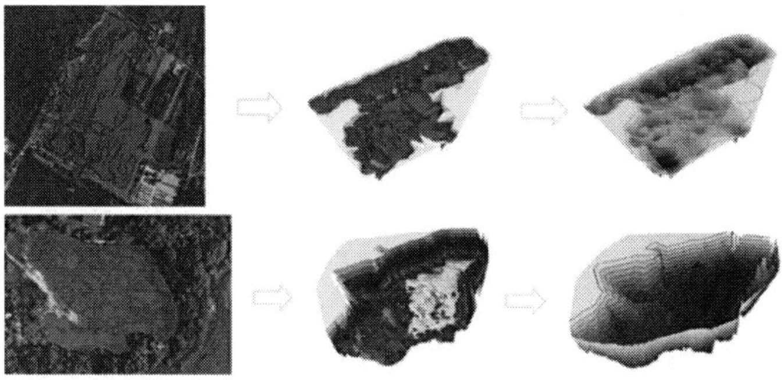

Fig. 4. Process of DTM formation and 3D vizualization of the bottom surface of the water reservoir in ArcMap

In order to analyze the conducted hydroacoustic survey in comparison with the direct depth measurements by using line with a centimeter scale and a weight at the end, surveys were made along three projected profiles through the reservoir. Measurement lines were signaled by a specially prepared rope, which was marked at every distance of 1 m. The beginnings and ends of the projected profiles were measured by the tachymetr. On each profile, the depth was directly determined every 2 meters by using a measuring line and tied stone and its end. Simultaneously, the value of the depth, which was displayed on the echo sounder's screen, was written down. The pairwise observation method was used to determine the accuracy of the measurements. The results of which are shown in Table 2. Data analysis showed an average error of both measuring methods about 0.14 m.

Table 2. Analysis of the measurements by the method of pair of observations

Profile	Mean difference error $d_{śr}$ [cm]	Mean unit error m_0 [cm]
AA′	12,5	8,9
BB′	20,2	14,3
CC′	25,9	18,3
Mean error	19,5	13,8

5 Interpretation of bathymetric and geodetic data by the example of lake Dwudniaki and Maly Zakrzowek

The inventory surveys of Dwudinaki Lake located in Wierzchosławice were divided in two parts: the first one (shown in the Figure 5 in green was made in May 2016) and the second one (marked in violet in August 2016). The first hydroacoustic survey was made with a good visibility of the bottom of the reservoir, and the second was already done with a large amount of plants hovering at the surface of the water or in the water. After validating and sorting the obtained database, it turned out that a lot of measured depth points were discarded because of the wrong depth value (for example zero value). Figure 5 also depicts the comparison of the flow path of the measuring unit and the number of points that, after validation, were suitable for use during the reservoir's bottom interpolation. Nevertheless, the received data was sufficient enough to carry out the entire process of acquiring, introducing, managing, analyzing and visualizing the bathymetric data.

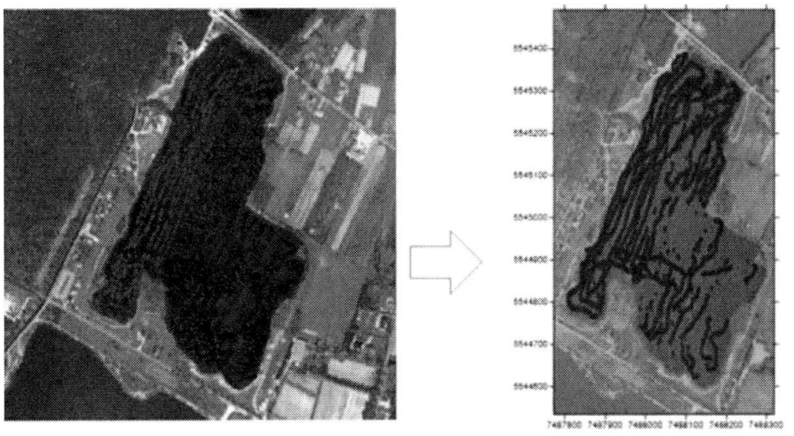

Fig. 5. Validation of depth points taken into account during analyzing the bottom shape of the Dwudniaki water reservoir

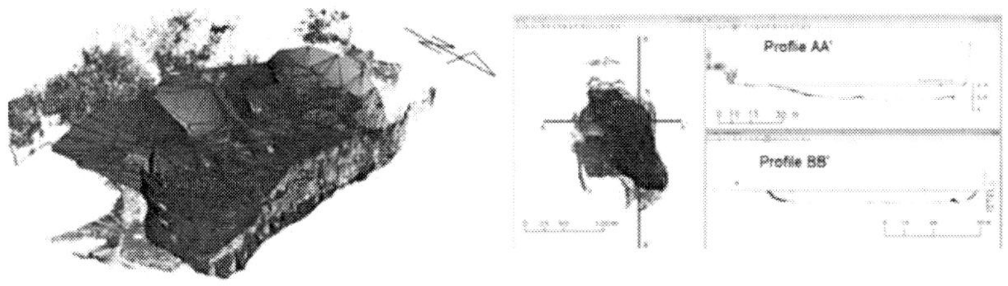

Fig. 6. Integration of bathymetric and geodetic data

For accurate inventory of the water reservoir and its surroundings, all obtained measured data should be combined. An integration of acquired information from bathymetric measurements (the shape of the bottom) and geodetic surveys (terrestrial laser scanning – visualization of the excavationn's walls, static GNSS measurements and tachymetry) consisted of putting every kind of data into the same reference global system (PL-2000, zone 7, EPSG 2178). Therefore, the data integration problem was mainly focused on transforming the depths so they correspond the elevation of geodetic measured points. The combination of researches carried out in two environments (on land and water) allowed to create the comprehensive and accurate model of the whole excavation and to make cross sections (Fig. 6).

6 Processing of bathymetric data acquired from measurements at operating open pit mine Kunice IV

The data obtained from the bathymetric measurements at the Kunice IV open-pit mine at different times allowed to visualize the progress of the underwater mine exploitation. Figure 7 depicts the visualization of bathymetric surveys at all measuring cycels, so all recived bathymetric points and the DTM created based on them. The deposits extraction proceeds in the north-east direction.

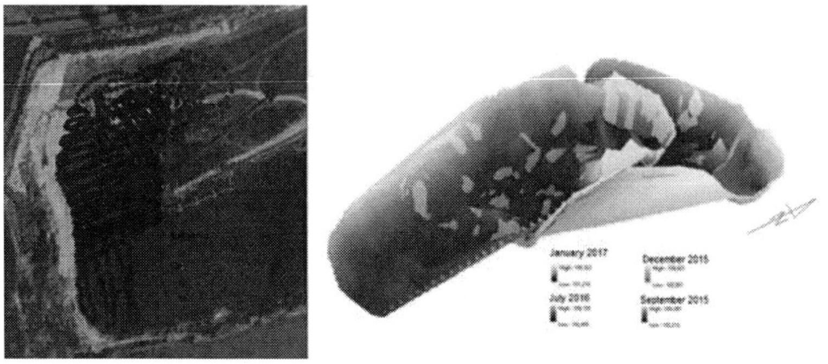

Fig. 7. The visualization of bathymetric surveys at all measuring cycels

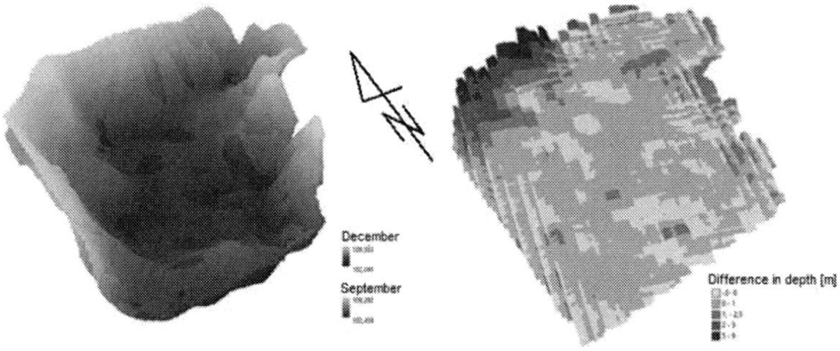

Fig. 81. DTM created from data obtained from measurements made by using SonarMite echosounder in September (orange colour) applied to DTM created from data acquired from surveys made by using AIRMAR B164 echosounder in December (blue colour) and the depth difference map

At the next stage of the study, the attention focused on the part of the water reservoir where measurements were made twice – in September 2015 by using SonarMite echo sounder and December 2015 with AIRMAR B164 echo sounder. In order to more accurate analysis of the received data, DTMs in raster format were created. Before this process, it was necessary to determine the appropriate area for the analysis, which included the same range of measurements performed at both dates. It allowed to make raster analysis in selected environment. Additionaly, apart from classical cross sections through DTMs, the method of comparison of the measured depths is shown in the Figure 8. There is the imposition of the models on each other – the orange shades indicate surveys in December, the blue shades in September – and the depth difference map showed. The minus results indicate a greater depth for the September's model. The maximum difference in height was 8.71 m, while the average was 0.55 m. The differences were mainly due to the fact of carrying out mining works. What is more, the volume of the two models was also calculated. The calculated difference in volume of the two models may determine the amount of excavated deposits in the analyzed area during 3 months, it's about 7860 m^3.

7 Conclusions

Acquiring information about depth or bottom shapes of water reservoirs or identification and verification of underwater objects are complex processes which require comprehensive elaboration of collected data. The more interdisciplinary character of research is, the more effective cooperation between scientists and engineers from different fields is needed. Modern measuring techniques and development of GIS systems bring new opportunities for data processing. Bathymetric measurements in the mining industry are mainly used to determine the change of bottom shape of water reservoirs due to exploitation, its speed and quantity of extracted deposits. The inventory of post-mining reservoirs in order to determine the direction of their reclamation and its subsequent use is also a significant issue. More often, bathymetric data is presented in the form of three dimensional models and then used in designing and vizualizing reclamation.

The researches showed that the accuracy of the bathymetric measurement depends on many factors, and the measurement noises. These are mainly the result of the occurrence of underwater vegetation, changes in water density or the presence of fish, significantly disturb the process of obtaining and processing the data. The integration of geodetic and bathymetric data is a complex problem that requires detailed elaboration of collected data. However, it allows better visualization of measurement data.

Acknowledgements

Bathymetrical data from open pit mining Kunice IV obtained from mine surveyor – eng. Piotr Szajt. Article was developed in the framework of the AGH statutory no 11.11.150.195.

References

1. E. Gurzęda, *Raport oddziaływania na środowisko projektowanej eksploatacji kruszywa naturalnego ze złoża „WITOWĄŻ I"* (Gdynia, 2009)
2. K. Różkowski, K. Polak, M. Cała, *Wybrane problemy związane z rekultywacją wyrobisk w kierunku wodnym* (Górnictwo i Geoinżynieria, 34/4, 2014)
3. K. Nazan, *Reclamation of Degraded Landscapes due to Opencast Mining* (Advances in Landscape Architecture, InTech, 2013)
4. D. Madusiok, *Analysis of the possibilities for reclamation after aggregate exploutation in the Cracow's quarry of Brzegi by using remote sensing and geoinformatics* (Geoinformatica Polonica, nr 15, Kraków, 2016)
5. R. Chesnaux, M. Llambert, J. Walter., U. Fillastre, M. Hay, M.A. Rouleau, R. Daigneault, A. Moisan, D. Germaneau, *Building a geodatabase for mapping hydrogeological features and 3D modeling of ground water systems*: Application to the Saguenay–Lac-St.-Jean region (Computers & Geosciences 37 (11), p. 1870, 2011)
6. B. Pączek, *Współczesne Metody prowadzenia prac hydrograficznych* (Przegląd hydrograficzny, No. 1, Gdynia, 2016)
7. Y. Kim, N. Koo, M. Riedel, H. Namgoong, J. Lee, S. Cheong, Y. Joo, D. Yoo, J. Chun, H. Lee, *A case study on swell correction of Chirp sub-bottom profiler (SBP) data using multi-beam echo sounder* (MPES) data (Journal of Applied Geophysics, vol. 145)
8. D.G. Simons, M. Snellen, *A Bayesian approach to seafloor classification using multi-beam echo-sounder backscatter data* (Applied Acoustisc, vol. 70, pp. 1258–1268)
9. S. Innangi, A. Bonanno, R Tonielli, F. Gerlotto, M. Innangi, *High resolutron 3-D shapes of fish schools: A new method to use the water column bachscatter from hydrographic MultiBeam EchoSounders* (Applied Acoustics, vol. 111)
10. R.A. Altes, *Wide-band, proportional-bandwidth Wigner-Ville analysis* (IEEE Transactions on Acoustics, Speech and Signal Processing, vol. 38, pp. 1005–1012)

Defining the Cubature Changes of Historic St. Kinga Chamber in Bochnia Salt Mine, Using Laser Scanning Technology

Anna Szafarczyk[1,*], and *Rafał* Gawałkiewicz[1]

[1] AGH University Science and Technology, Faculty of Mining Surveying and Environmental Engineering, A. Mickiewicza Av. 30, 30-059 Krakow, Poland

Abstract. In Poland, there are many mining enterprises, of historic character registered in the UNESCO World Heritage List. One of the oldest mining enterprises in Poland is the Salt Mine in Bochnia. The processes inside the rock mass require that surveying services carry out regular geometric control of the cavities. A particular attention should be paid (due to its sacral function) on St. Kinga Chamber, located 195 metres below the surface, on the mine level "August". So far measurement technologies have been connected with the studies on changes in the geometry of cavities and based on linear bases used to measure convergence. This only provides discrete information (in a point) and not always presents a real state of deformation. In the scanning method, in practice a three dimension image of changes (structural deformations) is obtained, impossible to determine with the application of measurement methods, applied to measure the value of linear convergence (the method with a limited number of bases). Laser scanning, apart from determining the value of volume convergence, gives also the possibility of the visualization of 3D cavern. Moreover, it provides direct information to update mining numerical maps and make it possible to generate various cross-sections through the cavern. The authors analysed the possibility of the application of laser scanning (scanner Faro Focus 3D), as a modern tool allowing the measuring of the value of volume convergence. **Key words** – cubature changes, volume convergence, historic mine, laser scanning.

1 Introduction

Salt is a substance necessary for a human organism to function and develop. The search for salt has been carried out since the pre-historic times. The beginning of salt mining in Bochnia (Poland) goes back to 1248 [1]. Now the Salt Mine in Bochnia is available for tourists, and the tourist route including the one localized on the route going through St. Kinga chapel was in 2013 written into the World Heritage UNESCO List, which makes it necessary to take a particular care about this object.

To make sure that tourists are safe, the mine surveyors periodically make linear measurements of convergence in the St. Kinga Chapel with classical measurement methods meaning periodical measurement of the length of the section between the assembled pairs

* Corresponding author: szafarcz@agh.edu.pl

of points making measurement bases. These measurements are characterized by high accuracy; however they refer to selected points of the chamber.

In this paper, the possibility of the application of modern measurement technology, which is laser scanning to determine the volume convergence of the chamber. Scanning allows obtaining extensive spatial information on the object being measured, thus some mines in Poland e.g., KGHM-Polska Miedź S.A. apply this technology on daily basis for the inventory of cavities.

1.1 Geological Structure of the Study Area

Salt deposit in Bochnia has the shape of a lens extending over the length of 10 km and width up to 200 m. The depth of the salt bed ranges from 10 to 80 m. It was formed in the form of two main folds (anticlines): the anticline of Bochnia, going directly under the town and smaller anticline of Uzbronia, lying south of the town [1, 2]. A proper salt deposit series (Fig. 1) is built from the layers of anhydrite, siltstone and halite.

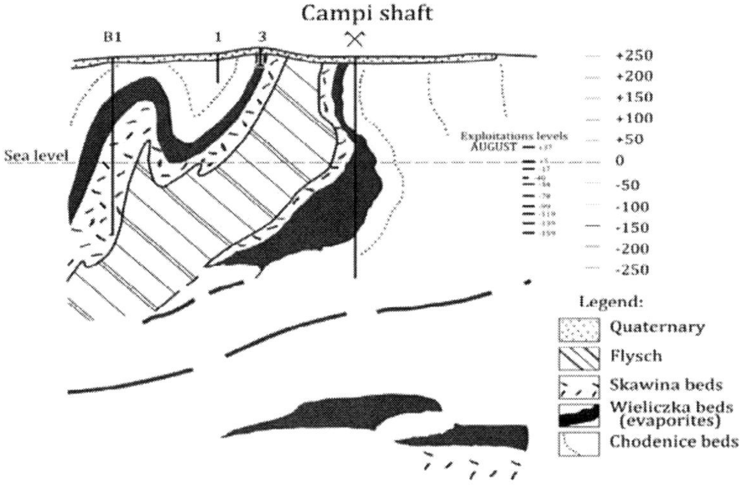

Fig. 1. Cross-section through the rock mass in the neighbourhood of shaft Campi (based on Garlicki A. 1968)

1.2 The Characteristics of the Objects

The inventoried chamber excavation is located 195m under the surface, on the mine level August, in the distance of about 600 m west of shaft Sutoris and 400 m east of shaft Campi, on the way of level August. This was made manually, its shape was semi-elliptic, and the area of the roof support was 5–10 per cent. The chapel was made in the northern halite, consisting of grey fine-grain silt-sand salt, containing some loam and anhydrite. At present, geometric parameters of the chamber excavation, based on the inventory chart and [2] are:

- volume: 800 m^3,
- floor area: 312 m^2,
- maximal length: 19 m, mean 16.0 m
- maximal width: 15.0 m, mean 12.0 m,
- maximal height: 6.0 m, mean 4.0 m.

The inside of the chamber with the historic altar making the integral part of the excavation was presented in Figure 2.

Fig. 2. The inside of St. Kinga Chamber in the Salt Mine Bochnia

1.3 The Convergence Phenomenon in the Salt Rock Mass

The phenomenon of the convergence of the cavities of the salt rock mass is usually a long process, which is a result of its specific physical properties. Its characteristic feature includes rheological properties, which are seen in the phenomena of creeping and relaxation. These phenomena show significant influence of time on the behaviour of the rock mass under the influence of load [3]. Laboratory studies on halite showed a linear connection between hydrostatic state of stress and volume strains. In salts permanent strains already occur at smallest changes in strains [4]. Floating of salt is caused by strains and is characterised by the increase of strains in time. The characteristic feature of the halite rock mass convergence is usually the preservation of the continuity of strains and consistence in time. Significant factors influencing the course of the process of the convergence of cavities found in the halite rock mass are temperature and humidity. The growth of the temperature of the rock mass increases the susceptibility of the rock masses on creeping [5]. This process is directly connected with the increase of the depth on which the excavation is located. This relationship can also be observed in periodical measurements of dislocations. In the spring-summer period a greater speed of dislocations was observed than in winter. A significant influence on the value of convergence has also lithological structure of the rock mass, especially if its layers are different in terms of susceptibility to creeping [6]. The presence of the rocks of o low susceptibility to creep in the floor and roof layers for the seam of halite, decreases the value of convergence, both in horizontal and vertical direction.

The paper presents the values of convergence occurring in the St. Kinga chamber in the Bochnia Salt Mine, based on the values of direct linear measurements and the model obtained as a result of the laser scanning of the cavity.

2 Convergence Measurements in St. Kinga Chapel

2.1 Archive Measurements

Convergence measurements have been made in St. Kinga Chapel since 1993. They are realized two times a year (spring and autumn time). Inside the chapel there is a well developed network of the control points (Fig. 3) used mainly to define the convergence value in time. First measurement points of this network were established in 1987.

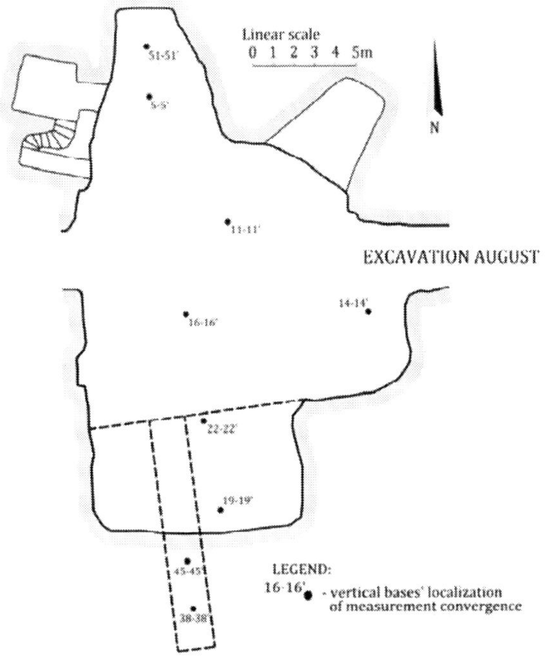

Fig. 3. Distribution of measurement marks in St. Kinga Chapel in horizontal projection (on the basis of materials from the surveying department of the Bochnia Salt Mine)

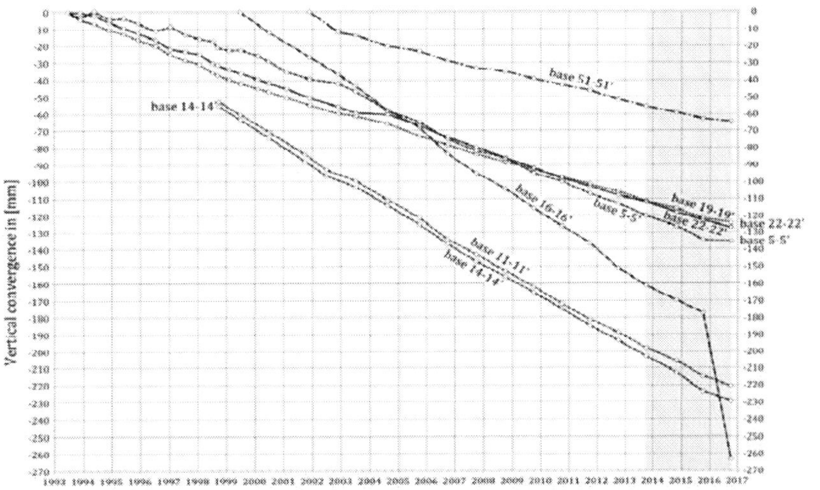

Fig. 4. Vertical convergences in St. Kinga Chapel in 1998–2014

As a result of cyclic measurements the value of the lengths of bases localized inside the chamber are defined. In case of the damage of the base, a new base is established and adjusted to a general measurement concept [4]. The results of the measurements in vertical bases were presented in Figure 4.

In the discussed period of time vertical bases located in St. Kinga Chapel were getting shorter with changeable speed, which was presented in Table 1.

Table 1. The values of the changes in the lengths of vertical bases in the Chapel of St. Kinga from October 2013 until October 2016

Base	Changes in the lengths w [mm]
5–5'	–15.5
11–11'	–22.8
14–14'	–26.2
16–16'	–102.1
19–19'	–12.3
22–22'	–14.5
51–51'	–9.1

Based on these measurements, the speed of roof subsidence can be defined, which ranges between –0.25 and – 2.8 mm/month and is relatively uniform in time except base 16–16', where local changes of anomalous convergence have occurred. The greatest values are recorded in the highest central part of the chapel and in its southern and eastern part. In these areas changes on the level –2.8 mm/month were recorded. The smallest changes occur in the northern part of the altar and are about –0.25 mm/month. Smaller value of the rate of subsidence is noted for the floor. The subsidence rate is the smallest in the middle of the chamber, and the biggest in the southern zone of the verge of the chapel's floor opposite to the altar. Vertical speed of convergence in the chamber ranges between –0.4 and 1.2 mm/month achieves the highest values in the middle of the chamber and diminishes towards the walls of the excavation, which allows favourable assessment of the stability in the whole chamber. The registered horizontal convergence speeds range between –0.3 and – 0.8 mm/month. Despite relative and absolute movements of the whole region, the outline of the chamber shows great stability [7].

2.1 Present Measurements

In the framework of integrated measurements carried out within this study, the technology of laser scanning was applied, due to the fact that, in a short time, allows us to obtain incomparably higher amount of information on the studied object in a numerical version, combined with traditional measurement methods. For the measurement laser scanner FARO FOCUS 3D and reflectorless total station TRIMBLE 5503 DR were applied, and the measurements were carried out in two series in 2013 [8] and in 2016. Every time the co-ordinates of the scanner in the mine system were marked based on the angle-linear measurement, in reference to the same two points of the mine control line; located in the neighbouring gallery cavities. Scans were made in several places and spatially bound based on reference shields used to give geo-reference to the point model. Co-ordinates XYH of the shields in the mine system were marked using the method of spatial indentations, with the application of reflectorless total station Trimble 5503 DR in two series, to increase the accuracy of marking their position.

2.1.1 Measuring the Subsidence Value of the Chapel's Roof

Based on the selected points initiated from the cloud, with the kriging method in Surfer program, two roof models were generated. The difference between roof models makes the value of subsidence over the period of three years between the carried out measurements was measured and presented in Figure 5. Data referring to the roof subsidence in the period of 2013–2016, despite relatively short time interval, should be recognized as very accurate (from 2 mm up to 1cm for unfavorable geometric conditions).

Because of the technology that was applied to obtain them. Generally, the roof of the chamber subsides, and these values uniformly grow in the function of time. In south-eastern part of the chamber there is the area, where uplift was presented (marked yellow) (Fig. 5), however this phenomenon is connected with the occurrence of wooden cribs in this part of the chamber.

2.1.2 Measuring the Volume Change in the Chapel in 2013–2016

Using the model of the point cloud in St. Kinga's chamber, generated based on data obtained from laser scanning, made in 2013 and the measurement in 2016, the value of the volume convergence occurring over the 3 years, was measured. To achieve this, the volume of chamber in 2013 and 2016 was calculated in program CloudCompare. The points from both models were given coordinates in local system KS Bochnia (Bochnia Salt Mine), which allowed the isolation from the modes, sets of points covering the same ranges of data. Then, based on selected data, using the option of calculating volume, the values of chamber volume from both measurements were calculated (Table 2).

Table 2. The volume and volume convergence of St. Kinga Chapel

Volume in 2013 [m^3]	Volume in 2016 [m^3]	Volume Convergence [m^3]	Volume Convergence ‰	Volume Convergence ‰/year
1051.87	1044.12	7.75	7.4	2.5

Over the period of 2013 and 2016 the volume of chamber was decreased by 7.75 m^3, which corresponds the value of volume convergence equalling 2.5 ‰/year, what was done for the first time. Comparison of the used methods (linear convergence and volume convergence obtained using TLS) could not be done in a simply way because of the different character of the obtained data. The accuracy of obtained liner convergence is 2 mm. The accuracy of volume convergence is compilation of several factors such as distance from the object, the angle of laser beam incidence and surface type (roughness and colour) and oscillates between 2 mm up to 1 cm for inconvenient geometrical conditions.

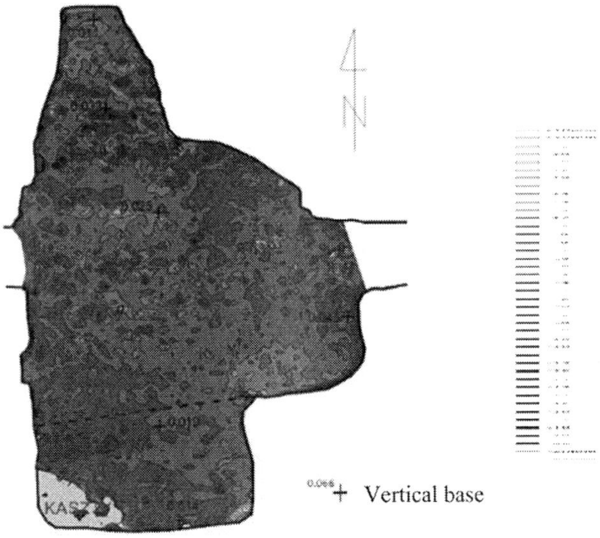

Fig. 5. Subsidence of the roof in St. Kinga Chapel in the period of 2013–2016 (subsidence values given in [mm])

3 Conclusions

Laser scanning technology is very useful in the inventory and visualization of historic salt chambers [9, 10]. The implementation of these measurements does not require good light and is carried out automatically. At great resolution of the measured points, scanning technology allows making a quasi-continuous model of the object. Numerical spatial model, obtained as a result of the point clouds obtained from the measurement, gives many extra possibilities in the analyses of the studied object.

The phenomenon of convergence in Salt Mine Bochnia is very small, thus to assess its scale, one has to apply the measurement method guaranteeing high accuracy. The accuracy of measuring the volume of objects with laser scanning, the biggest (fundamental) influence among all the components of situation error in the cloud point is in distance measurement. According to the producer's declaration, assuming distance measurement error as ± 2mm multiplied by the double floor area (625 m^2) and chamber volume measurement error is ± 1.25 m^3. The difference in volume determined with two measurements is 7.75 m^3, which, in case of volume error measurement ± 1.25 m^3, should be considered significant value from the statistical point of view.

In case of halite excavations, the method of laser scanning fulfils the accuracy requirements connected with the definition of anomaly zones and the change in the geometry of selected cavern elements, e.g., the roof.

Laser scanning is a useful method of observing the phenomenon of convergence, due to the fact of holistic (3D) observations of the character of this phenomenon. One should remember that it has certain disadvantages compared to classical linear convergence measurements. First of all, this technology is significantly more expensive and more time-consuming and complicated. Despite this, it provides incomparably larger amount of information on studied objects, not only in terms of convergence (linear in any direction or volume convergence), but also other parameters, such as e.g., geometry, volume, etc. This is the only method allowing, based on the comparison of two models obtained measurements from different periods of time, defining deformations of any fragment of object, or the whole object in a comprehensive way. Thus, in a direct manner (by contrast to classical method) it allows measuring the value of volume convergence.

In the measurements carried out in one, three-year period, measurable significant results were found, showing the uniform movement of the rock mass in the zone of the localization of the void. The phenomenon of convergence has occurred since the formation of the chamber and will only stop when the excavation is completely converged, thus the use of complex technology (laser scanning) seems to be fully justified in this case.

Acknowledgements

Article was financed from the grant Badania Statutowe 11.11.150.195.

References

1. J. Freudenheim, T. Migdas, *Restrukturyzacja Kopalni Soli "Bochnia"*, Warsztaty z cyklu: Zagrożenia naturalne w górnictwie [Natural threat in mining], Mat. Symp. s. 597–603 (2006)
2. G. Kortas, T. Pałys, R. Dudek, *Convergence measurement in the Blessed Kinga Chapel in the Salt mine of Bochnia* (1993
3. A. Maj, *Opracowanie modelu konwergencji wyrobisk w podziemnych kopalniach* soli (2009)

4. G. Kortas, A. Maj, *The assessment of the efficiency of filling the cavities in Salt Mine Bochnia in the context of the protection of the surface and underground historic objects* (2013)
5. M. Cardenas, C. LeCompte and G. Cua, *Estimation of Permeability of Disturbed Halite Using Inverse Modeling*, Groundwater, Vol. 37, Issue 4, pp. 539–545, (1999)
6. G. Wang, K. Guo, M. Christianson, H. Konietzky, *Deformation characteristics of rock salt with mudstone interbeds surrounding gas and oil storage cavern,* [In:] International Journal of Rock Mechanics and Mining Sciences, Vol. 48, Issue 6, pp. 871–877, (2011)
7. J. Chmura, T. Migdas, *The assessment of the stability of the tourist route and sanatory chambers in the salt mine of Bochnia, with the project of the measurement of new tourist objects* (2005)
8. J. Maciaszek, D. Madusiok, W. Matwij, *The comparison of the results of laser scanning and classical surveying methods of measuring the area and volume in the chapel of St. Kinga in the salt mine of Bochni*a (2013)
9. R. Kukutsch, V. Kajzar, P. Konicek, P. Waclawik, J. Ptacek, *Possibility of convergence measurement of gates in coal mining using terrestrial 3D laser scanner*, In Journal of Sustainable Mining, Vol. 14, Issue 1, pp. 30–37, (2015)
10. T. Lipecki, W. Jaśkowski, W. Gruszczyński, K. Matwij, W. Matwij, P. Ulmaniec *Inventory of the geometric condition of inanimate nature reserve Crystal Caves in „Wieliczka" Salt Mine*, Acta Geodaetica et Geophysica, Vol. 51, Issue 2, pp. 257–272, (2016)

Research for Forming up the Controlling Diagram Utilizing the Connection of a LV Resistor on Voltage Transformer's Open-Triagle Coil to Reduce over Voltage Caused by Earth Fault in 6 kV Grid of QuangNinh Underground Mines

Ho Viet Bun[1,*], *Le* Xuan Thanh[1]

[1] Department of Electrification, Electromechanic Faculty, Hanoi University of Mining and Geology, Viet Nam

Abstract. Single phase earth-fault in MV grids usually causes overvoltage that harm to human being and electric equipment. If the magnitude of over voltage is so great, many grids' eco-technical parameters will be affected. The paper analyzes all possible consequences of over voltage occurred in 6 kV grid of QuangNinh underground mines. Base on the analysis, a controlling diagram utilizing the connection of a LV resistor on voltage transformer's open-triangle coil to reduce over voltage is recommended. The simulation results of the diagram are used to prove the effectiveness of solution: the over voltage magnitude is only in the range of (2,1–2,4).Uf. Other advantage that solution brings to relay system will be pointed out. **Key word** – earth fault, over voltage, open-triangle coil, 6 kV grid.

1 Introduction

Earth faults are very popular phenomenon in 6 kV grids of QuangNinh underground mines. The ratio of this fault is about 55 to 73% [2] of the total faults in power system. It usually occurs with arc, especially in storm season. Some cases, earth fault happened with continuing arcing over voltage which is the most dangerous phenomenon (in the field of appearing and distinguishing the arc), [3–5]. The first damage of arcing over voltage is the insulation destroying.

This kind of over voltage depends on some main matters: quantity and property of the circuit, recovery voltage of grid right after the fault. Moreover, the transient of arcing over voltage depends also on contacting resistor, the contacting momentary and the location of earthing. According to researches presented in [3–5], the value of voltage on unfaulted phases is vary from 2,1 Uphase to 4, 6 Uphase.

* Corresponding author: hovietbun@gmail.com

Base on the above analysis, in 6 kV grid of QuangNinh underground mines, it is necessary to form up the model for determining correctly the earthing moment in order to connect a LV resistor to reduce arcing over voltage.

2 The application of Microprocessor on determining the earthing momentary in order to connect a LV resistor to reduce arcing over voltage

2.1 Diagram of connecting LV resistor to reduce arcing over voltage

There are many solutions to reduce over voltage; however, because of strict safety regulation in underground mines, connecting a LV resistor is optimal and effective solution that does not change the property of neutral point [1]. To implement the connection, a special transformer is arranged on 6 kV bus bar, the resistor is connected to its open-triangle coil. The connection diagram is shown in Figure 1. In this diagram the special transformer is connected to the bus bar through circuit breaker B4, LV resistor RH is controlled by microprocessor corresponding to the value of 3U0.

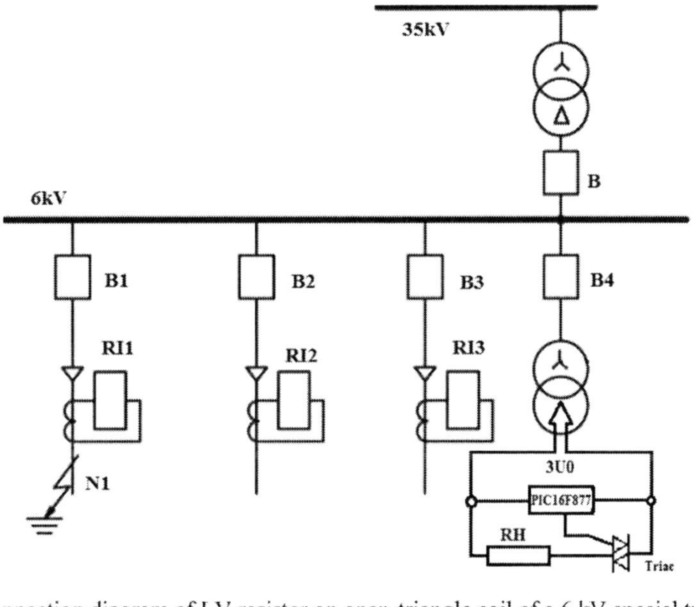

Fig. 1. The connection diagram of LV resistor on open-triangle coil of a 6 kV special transformer

2.2 Block diagram of controlling method utilizing microprocessor

In Figure 1, microprocessor PIC16F877 is used to generate controlling signal for connecting or disconnecting RH on the LV side of transformer. The requirements for over voltage relay tripping time must be lower than 5 ms [2]. Block diagram expressing the fault determining process and controlling operation of Triac KS200A is shown in Figure 2 and 3.

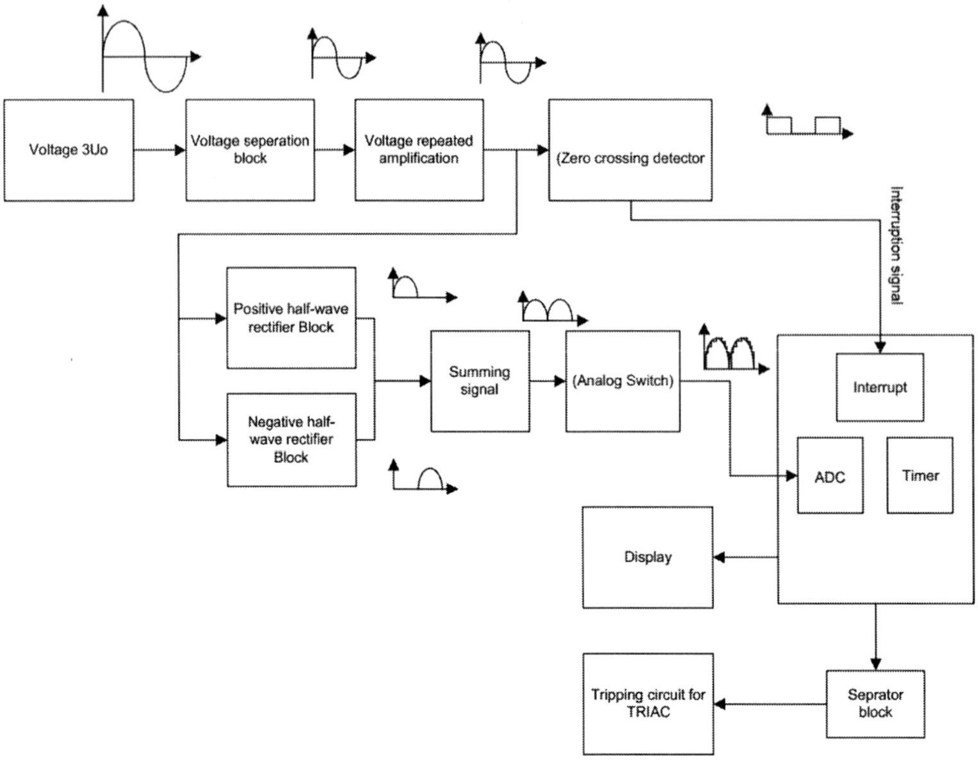

Fig. 2. Block diagram of controlling process

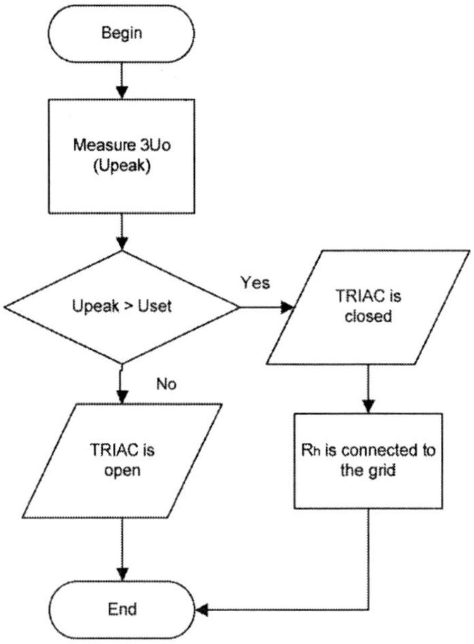

Fig. 3. Mathematic diagram of PIC 16F877 controlling process

2.3 Calculation, simulation and results

Using Proteus software to form up the controlling process, the diagram is shown in Figure 4.

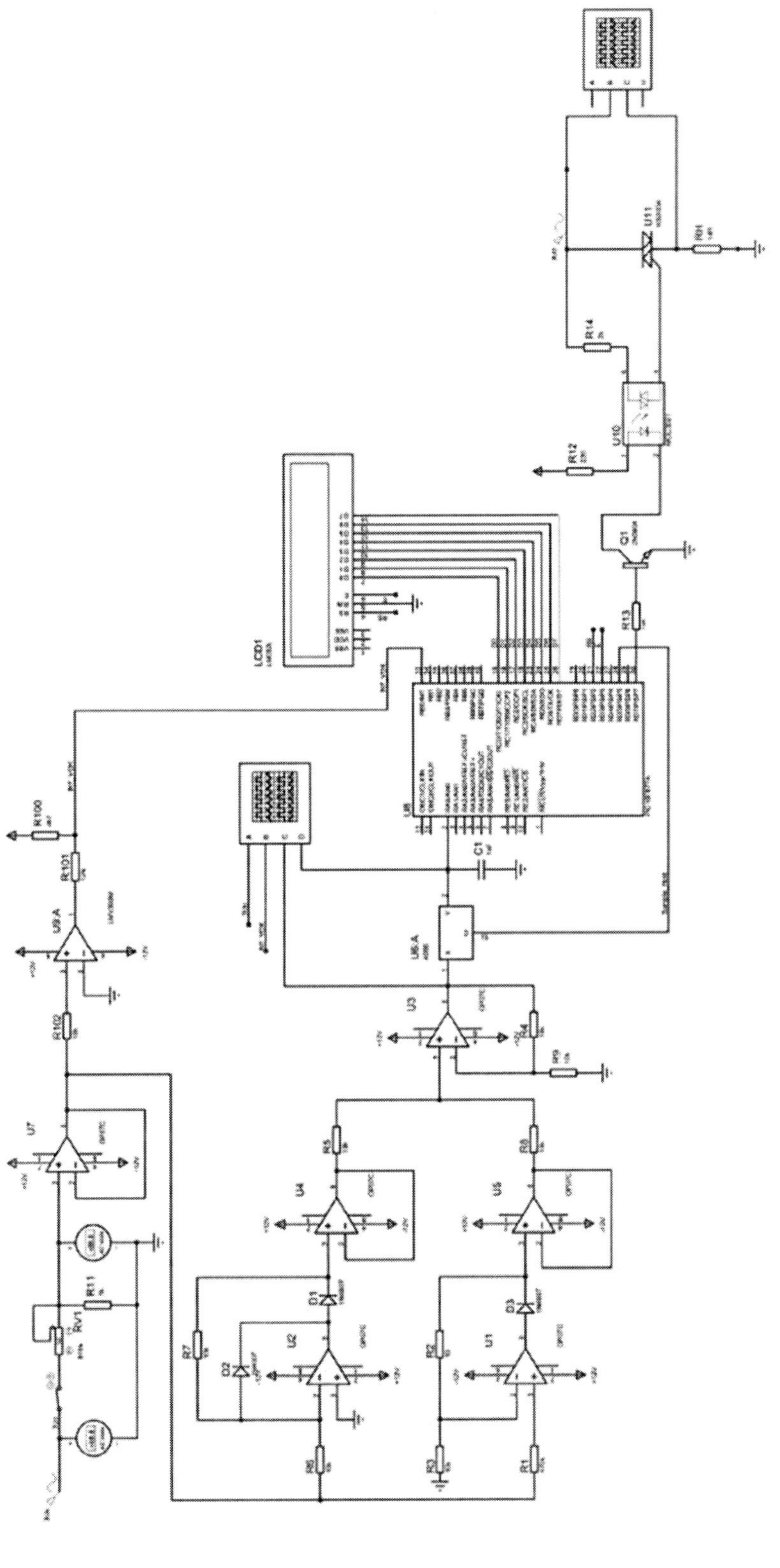

Fig. 4. The diagram of controlling circuit that act upon LV resistor RH

 https://doi.org/10.1051/e3sconf/20183505001

The simulation results are shown in Figure 5 to 9.

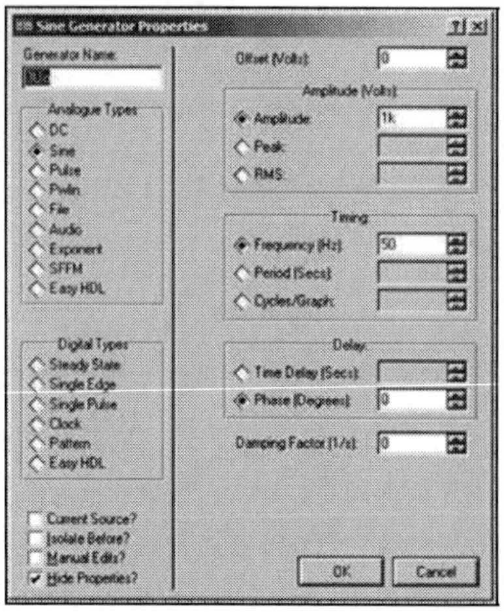

Fig. 5. Set the 3U0 threshold as 1000 V

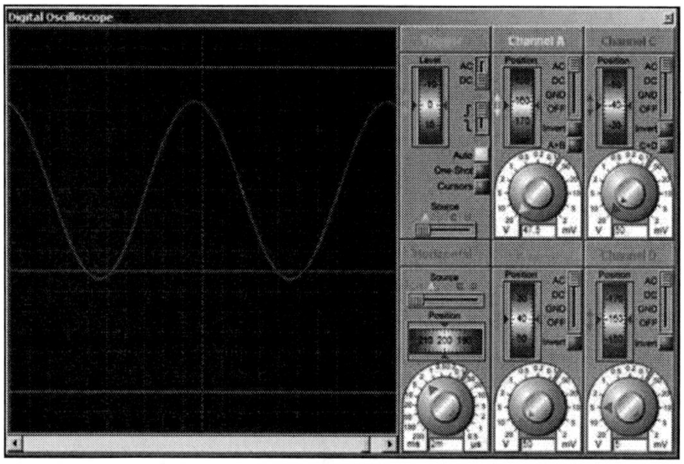

Fig. 6. 3U0 = 9 V (cyan) and voltage across LV resistor (violet)

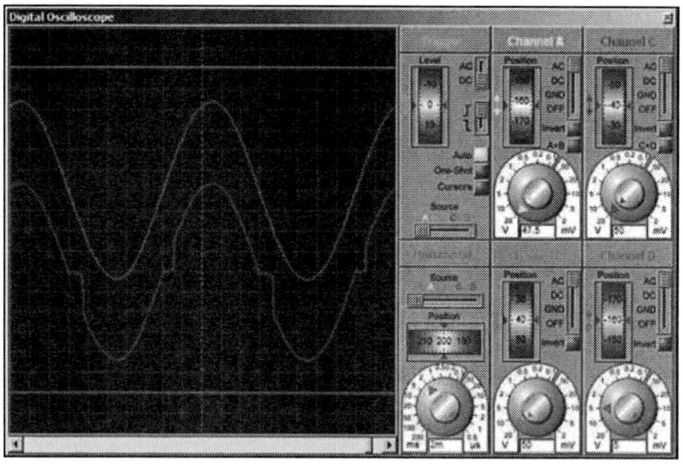

Fig. 7. Value of 3U0 = 1 kV (cyan) and voltage across LV resistor (violet)

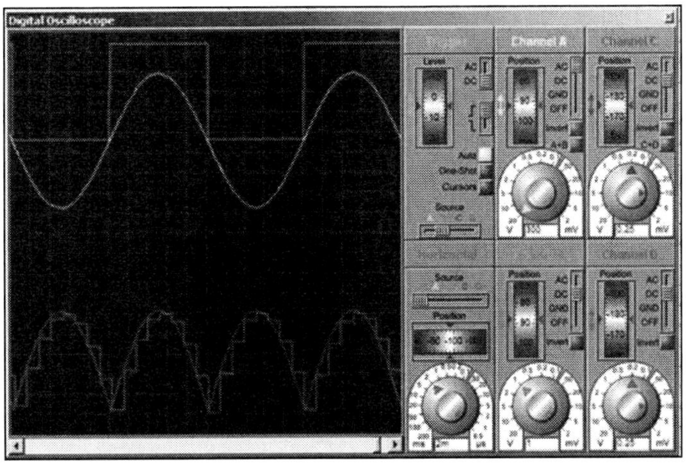

Fig. 8. Value of 3U0 = 1 kV (cyan) and voltage across LV resistor (violet)

Base on datasheet and the simulation results, the response time of OPAM is 1,3 μs, opto MOC3021 time (1/200 kHz) is 5 μs, switching time 4066 is (1/40 MHz) , sampling time and calculation procedure 500 μs and Triac KS200A response is 1/100 kHz = 10 μs. Therefore total response time of circuit is ≤ 1 ms.

The proof of using LV resistor in earthing process is shown in Figure 9.

a)

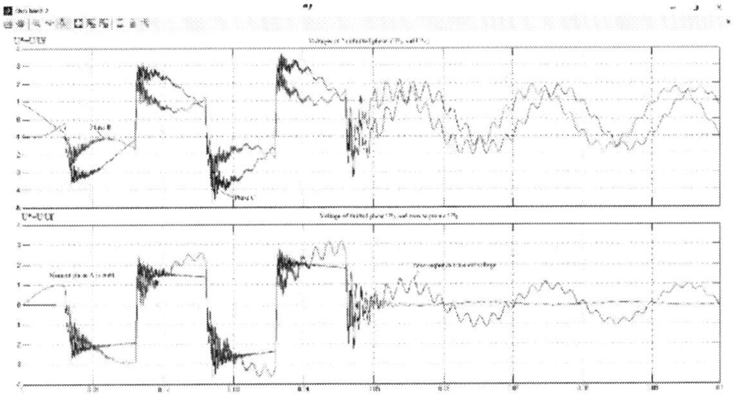

b)

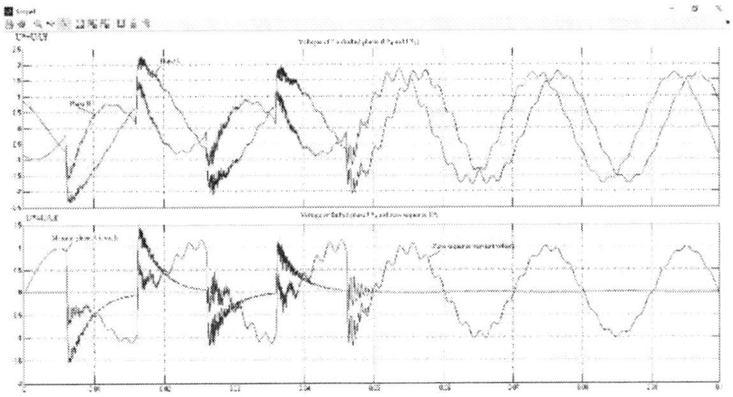

Fig. 9. Waveform of voltage transient when there is earth fault; a) without RH b) with RH is connected right the moment of earthling

3 Conclusion

The paper presented a very suitable method to reduce the over voltage when there is an earth fault in mining grid. The suggestion methodology using a LV resistor to connected to 6 kV bus bar through a special transformer does not violate the hazard regulation in underground mining, but it reduces significantly the magnitude the over voltage (as shown in Fig. 9). This over voltage value is only in the range of (2,1–2,4).Uf. The range is falled within the accepted one rated in National regulation applied for electric equipment [1].

References

1. Ministry of Industry and Trade, *National technical code of under ground mining*, QCVN 01:2011/BCT, (2011)
2. Hồ Việt Bun, *Reasearch and evaluate overvoltage transience caused by single phase-ground failures in Cam Pha-Quang Ninh underground mines' 6 kV grid*, International conferrence of Advanced in Mining and Tunelling 21–22 October 2014, Vung Tau, Viet Nam, pp. 409–411, (2014)
3. Б.И. Базылев, А.М. Брянцев, А.Г. Долгополов и др, *Дугогасящие реакторы с автоматической компенсацией емкостного тока замыкания на землю*, СПб.: Изд. пэипк, (1999)
4. Г.А. Евдокунин, С.С. Титенков, *Внутренние перенапряжения в сетях 6-35кВ*, СПб: Издательство Терция, (2004)
5. Р.А. Вайнштейн, Н.В. Коломиец, В.В. Шестакова, *Режимы заземления нейтрали в электрических системах*, Томск: Изд-во ТПУ, (2006)

Analysis the Transient Process of Wind Power Resources when there are Voltage Sags in Distribution Grid

Do Nhu Y[1,*]

[1] Hanoi University of Mining and Geology, Department of Electrification, Electromechanic Faculty, Viet Nam

Abstract. Vietnam has many advantages of wind power resources. Time by time there are more and more capacity as well as number of wind power project in Vietnam. Corresponding to the increase of wind power emitted into national grid, It is necessary to research and analyze in order to ensure the safety and reliability of win power connection. In national distribution grid, voltage sag occurs regularly, it can strongly influence on the operation of wind power. The most serious consequence is the disconnection. The paper presents the analysis of distribution grid's transient process when voltage is sagged. Base on the analysis, the solutions will be recommended to improve the reliability and effective operation of wind power resources. **Key word** – wind power, transition process, reliability.

1 Introduction

Wind power is renewable and green energy in Vietnam. According to accounting data, Vietnam has rich wind power capacity with 8,6% area that is applicable to build up big wind power stations, among that 41% area in urban area is suitable to build small stations. Total capacity of wind power is about 513.360 MW [1]

Voltage sag is a popular phenomenon in power system that makes wind power parameters changed. Therefore analyzing the transient of wind power sources at voltage sag moments, in order to recommend suitable operation for power system containing wind power, plays an important meaning.

2 Research method

Nowadays, most of wind power stations utilize Double-feed-Induction-Generator (DFIG) because they have the prior advantages when stator is directly connected to the grid whereas rotor is connected to the grid through controllable power electronic devices. Because the controller in rotor, therefore its power is much smaller than the power of generator; power flows directly from stator to grid, especially with big capacity generator. In the limit range

* Corresponding author: donhuy@humg.edu.vn; donhuy.humg@gmail.com

of speed, the power of converter is only 30% grid transmitting power. The structure of wind power generator DFIG is shown in Figure 1 [3].

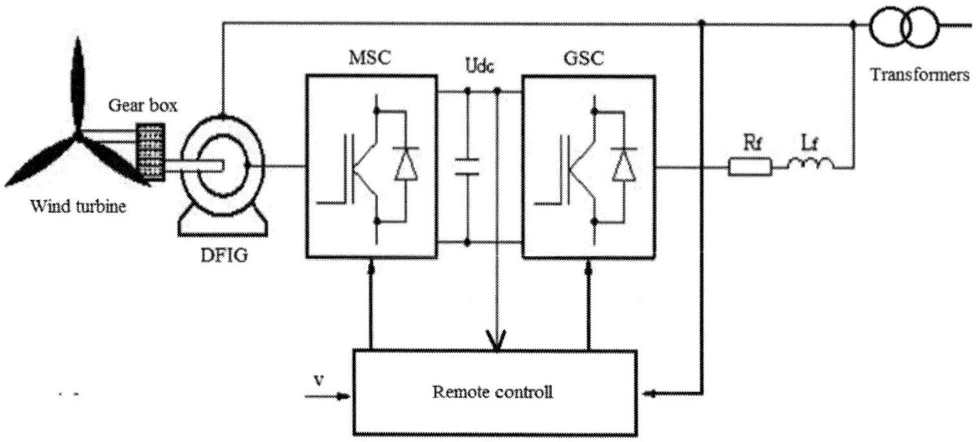

Fig. 1. Structure of wind power DFIG

In Figure 1 (MSC) is the converter on generator side used to control generator's speed in order to get maximum power. Input signals of speed controller is deducted from wind velocity. Output signals of speed controller is the ref signal for power loop controller. Grid source converter (GCS) controls DC voltage (U_{dc}). It is also used to control reactive power Q.

To analyze the transient occurred in wind power when there is voltage sag, power system is simulated on Matlab-Simulink. Power generator DFIG is simulated on dq axis as Figure 2.

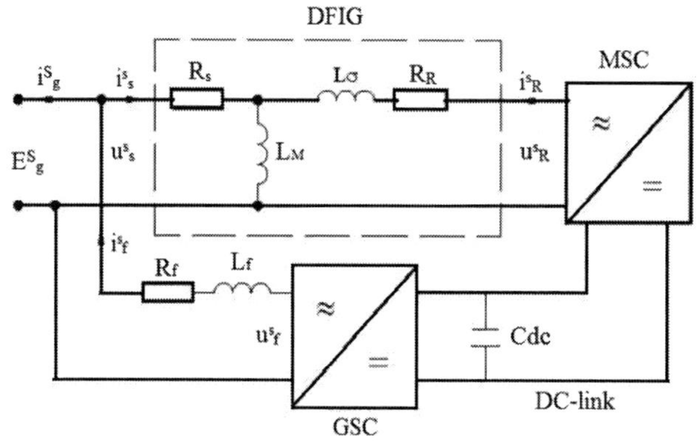

Fig. 2. Simulation of DFIG generator on space vectors

Replacing equivalent model of DIFG is expressed in stator's space vectors as shown in Figure 3.

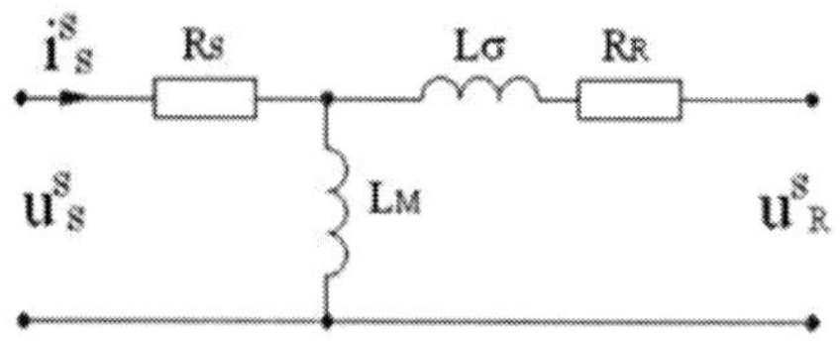

Fig. 3. Equivalent diagram of DFIG on space vector

Where: R_s – stator winding resistor; L_M – Magnetic induction; L_σ – Induction of statorand rotorconverted into stator side; R_R – Rotor resistor converted to stator. Letter s means space vectors ref on statorof DFIG.

According to [2], Kirchhoff equations for loop circuits:

$$\begin{aligned} u_S^s &= R_S i_S^s + \frac{d\Psi_S^s}{dt} \\ u_R^s &= R_R i_R^s + \frac{d\Psi_R^s}{dt} \end{aligned} \tag{1}$$

The fields of stator, rotor, and electromagnetic torque are computed by the following equations:

$$\begin{aligned} \Psi_S^s &= L_M \left(i_S^s + i_R^s\right) \\ \Psi_R^s &= L_\sigma i_R^s + L_M \left(i_S^s + i_R^s\right) \\ T_e &= 3 Z_p I_m \left[\Psi_s i_R^*\right] \end{aligned} \tag{2}$$

Where: L_M – Magnetizing induction; L_σ – Convert induction

$$\begin{aligned} L_S &= g L_{sl} + \lambda^2 L_{rl} \\ \lambda &= \frac{L_{sl} + L_M}{L_M} \end{aligned} \tag{3}$$

Where: λ – Convert factor; $L_{sl}; L_{rl}$ – Induction ofstator, rotor; Z_p – generator pole pair.

Mechanical equation of system is expressed by:

$$\frac{J}{Z_p}\frac{d\omega_r}{dt} = T_e - T_s \tag{4}$$

Where: J – Inertia torque; ω_r – rotor rotating velocity; T_e – electromagnetic torque; T – mechanic torque on rotor's shaft.

Maths expression of filter on grid side includes R_f and L_f, replace models is shown in Figure 4.

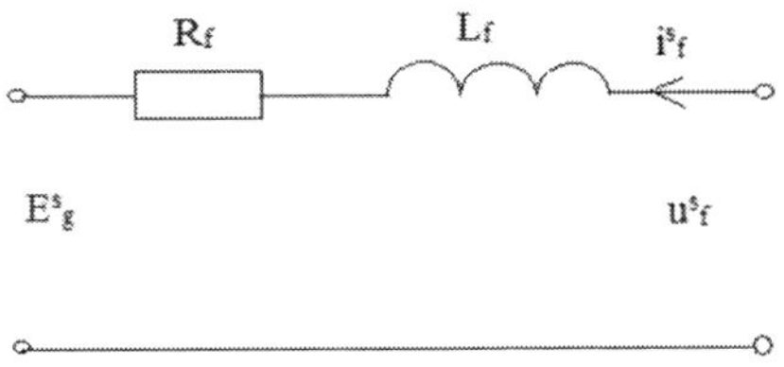

Fig. 4. Model of filter on grid side in space vector

According to [2], Kirchhoff equation on dq axises:

$$E_g^s = -R_f i_f^s - L_f \frac{di_f^s}{dt} + u_f^s \tag{5}$$

Maths expression of DC – Link, replaced model is shown in Figure 5.

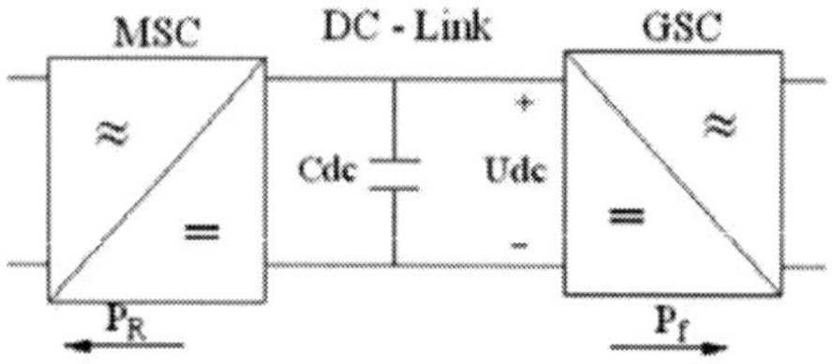

Fig. 5. Model of DC – Link

Energy W_{dc} is deducted from capacitor C and is computed by the following equations:

$$\begin{aligned} W_{dc} &= \frac{1}{2} C_{dc} U_{dc}^2 \\ \frac{dW_{dc}}{dt} &= \frac{1}{2} C_{dc} \frac{d}{dt} u_{dc}^2 = -P_f - P_r \\ C_{dc} u_{dc} \frac{d}{dt} u_{dc} &= -P_f - P_r \end{aligned} \tag{7}$$

Output power of wind generator is calculated by:

$$P_m = C_p(\lambda, \beta) \frac{\rho A}{2} v^3 \tag{8}$$

Where: P_m – Output power of wind turbine; $C_p(\lambda, \beta)$ – power conversion factor (ratio of wing velocity λ and wing angle β); A –cross section of; ρ – air density; v – wind velocity.

Power conversion factor C_p is determined as:

$$\begin{aligned} C_p(\lambda, \beta) &= 0,5176 \left(\frac{116}{\lambda} - 0,4\beta - 5 \right) e^{\frac{21}{\lambda_i}} + 0,0068\lambda \\ \frac{1}{\lambda_i} &= \frac{1}{\lambda + 0,08\beta} - \frac{0,035}{1 + \beta^3} \end{aligned} \tag{9}$$

Ratio of wing velocity and wind velocity is determined by:

$$\lambda = \frac{R\omega}{v} \tag{10}$$

Where: ω – Turbine rotating velocity; R – Turbine radious

On other hand wind turbine can be operated by different controlling rules base on wind velocity. Figure 6 is the expression of relation between wind turbine's power and wind velocity.

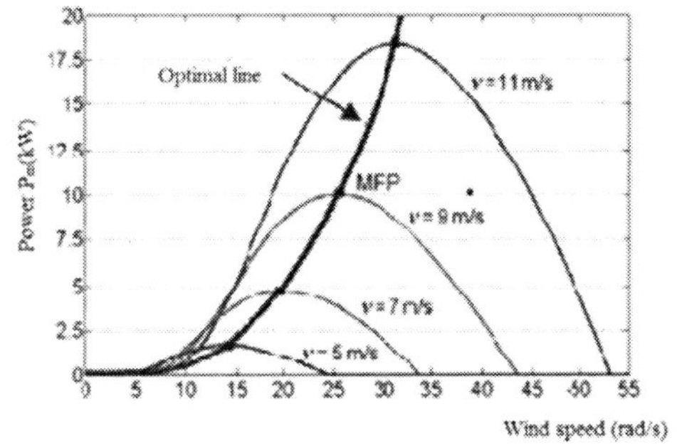

Fig. 6. The relation of wind turbine's power and wind velocity

Utilizing the simulation on Matlab-Simulink, above mentioned equations are expressed in Figure 7 [2].

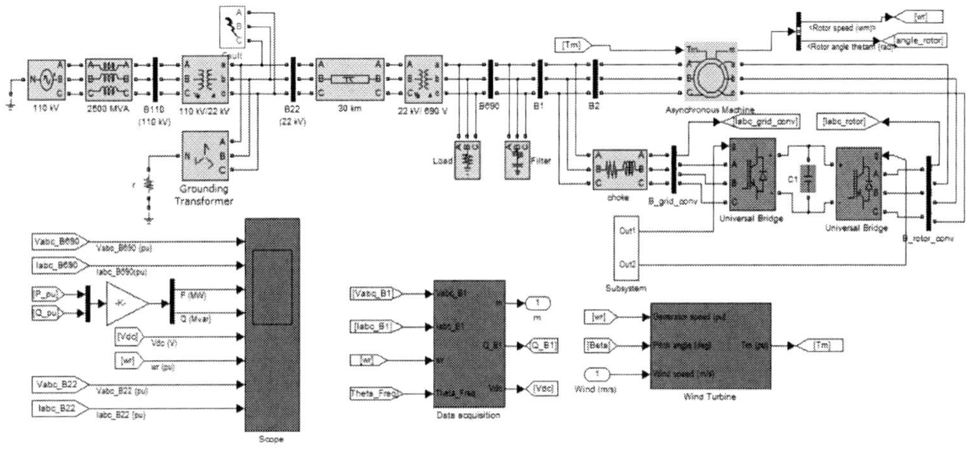

Fig. 7. Simulation diagram

3 Results

Simulation results are implemented on system with the following parameter: output voltage 690 V, step-up transformer is 690 V/22 kV, the wind power system is connected with power system on 22 kV bus bar. In the operation of the system, there is a voltage sag, consequently voltage on 22 kV bus bar is reduced (as shown in Figure 8 The transient process of output voltage of wind power is shown in Figure 9, the current changing is presented in Figure 10 and the transient velocity of motors is on Figure 11.

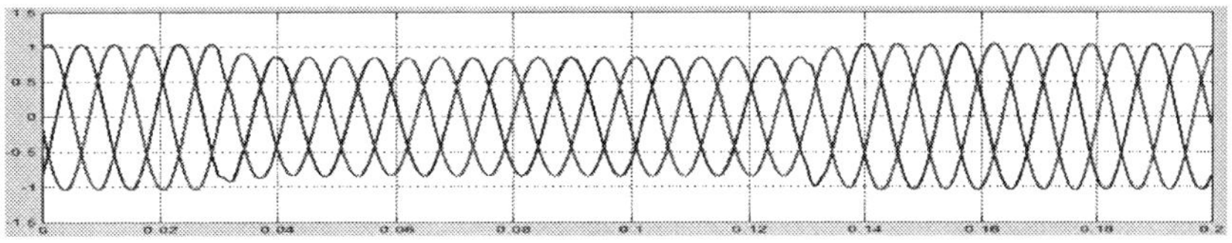

Fig. 8. Voltage on 22 kV bus bar (B22)

https://doi.org/10.1051/e3sconf/20183505002

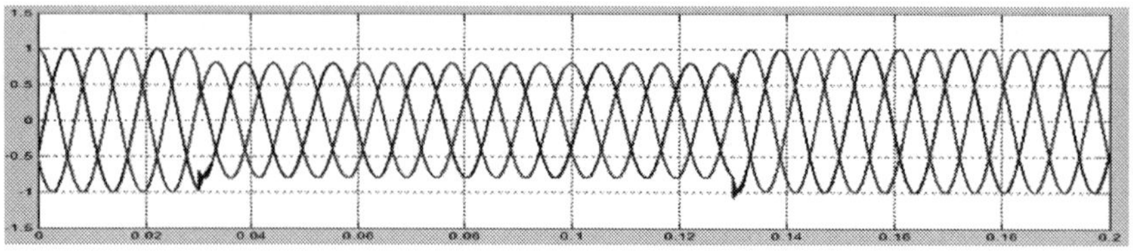

Fig. 9. Response voltage of wind power

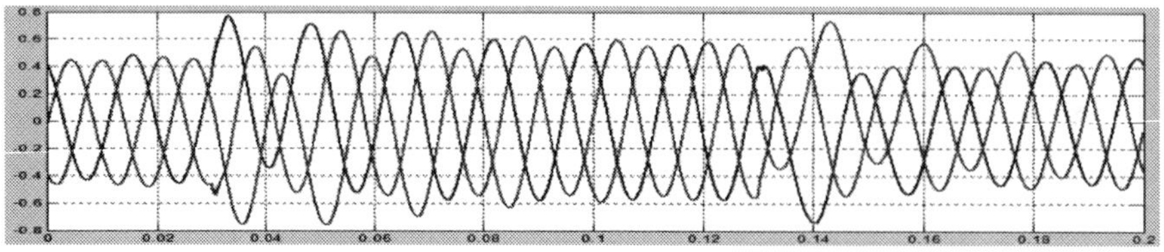

Fig. 10. Response current of wind power

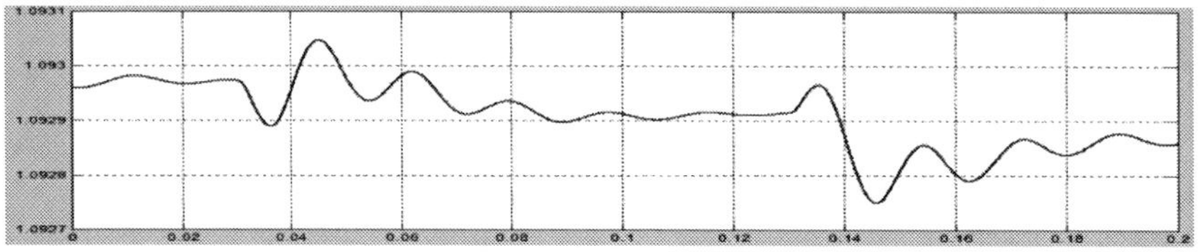

Fig. 11. The transient of generator velocity

The above simulations show that, when there is voltage sag on distribution grid (22 kV bus bar) output voltage of generator is also down (Fig. 9), consequently current of generator is increased (Fig. 10). If the process is long lasting can cause over load or destroy the generator. At the moment of voltage recovery, wind power voltage is also recovered, but it is transient much over rated voltage. This cause to the transient of generator velocity (Fig. 11).

4 Conclusions

Wind power is increased rapidly in both quantities and capacity. The Power system containing wind power resources needs a stable operation because its voltage sag could decrease the output voltage of generator. Consequently, generator's current will increase. The long lasting increasing of current could lead to over load or destroy the generator.Furthermore, at the moment of votage recovery on distribution grids, voltage is also fluctuated rapidly. The simulation results express the transient of wind power energy at the moment of voltage sag in distribution grids. The results will be used to recommended solutions for improving the stabilities, effectiveness of wind power.

References

1. Do Nhu Y, Le Xuan Thanh, *Improvements to the opreation of wind power genertors in Vietnam*, Journal of Vietnam Environment (ISSN-2193-6471), Vol. 8, pp. 173–178, (2016)
2. Do Nhu Y, *Research using SVC to Improve the Operation of Wind Power generators in Vietnam*, European of technology Journal and design, Vol. 14, pp. 159–163, (2016)
3. Lai Khac Lai, Duong Quoc Hung, Tran Thi Thanh Hai, *Design the grid synchronizing controller for wind power using Double-Feed Induction Generator DIFG*, Journal of Science and Technology, Vol. 10, pp. 219–225, (2010)

Experience in the Education of Engineers from Vietnam in the Faculty of Mining and Geoengineering AGH

Marek Cała[1], and *Marek* Borowski[1, *]

[1] AGH University Science and Technology, Faculty of Mining and Geoengineering, A. Mickiewicza Av. 30, 30-059 Krakow, Poland

Abstract. The AGH University of Science and Technology collaborates closely with other universities, economic units, governmental and local administrative bodies. International cooperation plays a very important role in the academic research. The AGH University of Science and Technology has signed many collaboration agreements. They aim at multidimensional cooperation in the fields of education and academic research. AGH UST has always focused on collaboration with business and industry. In recent years, the global economy is undergoing massive transformations, what creates new challenges to companies and educational institutions that cater to the needs of industry. The expansion of business enterprises is largely dependent on their employees' expertise, skills and levels of competence. Certified engineers are provided by universities. Therefore, the qualifications of the graduates are determined by the curriculum and teaching methods, as well as the available educational and research facilities. Of equal importance is the qualified academic staff. Human activities in the field of engineering require finding solutions to problems of various nature and magnitude. An engineer's work consists in the design, construction, modification and maintenance of useful devices, processes and systems, using scientific and technical knowledge. In order to design complex engineering solutions, an engineer uses his imagination, experience, analytical skills, logical reasoning and makes conscious use of his knowledge. At the Faculty of Mining and Geoengineering of the AGH University of Science and Technology in Cracow, 15 engineers from Vietnam are studying Mining and Geology at the second-cycle studies (specialization: mine ventilation). The solutions proposed in the field of the engineers' education guarantee that foreign students gain both engineering knowledge and problem-solving skills. Therefore, the study programme was complemented by a series of practical aspects. **Key words** – engineers' education, international cooperation.

1 Introduction

International cooperation is regarded as one of the key elements in the educational and scientific development of the AGH University of Science and Technology and a way to boost

* Corresponding author: marek.borowski@agh.edu.pl

its international prestige. As a university of applied knowledge, the AGH UST has always been open to cooperation with various industries.

The university concluded more than 400 agreements about close cooperation with foreign universities in Europe, the Americas and Asia. In those agreements, the parties promise to join their efforts in many fields to promote education and scientific research. In addition, the AGH UST participates in numerous educational initiatives, including the Erasmus LLP programme. As part of its cooperation with other educational units, the university undertakes various organizational challenges to establish research centres, technology development centres and intercollegiate laboratories, to name just a few. The AGH UST strives to develop intercollegiate courses, create new interdisciplinary faculties and move studying to an increasingly international level. These activities are consistent with the university's priorities, which include making university education international, a challenge faced by potential partners worldwide, especially in the developing countries.

In particular, the longest-standing partners of the AGH UST are Asian countries, including Vietnam, with which the university collaborates on educating geologists, mining engineers and specialists in other fields.

In 2012, Professor Tadeusz Słomka, the rector of the AGH UST, and Le Minh Chuan, the President of VINACOMIN, signed a general agreement. One of its objectives was to start masters courses for a group of Vietnamese students. It was not until September 2014, however, that a delegation of the VINACOMIN Group paid a working visit at the Faculty of Mining and Geoengineering. As a result, a detailed agreement was concluded, and 15 engineers from Vietnam started a second-cycle master's programme in 2015. The agreement specified details concerning the financing, accommodation and other issues related to the stay of Vietnamese students in Poland. Based on the agreement, a detailed syllabus of the specialization was elaborated and the relevant legal proceedings were initiated to register the new unit. It should be emphasised that a special curriculum was designed to meet the specific requirements of the VINACOMIN Group. In the curriculum, emphasis was placed on gaining relevant knowledge and practical skills in the fields of mine ventilation systems and preventing natural hazards, especially endogenous fire hazard.

2 Aspects of educating engineers

In recent years, the global economy is undergoing massive transformations, which creates new challenges to companies and educational institutions that cater to the needs of industry. The expansion of business enterprises is largely dependent on their employees' expertise, skills and levels of competence. Certified engineers are provided by universities. Therefore, the qualifications of the graduates are determined by the curriculum and teaching methods, as well as the available educational and research facilities. Of equal importance is the qualified academic staff. The paramount importance of those issues results from the challenges and transitions that are taking place worldwide. It is possible to identify the following key factors [2]:

- technological advances, meaning that without ongoing education of engineers the economy will lose its competitive edge. For this reason, it is vital to bear in mind that knowledge becomes outdated and new fields appear, which calls for an interdisciplinary approach. Consequently, engineers are expected to be proficient not only in their chosen specialization, but also in a range of other fields.
- globalization of markets, which removes limitations to communication and trade, allowing businesses to expand their activities. New markets and new ways to exchange information and acquire knowledge emerge. For this reason, it is necessary to have a knowledge of various cultures and to adjust the strategies of human resources management accordingly, as well as educational opportunities.

- the dissemination and recording of information in today's world takes place electronically. As a result, a huge amount of information has become easily accessible. In the myriad of information, however, it is extremely difficult to identify the useful bits and to verify, select and integrate information.
- threats to the natural environment caused by the exploitation of limited natural resources, pollution and the resulting degradation of the ecosystem. Therefore, human activity should employ methods that protect the environment.
- increasing awareness and social responsibility means that apart from improving the standards of living, technology has a detrimental effect on human health because of pollution, its impact is revealed in environmental disasters, it causes the depletion of non-renewable resources, or facilitates the spreading of diseases and epidemics.
- the pace of changes caused by technological advances accelerates the evolution of tastes and consumer needs. In fact, the requirements of the industry are shaped by consumers.

In order to meet all these challenges, universities continuously adjust their curricula, in some cases designing new educational programmes, like in the case of the cooperation of the AGH UST with the VINACOMIN Group. At the same time, university students are made aware of the necessity to adapt to the changing reality and to continue education throughout their careers. Learning this skill is one of the objectives of university education. Moreover, any company – regardless of the technological solutions it employs – is only as efficient and effective as its employees. It is believed that the crucial asset of every organization, which determines its strategic potential, are intellectual resources, i.e. knowledge, skills, experience, culture, hierarchy of values, and a will to cooperate and share knowledge. At our university, we strive to foster the awareness of the above in our students. It can generally be stated that intellectual capital is a vital factor that affects the ability to stay competitive, win over new customers, or increase one's share of the market. The key element of the intellectual capital is knowledge, which becomes useless unless it is continuously updated. What follows is that effective management of an organization that is expected to develop provides opportunities to revise, update and expand knowledge. Adopting such a policy by the VINACOMIN Group should be appreciated. In addition, the choice of the AGH UST as a partner who will equip Vietnamese engineers with necessary qualifications was in all respects the right decision. The faculty set the following objectives of students' education:

- to gain knowledge (terminology, procedures),
- to learn about basic principles, theories and generalizations,
- to learn how to apply the acquired knowledge to problem-solving,
- to develop specialist skills, decision-making,
- to develop creativity, to provide opportunities for personal development, to learn how to be articulate in communication,
- to acquire analytical skills, to learn how to assess ideas and express opinions,
- to understand the necessity of updating one's knowledge and to learn how to do it.

3 The competences of an engineer today

Human activities in the field of engineering require finding solutions to problems of various nature and magnitude. An engineer's work consists in the design, construction, modification and maintenance of useful devices, processes and systems, using scientific and technical knowledge. In order to design complex engineering solutions, an engineer uses his imagination, experience, analytical skills, and logical reasoning, and makes conscious use of his or her knowledge. An engineer's knowledge comprises theory and relevant information, regarded as appropriate qualifications [1]:

- factual knowledge – "know-that", or facts, principles and rules,

- procedural knowledge – "know-how", or algorithms and methods enabling us to apply theoretical knowledge,
- conceptual knowledge, gained through understanding, reflecting, and reflective learning
- metacognitive knowledge, i.e. knowledge about one's own cognitive processes and our ability to control them.

The key problem of knowledge transmission is to decide which type of knowledge should be emphasised for the education to be effective while maintaining a high quality of education.

The definitions of competence vary. The traditional notion of competence as a combination of knowledge, skills and capabilities seems to be outdated. Currently, a broader interpretation is preferred and the term is defined as 'all the qualities of an employee which are applied and developed during work, and are conducive to achieving results consistent with the strategic objectives of the organization' [4].

During the process of education, it is possible to develop: the competences of an employee or a group regarded as professional competences, or the competences of a team of employees, taking into account the needs of the company. As a rule, an engineer's curriculum focuses on professional competences, especially in the field of engineering. Nevertheless, the cooperation with the VINACOMIN Group made it necessary to modify the curriculum in such a way that the Vietnamese engineers were able to acquire the competences of a professional team.

Also, it should be observed that an engineer's professional profile is not limited to knowledge. Apart from providing would-be engineers with the necessary qualifications, the AGH UST makes efforts to teach its students to observe ethical standards and to develop a correct attitude towards their work, the community and the environment. These qualities are in a way hidden and are instilled from birth by educators and other people. It should be emphasised that engineers from Vietnam are characterized by high standards of behaviour in their approach to work, studying, collaborators and teachers. Figure 1 presents the qualities of a professional engineer, including knowledge and other aspects.

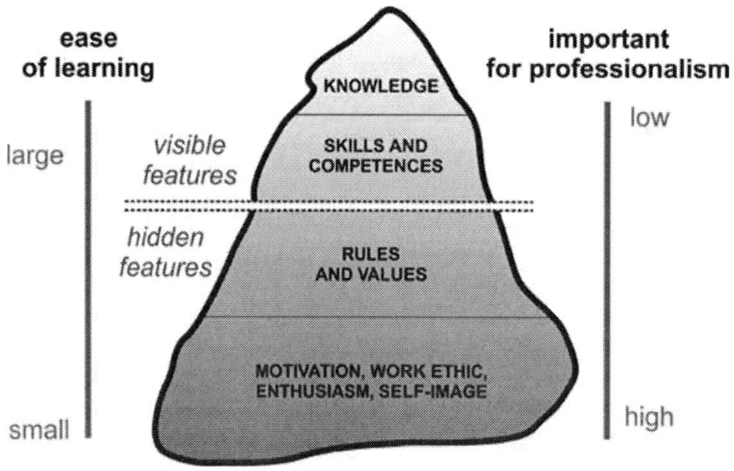

Fig. 1. The characteristics of a professional engineer [5]

Nevertheless, it is primarily the quality of education in the field of engineering that is decisive for an engineer's level of competence. Therefore, global organizations define educational standards for various fields, including of course engineering studies. The most widely known include ABET (Accreditation Board for Engineering and Technology), EUR-ACE (EURopean ACcredited Engineer project), IEA (International Engineering Alliance), or JABEE (Japan Accreditation Board for Engineering Education) [3].

In Poland, there are standards for every field of study that are supposed to be met by the authors of curricula. Educational standards are based on the Bologna process, which in turn was largely based on the EUR-ACE system. The standards specify the minimum number of semesters, classes, core curriculum classes or majors, and the number of ECTS points. However, universities and faculties have sufficient autonomy to design unique curricula for their master's courses. Such unique curriculum was elaborated in the case under discussion to accommodate the requirements communicated by the VINACOMIN Group.

4 The Vietnamese students' education

In accordance with the agreement concluded between the AGH UST and the VINACOMIN Group, the engineers from Vietnam were to be educated in Polish. Therefore, in Vietnam the Group selected a group of 15 engineers who took a course in the Polish language in the VINACOMIN High School of Business Management in October 2014. From April 2015 onwards, the Vietnamese group started an intensive Polish language course at the Department of Foreign Languages at the AGH UST in Cracow. The classes lasted until the end of July, eight hours a day and five days a week. During this time, integration meetings with Polish students were organized for the participants. Also, the Vietnamese were taken on trips enabling them to acquire and practice elements of the Polish language.

In September 2015, the would-be students passed an exam in the Polish language at the B1 level. On 1 October 2015, during the matriculation ceremony held at the Faculty of Mining and Geoengineering, the Vietnamese engineers took a solemn oath and became fully-fledged students of the AGH UST.

From that moment on, the four-semester education process started. The curriculum comprised three core curriculum subjects, such as statistics, modern physics, and Polish, 25 vocational subjects and two non-vocational (humanistic) ones. The total number of hours was 1440, including 660 hours of lectures, 450 hours of tutorials, 135 hours of laboratory work and 195 hours of project work. In consideration of the educational standards, a student's workload was calculated at 30 ECTS points per semester. This number corresponds to 750 hours in a semester, half of which were classes and the other half was a student's own work. Also, four weeks of vocational training in a hard coal mine were included in the schedule.

Fig. 2. Before entering the Borynia–Zofiówka – Jastrzębie, Part Borynia Coal Mine

In order to provide the students with the competences required by the VINACOMIN Group, they have been taken on a number of technical excursions during the four semesters. As for the remaining four semesters, other technical excursions are planned as well; they are expected to supplement the students' knowledge about the most up-to-date technological solutions and the standards and procedures applied in hard coal mining.

In July 2016, a group of 15 mining engineers from Vietnamese mines, studying the specialization Mine Ventilation Systems at the Faculty of Mining and Geoengineering in Cracow, completed a training period in the Borynia-Zofiówka-Jastrzębie, Part Borynia Coal Mine (Fig. 2).

In July 2017, Vietnamese students defended a master's diploma dissertation. Figure 3 presents students just after their defence of master theses, taken in the traditional place – on the stairs of the building A-0.

Fig. 3. Traditional picture taken after the master theses defence

5 Conclusions

In summary, it must be emphasised that the development of an organization depends to a great extent on the knowledge, skills and competences of the employees. Therefore, the AGH UST strives to design a curriculum that will equip the students with both general professional competence and specific competences needed by their company. In order to satisfy the requirements, a special curriculum was designed. Also, matching methods of teaching were selected and adequate educational facilities (specialist laboratories) were provided. Specially prepared academic staff employed at the faculty was appointed to the task. It must also be added that the Faculty of Mining and Geoengineering has the necessary organizational potential and can meet space requirements. Apart from mining, the faculty is also capable of educating Vietnamese students in the field of construction (with special emphasis on geo-mechanics) and environmental engineering.

The solutions implemented in order to rise to the challenge of educating Vietnamese students justify the conclusion that its objective has been attained. The students from Vietnam have completed the four semesters of a second-cycle master's programme with flying colours. The experience gathered so far raises hopes for the development of scientific and economic collaboration with Vietnamese companies that are interested in educating specialist staff. The activities pursued by the faculty are consistent with the priorities of the AGH UST, which include making university education international, a challenge faced by potential partners worldwide, especially in the developing countries.

References

1. L.W. Anderson, D.R. Krathwohl et al.: *A Taxonomy for Learning, Teaching, and Assessing*: A Revision of Bloom's Taxonomy of Educational Objectives, Longman, New York, (2000)
2. A. Kijewska, H. Przybyła: *Społeczno-techniczne aspekty kształcenia inżynierskiego na potrzeby górnictwa*, Wyd. Śląsk, Katowice. s. 139, (2013)
3. A. Kraśniewski: *Ocena sytuacji w szkolnictwie wyższym w Polsce w zakresie opracowywania efektów kształcenia związanych z wprowadzeniem Krajowej Struktury Kwalifikacji (Krajowych Ram Kwalifikacji).* Dokument elektroniczny. http://www.nauka.gov.pl/g2/oryginal/2013_05/ffb483b704698fe6493477de4dfcd4a3.pdf
4. T. Oleksyn: *Zarządzanie kompetencjami. Teoria i praktyka*, Kraków, Oficyna Ekonomiczna, s. 260, (2010)
5. Słowiński B., *Dobro w działaniach inżynierskich – teoria i praktyka.* Dokument elektroniczny. www.broneks.net/category/wyklady

Knowledge Exchange between Poland and Vietnam in Mining and Geology – the Status Quo and Future Development

Nga Nguyen[1,*], and *Nguyet* Pham[2]

[1] Mining Management Department, Faculty of Economic and Business Administration, Hanoi University of Mining and Geology, 10000 Hanoi, Vietnam

[2] Informatics for Economics Department, Faculty of Informatics, Hanoi University of Mining and Geology, 10000 Hanoi, Vietnam

Abstract. From the beginning of the 21st century, knowledge exchange between Poland and Vietnam in mining and geology has been focusing in technology, education and training. Since years, Polish academic and commercial partners have been developing a close collaboration with Vietnam National Coal – Mineral Industries Holding Corporation Limited. Major outcomes of the collaboration are installations and operation of mining equipments and machines in Vietnamese mining companies, and excellent training programs for graduate and post graduate students and mining staff for both countries, etc. From aspects of knowledge management in globalization, the article highlights the outstanding outcomes of knowledge exchanges between the two countries, outlines cultural and economic challenges for the exchange and proposes some improvement in the future. **Key words** – knowledge exchange, mining knowledge, Polish – Vietnamese corporation.

1 Introduction

The collaboration of Poland and Vietnam has been being developed since 1950s. Since then, Poland has received thousand researchers and students from Vietnam for research and educational programs. During the time 1960–1980s, scientific collaboration was mainly at state levels. To date, this partnership has been promoted at different levels of institutes, universities and organizations, and private activities. In the 21st century, knowledge exchange in mining and geology has been significantly changed by the engagement of Polish companies with mining companies in Vietnam, and internship programs of Polish students in geology and mining disciplines in Vietnam universities and institutes. Publications are increasing in the last ten years in many scientific journals and conferences. Education programs for a new generation of qualified scientist with master and doctorate degrees are launched by different organizations. A revision of achievements and look back to its disadvantages is supportive for overcoming the obstacles of a better knowledge exchange in the future between the two countries.

* Corresponding author: nguyenthihoainga@humg.edu.vn

2 The status quo of knowledge exchange in mining and geology – collaboration of Poland and Vietnam in the 21st century

2.1 Mining and geology knowledge exchange

New knowledge is created from mutual conversion of tacit knowledge and explicit knowledge in the individual or organization [1]. Tacit knowledge is rooted in action, procedure, routines, commitment, ideals, values and emotions. Tacit knowledge can be accessible through consciousness if it leans towards the explicit side of the continuum. On the contrary, explicit knowledge has a universal character supporting the capacity to act across contexts, for example drawings and writing, etc.

Mining and geology knowledge are specific knowledge, being accumulated from the development of the mining industry, and transferred from countries to countries and organizations to organizations. Mining and geology knowledge are not the truth or the belief of this industry, but also mental models, methods and skills, and especially know-how as well.

Table 1. Activities of mining knowledge exchange between Poland and Vietnam since 2010

No.	Activities	Year	Outcomes
From Poland			
1	International conference Pol–Viet (1)	2014	International conference with results of research between Poland and Vietnam, attracted more than 100 scientists from education organizations and research institutes from both countries. 40 presentations in geology and mining, and other related fields were on schedule of the conference. An outstanding outcome of the conference is an overall review of education and scientific research from University of Sciences and Technology AGH Krakow and Vietnamese education organizations and research institutes.
2	Conference on Polish mining equipment and machinery – modern technology, creative solutions	2014	Organized by the Polish Commerce Chamber in cooperation with Vietnamese Association of Mining Sciences and Technology, the conference attracted participants of senior staff of the Chamber and professors of AGH Krakow as well as representatives of 14 leading mining companies in Poland. Coming from Vietnam is representatives of Vietnam National Coal – Minerals Industries Holding Corporation Limited (Vinacomin), Quang Ninh (the largest mining area in Vietnam) and other representative from mining and machinery subsidiaries under Vinacomin, and leaders of mining universities and research institutes.
3	Conference on "Stimulation of export on selected market"	2015	In frame of a commercial promotion program by Polish Ministry of Commerce and Industry, mining companies from Vietnam visited some outstanding mining and machinery companies for further cooperation in supplying modern equipment and technology for deeper mining in Vietnam.
4	Sending students and staff to Vietnam for studying and internship	2010 – 2017	Host organizations are Vietnamese research institutes and universities in mining and geology. After finishing studying and internship, either they came back to Poland or staying in Vietnam. They play an important role in the development of the mining and geology industries in Vietnam and the cooperation of the two countries.

Table 1. cont.

No.	Activities	Year	Outcomes
From Vietnam			
1	International conference Viet – Pol (2)	2015	The conference brought excellent opportunities for Vietnamese scientists, especially the junior ones in fields of geology, geomatics and mapping, geophysics, mining and electro-mechanics.
2	International conference Viet – Pol (3)	2016	The conference is integrated in the conference on Earth sciences and sustainable geo-resources development ESASGD 2016 in frame of activities for the 50th anniversary of Hanoi University of Mining and Geology (HUMG). Participants of the conference came from many countries such as Poland, France, Germany, Russian, China, Laos, Thailand, Myanmar, Japan, India and USA. 391 papers are on themes of 8 different sessions of mining, petroleum, geology, geodesy, education and training, economic and administration in mining industry, mining electric and mechanic engineering, and mining automation etc.
3	Institute of Mining Sciences and Technology (Vinacomin)	2016	Corporation with Carboautomatyka Company in knowledge transfer in mining machinery production, mud management technology, and coal quality management.
4	Institute of Mining Sciences and Technology (Vinacomin)	2016	Corporation with Polish mining companies in knowledge transfer regarding methane gas control, automatic warning system, mining machinery components production, etc.
5	Sending junior engineers and staff to universities in Poland	2010 – 2017	Hundreds of students and staff from universities, research institutes, state officers and mining companies have been sent to Poland during this time. Vinacomin strengthens the cooperation with Polish universities to send post graduate students for further studying. Those who come back to Vietnam can obtain the knowledge and develop it at work, as well as be a bridge for the mutual relationship of the two counties

The specific knowledge is stored in various kinds of documents such as books, videos, audio, pictures and paintings and etc. Besides, mining and geology knowledge is also experiences of individuals, which is very difficult to express in documents but can be shared over time. In this article, mining and geology knowledge are mentioned in activities of collaboration programs such as education and training, scientific conferences, equipment installation and knowledge transfer, etc.

From these activities, it can be seen that knowledge exchange between Poland and Vietnam are mainly in research and technology transfer. Short courses and visiting lecturers/professors in universities and internship in mining companies are absent from knowledge exchange of the two countries. Without courses and lectures, knowledge cannot be multiplied or distributed, meanwhile, without practice, experiences and skills cannot be accumulated and knowledge cannot be upgraded or innovated.

2.2 Obstacle of knowledge exchange in mining and geology

Being known as one of the most important producers in Europe of coal (hard coal and lignite) and copper, Poland is also a reliable country for mining education to Vietnamese government and mining companies for decades. However, there are some obstacles for the knowledge exchange in mining and geology as of the differences between the two countries. Besides of technological conditions, which are difficult to overcome due to aspects of geology, mines and mining conditions, social – economic challenges and cultural challenges are worth being mentioned.

2.2.1 Socio-economic challenges

According to reports of World Bank in 2016, Poland has a population of about 38 million and Gross National Income per capita of nearly US$ 12,400. With this number, Poland has the largest economy in Central Europe [2]. Poland has achieved a development success milestone, moving from middle-income to high-income status in record time. Labor market conditions continued to improve in 2016, as employment increased and unemployment reached record lows (5.5% in the fourth quarter of 2016). In 2017, GDP of Poland is expected to increase, and driven by domestic demand [3]. Due to the relatively low potential of replacing coal with other energy sources, current forecasts show coal retaining its major role in the Polish energy mix for upcoming years [4]. In 2016, Poland Gross Domestic Product (GDP) reached 469.509 billion USD [5], two times larger than Vietnam's number of 202.616 billion USD [6].

With a population of more than 92 million [6], Vietnam is now one of the most dynamic emerging countries in East Asia region [7]. Mining still plays an important part in GDP of the country, taking about 7% in national annual GDP [8].

However, the two countries and their mining industries were deeply impacted by the economic downturns in recent years. The mining industries had to downsized to prevent from losses due to domestic and international markets difficulties, therefore, numbers of workers in mining sector, and consequently, those of student enrolment in mining universities have been being reduced [9, 10]. Trimming investment on education and training, heavier workloads and new job roles with higher working intensity, etc. are impact factors on knowledge transfer and its effects [11, 12]. Resigned and retired workers and staff without a contingence create a big loss of knowledge and experiences in mining and geology.

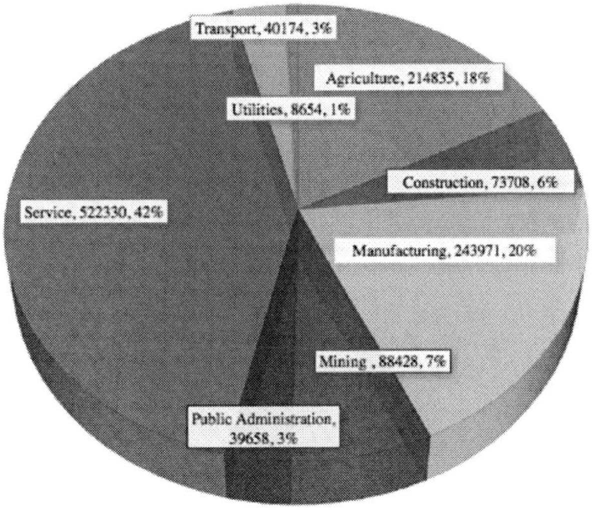

Fig. 1. Contribution of sectors in GDP of Vietnam 2016 (billion VND) (Source: Data from [8])

The economic gap between the two countries reflects also the investment gap of mining and geology, mining and geology education and research.

2.2.2 Cultural challenges

Although Vietnam and Poland has been building their mutual friendship since the middle of the 20th century, their mutual understand of languages, cultures and religions is still insufficient.

Languages seem to be the biggest challenges of the knowledge exchange, as the two countries are both non-English speaking countries and their native languages are not taught in education systems. Either using a third languages or interpreters/translators is an obstacle of knowledge exchange.

Other cultural differences come from cultural aspects, such as the way of presenting ideas, concept of punctuality, dealing with problems, ego and concept of individualism and collectivism, manners and behaviors [13] and safety competence [14]. These aspects impact on the collaboration, however, the impact can be improved by returning experts as these experts are bringing to their home countries not only new knowledge, experiences and skills but also new network and understanding of cultural aspects [14].

Besides of those differences, mining and geology have their own culture, which are quite common all around the world, such as: significant features of a dangerous, tough and monotone job, staff and workers come from typical faculty in typical universities and they are normally inspired in working in companies by their kinship and ancestors, from whom they get invisible influences and experiences, the proud of teamwork, solidarity and consensus, typical celebration and festivals, etc. [15]. This kind of culture can also be a barrier for people in adapting new knowledge, as they can create undernet with specific knowledge of particular groups (in an organization). Knowledge, therefore, can be easily exchanged in each group, but not easy in different groups.

Mutual trust between the two sides of knowledge exchange is also a significant issue, as knowledge sharers might not be sure that knowledge recipients understand and use knowledge appropriately, or knowledge recipients do not rely on the accuracy of knowledge [16]. This issue becomes more crucial when languages challenges cannot be resolved.

3 Perspective of knowledge exchange and further collaboration

Fig. 2. Staff of Vinacomin received their master degrees of ventilation from AGH Krakow in July 2017. Two of them will be chosen for doctoral program (Source: Vinacomin)

Being one of the biggest employers of Vietnam industry, Vinacomin aims to send staff to Poland, studying ventilation and safety management in master and doctoral degrees, to meet the requirement of going deeper and further underground in upcoming years. But Vinacomin is not just only one partner of knowledge exchange. There are other education and research institutes in Vietnam that can promote their activities with Poland. Besides of middle-term and long-term plans such as education in universities, as mentioned in the previous part, knowledge exchange in mining and geology between Poland and Vietnam can be broadened in numbers of exchange students and school staff, and developed in organizing short courses and small symposia. These academic activities cannot only be supportive for knowledge exchange but also can reduce the gap of culture between the two countries. Collaboration can be developed further under the technical supports of mining and geology schools such as Krakow AGH and Wroclaw from Poland, Hanoi University of Mining and Geology, Hanoi University of National Resources and Environment and other universities in from Vietnam, and financial supports from the two government as well as other organizations and programs such as Erasmus+. Members of Society of Mining Professors and other associations from Poland and Vietnam are also the bridges for the collaboration, because they can connect their networking for an expansion and an effect knowledge exchange. Funding should be found for guest lecturers and retired professors from Poland to come to Vietnam, giving lectures in a short course of 2–3 weeks, and junior lecturers from Vietnam to come to Poland for experiences exchange. In recent years these activities are developed in a program of AGH Krakow and Hanoi University of Mining and Geology. However, there is still a big demand from other mining and geology schools.

4 Conclusions

Advances in technology and management are two drivers for knowledge exchange, and further, knowledge management of organizations and countries. Sharing knowledge is not a loss of knowledge but a multiplication. Exchange of knowledge in mining and geology is an important aspect for the sustainable development of the mining industry in the forthcoming years. From perspective of a forerunner mining country, Poland can share the rich experiences to emerging mining counties like Vietnam. On the other hand, Poland can gain new experiences in mining and geology, filling gaps in research from emerging mining countries and supporting to investment and commerce. From perspective of an emerging mining country, knowledge exchange helps Vietnam to quickly obtain and control new technology, meeting the demand of production expansion in mining and geology, safely and effectively. Understanding of challenges in economy and culture in knowledge exchange in a specific sector of mining and geology, Poland and Vietnam can benefit from the collaboration in the upcoming years.

References

1. I. Nonaka, R. Toyama, T. Hirata, *Managing Flow*, Palgrave Macmillan (2008)
2. http://www.worldbank.org/en/country/poland/overview (accessible August 15th, 2017)
3. http://pubdocs.worldbank.org/en/832971492622358477/Poland-Snapshop13Apr2017.pdf (accessible August 15th, 2017)
4. https://euracoal.eu/info/country-profiles/poland/ (accessible August 15th, 2017)
5. http://data.worldbank.org/country/poland (accessible August 15th, 2017)
6. http://data.worldbank.org/country/vietnam (accessible August 15th, 2017)
7. http://www.worldbank.org/en/country/vietnam/overview (accessible August 15th, 2017)
8. https://tradingeconomics.com/vietnam/gdp-per-capita accessible August 15th, 2017)

9. https://www.statista.com/statistics/422202/number-of-employees-mining-quarying-sector-poland/ (accessible August 15th, 2017)
10. General statistics office of Vietnam, *Report of Labour force survey*, Q1 2017 (2017)
11. International Labor Organization – ILO, *Health and life at work: a basic human right, World Day for safety and health at work*, Geneva (2009)
12. International Social Security Association – ISSA, *The impact of crisis on the mental; health of workers*, Geneva (2010)
13. Y. Liu, Ost trifft West (2007)
14. J. Kretschmann, N. Nguyen, *Capacity building in emerging mining countries: Challenges in know-how transference*, Heliotopos (2013)
15. N. Nguyen, J. Kretschmann, *Communication in risk management in coal mining companies*, Technology and Science Publishing House (2012)
16. D. Dang, H. Nguyen, *Knowledge management in enterprises*, Communication and Broadcasting Publishing House (2016)

Constructing Indicators for Measuring Provincial Sustainable Development Index in Vietnam

Van Canh Truong[1], *Andrzej* Lisowski[1,*]

[1] University of Warsaw, Faculty of Geography and Regional Studies, Krakowskie Przedmieście Av. 30 00-927 Warsaw, Poland

Abstract. Sustainable development is zeitgeist of our age. It is one kind of development that in this trajectory humanity can create a stable and developed socio-economic foundations, conserve environment and therefore able to continue for a long time. Using indicators is one of the best ways to monitor and measure the progress toward sustainable development. In this paper we have proposed the way to create indicators for measuring provincial sustainable development index in Vietnam. We firstly made a framework of elements for economic, social and environmental component and compiled a list of indicators of 20 national and international agencies in the world. We then applied the SMART framework (Specific, Measurable, Achievable, Relevant, and Time-related) to choose indicators which will be relevant for Vietnam and put them back to the elements. We then have 39 relevant indicators with 12 indicators for economy, 17 indicators for social and 10 indicators for environmental component. Finally, we have established the way to determine the worst and best value for each indicator from available data for countries in the world. **Key words** – indicators; Vietnam; sustainable development; index.

1 The conceptualization of sustainable development

In the report of the United Nations Commission on Environment and Development, known widely by the name "Our Common Future" in the Brundtland Commission in 1987, a classic definition of the concept of sustainable development, one that was used for the next twenty-five years, was first introduced: sustainable development is development that meets the needs of the present without compromising the ability of future generations to meet their own needs. [1]

Over the years the concept of sustainable development (SD) has led to various definitions, understandings, instrumentations, evolved into a more practical approach, focusing less on intergenerational needs and more on the holistic approach linking economic development, social inclusion, and environmental sustainability, meaning that each of these dimensions is as important as the next one. [5]

SD became political objective of United Nations with several important summits. Some of key SD policy frameworks include comprehensive action plan Agenda 21, UN's Global Initiative "Millennium Development Goals (2001). The holistic approach of sustainable development was emphasized in "Report of World Summit on Sustainable Development"

---

* Corresponding author: trvcanh1712@gmail.com

in 2002 at the UN World Summit on Sustainable Development in Johannesburg, and again in "The Future We Want" at Rio+20 Summit in 2012. A plan of action for people, planet and prosperity showed in the report "Transforming Our World: The 2030 Agenda for Sustainable Development" after 2030 Agenda for Sustainable Development. [13]

2 Sustainable development in the case of Vietnam

Vietnam, officially the Socialist Republic of Vietnam, is the easternmost country on the Indochina Peninsula in Southeast Asia. With an estimated 91.7 million inhabitants as of 2015, it is the world's 14th-most-populous country, and the ninth-most-populous Asian country. It covers a total area of approximately 330,966 km2. Vietnam is divided into 58 provinces. There are also five municipalities, which are administratively on the same level as provinces. [4]

For sustainable development objectives to be realized in Vietnam, a series of policies have been adopted in the economic, social and environmental fields and with the implementation of international commitments on sustainable development to which Vietnam is a signatory. Vietnam has recorded remarkable achievements in economic, social and environmental sectors, all three pillars of sustainable development.

In the economic field, Vietnam has been recognized as one among the developing countries which have made notable achievements in economic reform toward growth and poverty reduction. Vietnam has jumped out of the group of poor countries into the group of middle-income countries. [9]

In the social field, initial encouraging achievements have been recorded in hunger eradication, poverty reduction, population, job creation, education and health care. Social security has been attended to so as to ensure stability for the people's livelihoods and productivity, especially given the high inflation and frequent natural disasters. Vietnam's gender equality index is higher than countries with a similar level of development and income. The HDI has also improved over the years. [9]

In the environmental field, the system of policies and laws on environmental protection has been in place in a sufficient and comprehensive manner. Many of the tasks regarding pollution prevention and control and biodiversity conservation have brought about heartening results. Integration of environment-related issues into strategies, schemes and plans right at the designing stage as well as into projects at the stage of investment preparation has helped limit and minimize environmental pollution. [9]

On the way to continued implementation of its national sustainable development objectives, Vietnam is faced with many inherent or emerging challenges, including: Climate change has made its presence increasingly visible, with an increase in natural disasters, causing huge human and material destruction to the Central coastal region and significant destruction to many others. Natural resources, especially water and biodiversity, have been seriously degenerated. In certain localities, land and minerals resources have suffered from exhaustive or highly wasteful exploitation. Environmental pollution, the prolonged degeneration of the environment as a consequence of the war (bombs, landmines, Agent Orange/dioxin) and the environmental pollution due to the present socio-economic development, constitute a very grave challenge to Vietnam in the sustainable development process. [9]

3 Indicator selection method

Indicators are simple measures related to something more complex of primary interest. To ensure that selected indicators are applicable, they must fit with themes of sustainable development considering that some of them are already mentioned in policy documents. The selection of indicators followed by the method which is presented below:

+ **First step:** *Make a framework of elements.* The research endorsed three dimensions of sustainable development: social wellbeing, economic opportunity, and environmental quality and agreed upon the elements necessary to ensure each [7]. The result is the elements necessary for sustainable development:

Table 1. The elements for three components of sustainable development

Components	Elements
Economic	Prosperity
	Employment
	Investment
	Competitiveness
Social	Education
	Living conditions
	Poverty
	Inequality
	Heath
	Safety
Environmental	Resources
	Water and sanitation
	Air quality
	Waste / Reuse / Recycle

Source: Authors' analysis

+ **Second step:** *Compile a list of indicators used for evaluating sustainable development in countries, institutes, organizations in the world.* In this study, researcher explore how a standard set of sustainable development indicators could be drawn or adapted from existing systems, the first steps towards creating the standardized system needed to understand national sustainable development status and trends. [7]

Following an exploration of the literature on indicators and indicator systems, the research began a broad analysis of the 20 identified indicator systems. While the list of 20 systems are not comprehensive, can be considered representative of the types of organizations that are active in the area. The 20 systems, shown below, fall into three broad categories, Institutional, NGO, and Governmental from 1998 to 2016.

Table 2. The 20 identified indicator systems

No.	Countries/Regions	Year	Economic	Social	Environmental	Total
1	USA	1998	13	11	16	40
2	UNEP- MAP*	1999	40	23	67	130
3	Finland	2000	23	28	31	82
4	Korean	2006	14	24	26	64
5	Asian Development Bank	2006	24	30	8	62
6	United Nations	2007	29	31	36	96
7	The MDGIs**	2008	19	31	10	60
8	Germany	2010	8	12	8	28
9	France	2010	16	21	17	54
10	Poland	2011	23	31	22	76
11	Israel	2011	7	7	12	26

12	Vietnam	2013	9	13	19	41
13	Australia	2013	16	26	14	56
14	Malaysia	2013	8	24	8	40
15	Taiwan	2013	24	30	44	98
16	Nordic Region	2013	4	7	12	23
17	United Kingdom	2013	8	11	16	35
18	OECD	2014	20	6	23	49
19	EU commission	2015	49	38	22	109
20	SDSN***	2016	17	39	23	79
			371	443	434	1248

* United Nations Environment Program- Mediterranean Action Plan
** The Millennium Development Goals Indicators
*** Sustainable Development Solutions Network

Source: Authors' analysis and calculations

+ **Third step:** *Apply the SMART framework to take relevant indicators.* The acronym SMART, put forth by the European Union and Statistical Institute for Asia and Pacific (2007) describes the qualities of a 'good' indicator. Identified during the literature review, the qualities (Specific, Measurable, Achievable, Relevant, and Time-related) were endorsed by the SUD Working Group at its August 2010 meeting. [7] In this paper, about S for specific in the SMART, the researcher also concern about the S for spatial.

The indicators in this study were chosen on the basis that they were, as far as possible, quantifiable and spatially specific. It was also important to check if indicators were capable of capturing change over time and territory, and thus provide information sensitive to change in a timely manner. Based on these concerns and the SMART framework, indicators were filtered based on assessment of each of the following questions:

- Does the indicator address the policy objectives and development priorities (i.e. overall priority themes) of the case studies?
- Are the indicators available and does the indicator enable assessment of the performance and dynamics of balanced territorial development (i.e. can it be mapped to illustrate spatial patterns)?
- Is the indicator regularly measured (i.e. are there reliable and regularly updated datasets available or monitoring arrangements in place)?
- Does the indicator effectively provide information sensitive to change to timely aid decision-making processes?
- Is the indicator well understood by planners and decision-makers (i.e. can it communicate the results in a concise and accessible manner)? [14]

So a number of indicators were measurable, but not achievable, meaning they asked for information that could be collected, but to do so would be prohibitively expensive or difficult. In the economic and social dimensions, in particular, many indicators were removed for referencing resource- intensive one-time surveys. Some indicators were also irrelevant, given the goals of the project. Interpreting the time-related criterion proved more of a challenge, but researchers liken it to timely, that is whether an indicator is based upon current, continuous data that can be tracked over time, as opposed to old surveys or databases. [7] If most criteria were met, the indicator was selected, if not – the indicator was discarded. If the indicator partially met criteria it was put on a list of additional indicators.

+ **Fourth step:** *Put the relevant indicator after applying SMART framework to each of elements.*

+ **Fifth step:** *Determining the worst and best value for each indicator.* We order data for each indicator from worst or min (0) to best or max (1 or 100). In some cases the highest numerical value on an indicator is "worst" (e.g. infant mortality rate) while for other indicators the highest numerical value is "best" (e.g. life expectancy). We are going to create the worst and the best value for each indicator. In most cases the best and worst score is the natural "perfect" and technically feasible target in line with the principle of "leaving no one behind" (e.g. zero extreme poverty, zero undernourishment, 100% school completion). [10] In some cases no such "perfect" target exists as the theoretical optimum may not be achievable or might be undefined (e.g. child mortality rate, physician density, traffic deaths, life expectancy...). We can take thresholds for indicators from data for the sample countries in the world. To determine the worst value for each indicator, we first remove the worst 2.5% of observations (mean the sample countries in the world) in order to ensure that our scoring is not overly influenced by outliers. We then identify the next-worst value on each indicator and apply this value to the bottom 2.5 percentile of the distribution. To identify the best value we can use average of the top 5 values in the sample of countries for that indicator. All countries that exceed the average of the best values are assigned the best value. [10]

4 Results and discussion

After applying the method to select indicators, researcher have constructed a list of 39 indicators with 12 indicators for economic component, 17 indicators for social component and 10 indicators for environmental component. All of them are met the criteria for selecting indicators and have been set up the best and the worst values as the threshold for the next step of measurement. The result of selected indicators is shown in Table 3.

Table 3. The core indicators for measuring provincial sustainable development in Vietnam

Component	Element	Indicator	N	Best (= 1)	Worst (= 0)	Source
Economic	Prosperity	GDP per capita (PPP USD)	177	96000	1184	WB, 2015
		Annual growth rate of labor productivity (%)	183	10.8	–22.3	WB, 2015
		Percentage of budget revenue and expenditure accounts per GDP (%)	107	25.7	–5.1	WB, 2015
		Urban population rate (%)	195	100	16.3	WB, 2015
		Proportion of GDP generated by service (%)	151	83.7	37.5	WB, 2015
	Employment	Unemployment rate at ages 15+ (%)	181	0.5	25.8	ILO, 2015
		Employment to population ratio (%)	169	85	37.2	WB, 2015
		Percentage of advanced trained employed workers at 15 years of age and above (%)	71	57.4	8.0	ILO, 2014–2015
	Investment	Expenditure on Research and Development as a percent of GDP (%)	67	3.7	0	UNESCO, 2015
		Expenditure on education as a percent of GDP (%)	112	8.2	1.6	UNESCO, 2015
		Expenditure on health services as a percent of GDP (%)	188	14.2	2.6	WB, 2014
	Competitiveness	Provincial Competitiveness Index		1	0	Optimum

Component	Element	Indicator	N	Best (=1)	Worst (=0)	Source
Social	Education	Adult literacy rate (%)		100	0	Optimum
		Population aged 5 and over ever or never attending school (%)	98	0	45.0	UNESCO, 2014
		Combined gross enrolment ratio (%)		100	0	Optimum

	Living condition	Proportion of household own permanent house		100	0	Optimum
		Percentage of households using electricity (%)		100	0	Optimum
		Proportion of households with broadband internet access (%)		100	0	Optimum
	Poverty	Poverty rate (%) - national poverty line	117	0	64.6	WB, 2009–2015
		Cereals per yield (ton/ha)	171	8.6	0.5	WB, FAO, 2014
	Inequality	Gini index	138	0.26	0.61	WB, 2013
		Female labor force participation rate (% male)	121	100	22.5	SDSN, 2010–2014
		Sex ratio at birth (male/100 female)	194	105	110	CIA, 2016
	Health	Full immunization coverage among children under 1 year of age (%)		100	0	Optimum
		Proportion of malnutrition of children under 5 (underweight) – %	129	0	33.0	2005–2015 UNICEF, WHO & WB (2015)
		Under 5 mortality rate (per 1.000 live births)	189	2.3	114.7	WB, 2015
		Average life expectancy at birth	188	83.0	52.0	WB, 2014
	Safety	Total persons convicted per 100.000 population	80	0	2775	2010–2014 UNODC, 2014
		Proportion of death due to traffic accident per 100.000 population	179	0	33.2	WHO, 2013

Component	Element	Indicator	N	Best (=1)	Worst (=0)	Source
Environ-mental	Resources	Change rate of forest areas 2005–2015 (%)	192	36.5	–22.6	WB, 2015
		Percentage of natural forest area (% of total forest area)		100	0	Optimum
		Change in agriculture land 2004–2014 (%)	191	0.2	–0.4	WB, 2013
	Water and sanitation	Proportion of household with access to improved sanitation (%)	179	100	12.1	WB, 2015
		Percentage of household access potable water (%)	183	100	51.5	WB, 2015
	Air quality	Proportion of households using solid fuels for cooking (%) – coal or biomass (such as dung, charcoal, wood, or crop residues)		0	100	Optimum
		Percentage of households using motorbike (%)	44	86	6.4	Pew Research Center
		Annual mean concentration of PM2.5 ($\mu g/m^3$)	191	4.0	71.0	WB, 2015
	Waste / Reuse / Recycle	The total amount of solid waste collected per capita (kg/person/day)	63*	0.1	0.8	GSO, 2015
		Proportion of solid waste that are treated according to national standards (%)		100	0	Optimum

Source: Authors' analysis and calculations

4.1 Economic component

Overall 17 key indicators related to economic dimension of sustainable development for Vietnam were selected. Among those are 5 indicators for assessment of prosperity of economic. The key indicators for employment are three and also three key ones for investment. Especially, we recommended using a special indicator to assess the business environment in Vietnam which is Provincial Competitiveness Index (PCI).

Several related tourism indicators were not included among key economic indicators. Economic dimension also did not included infrastructure and transportation related indicators. We know that this is so important to use those indicators for measuring sustainable development but the data for those ones is not spatial available by provinces.

4.2 Social component

Social dimension included 17 indicators. Not the same situation in economic and environmental dimension, the indicators for social dimension are quiet overwhelming. Among 17 indicators 3 education indicators, the role of education is clearly seen as priority among stakeholders. Two of the education indicators which are used so popular to overall see the knowledge of population are adult literacy rate and combined gross enrolment ratio. Several indicators are suggested to capture living condition. In the social dimension the research concentrated on the indicators for health gathering to children. Sex ration at birth indicator was included in the list of key indicators for social dimension. This is essential indicator in the context of Vietnam due to the preferring for male children.

Citizen satisfaction with life in Vietnam is not measured but is preferred. Growing importance of life quality and wellbeing means that this dimension will have to be addressed.

4.3 Environmental component

The data for environment is so limited, especially provincial data. Environmental dimension contains 10 indicators most of which are already used. Based on available indicators, environmental dimension emphasizes availability of resources, water and sanitation, air quality and reuse or recycle of waste. Among preferred an indicators that is included, but not yet used is annual mean concentration of PM2.5 ($\mu g/m^3$), but data for this indicator we will have from the satellite images. To evaluate air quality the researcher has concerned about two aspects: outdoor and indoor air quality. We have decided to take indicator like proportion of households using solid fuels for cooking (%) for indoor air quality and percentage of households using motorbike for outdoor air quality.

Waste management can also have profound impact on environment, indicators of waste management were selected as key indicators. These were – The total amount of solid waste collected per capita (kg/person/day) and Proportion of solid waste that are treated according to national standards (%).

5 Conclusion

Indicators reflect certain development priorities, but only some indicators are able to communicate complex relationships between phenomena in a simple way and in a manner, which is easily understood. Selecting indicators is not a straightforward task [14]. To measure the status of sustainable development of a country according to the spatial change, we must also base on the recent statistical data which we can have to use. The study has explored a list of indicators of 20 national and international agencies in the world. We then applied the SMART framework (Specific, Measurable, Achievable, Relevant, and Time-related) to choose indicators which will be relevant for Vietnam. We then have 39 relevant indicators with 12 indicators for economic, 17 indicators for social and 10 indicators for environmental component. Finally, we have established the way to determine the worst and best value for each indicator from available data for countries in the world. Although the study constructs the list of limited indicators, it finds the core indicators in all sustainable development dimensions which we can use the recent statistical data to measure. But a lot of essential indicators about transportation, economic, wellbeing, environmental quality and so forth are still missing in the list of core indicators due to the missing the data.

References

1. Brundtland Commission, UN, *Our Common Future* (1987)
2. D. Reid, *Sustainable Development: An Introductory Guide*, Earthscan in UK and US, (2013)
3. ESPON, KITCASP, *Key Indicators of Territorial Cohesion and Spatial Planning, Target Analysis, Final Report*, Scientific Report (2013), URL: http://www,espon,eu/main/Menu_Projects/Menu_TargetedAnalyses/kitcasp,html,
4. *General Statistic Office of Vietnam*, Statistical Yearbook of Vietnam (2016)
5. D. Jeffrey, Sachs, *The Age of Sustainable Development*, Columbia University Press (2015)
6. Jennifer A, Elliott, *An introduction to sustainable development (third edition)*, Published by Routledge Taylor & Francis (2006)
7. Penn Institute for Urban Research, *Sustainable Urban Development Indicators for the United States* (2011)
8. S. Bell and S. Morse, *Sustainability Indicators: Measuring The Immeasurable (second edition)*, Earthscan (2008)
9. *Socialist Republic of Vietnam, Implementation of Sustainable Development in Vietnam* - National Report in Rio+20 (2012)
10. Sustainable Development Solutions Network (SDSN), *SDG Index and Dashboards - Global Report*, New York: Bertelsmann Stiftung and Sustainable Development Solutions Network (2016)
11. Sustainable Development Solutions Network, *An Action Agenda for Sustainable Development* (2013)
12. Tatyana P, Soubbotina, *Beyond Economic Growth: An Introduction for Sustainable Development*, The International Bank for Reconstruction and Development, WB (2004)
13. UN, *Transforming our world: The 2030 Agenda for Sustainable Development* (2015)
14. V. Visvaldisa, G. Ainhoab, P. Ralfsc, *Selecting indicators for sustainable development of small towns: The case of Valmiera municipality*, URL: http://www.sciencedirect.com/science/article/pii/S1877050913012763, Published by Elsevier (2013)

Integrated Reporting as a Tool for Communicating with Stakeholders – Advantages and Disadvantages

Iwona Matuszyk[1], and *Bartosz* Rymkiewicz[1,*]

[1] AGH University Science and Technology, A. Mickiewicza Av. 30, 30-059 Krakow, Poland

Abstract. Financial and non-financial reporting from the beginning of its existence is the primary source of communication between the company and a wide range of stakeholders. Over the decades it has adapted to the needs of rapidly changing business and social environment. Currently, the final link in the evolution of organizational reporting, such as integrated reporting, assumes integration and mutual connectivity to both financial and non-financial data. The main interest in the concept of integrated reporting comes from the value it contributes to the organization. Undoubtedly, the concept of integrated reporting is a milestone in the evolution of organizational reporting. It is however important to consider whether it adequately addresses the information needs of a wide range of stakeholders, and whether it is a universal tool for communication between the company and its stakeholders. The aim of the paper is to discuss the advantages and disadvantages of the concept of integrated reporting as a tool for communication with stakeholders and to further directions of its development. The article uses the research methods such as literature analysis, the content analysis of the corporate publications and comparative analysis. **Key words** – organization reporting, integrated reporting, communication with stakeholders.

1 Introduction

Organizational reporting has been an essential tool for communication between the company and stakeholders since its inception. However, rapid social and economic changes lead to changes in the expectations of stakeholders. In order to meet these expectations organizational reporting is constantly evolving. Over the years, companies have been expanding the scope of disclosures by multiplying reports related to various spheres of the organization's activity. Among others, reports have been created on marketing, social, environmental, and sustainable development issues, as well as on compliance with the principles of corporate governance and many others. Excessive multiplication of corporate publications, often large in size, can sometimes lead to counterproductive effects. A decrease in the effectiveness of the information provided may lead to disruption of the organization's informational role of reporting. An international consultation has resulted in a modern concept of reporting, i.e., by integrating financial and non-financial data, with the latter being frequently scattered. One comprehensive,

* Corresponding author: bartosz.rymkiewicz@wp.pl

integrated, report appeared instead of many reports so far[1]. Integrated reporting is now the final link in the evolution of organization reporting and a milestone in its development. However, it should be considered whether it adequately meets the information needs of a wide range of stakeholders. As a consequence, doubts are raised whether such approach fulfils its fundamental role as a universal tool of communication between the company and its stakeholders.

The aim of this paper is to consider the advantages and disadvantages of integrated reporting as a tool for communication with stakeholders and to further development of the concept. The article uses research methods, i.e., literature analysis of the subject, analysis of corporate publications and comparative analysis.

2 The problems of the traditional organization's reporting

Economic and social reality are characterized by high levels of changes and dynamics. Increasing levels of globalization, opening of markets and technological development, as well as changes in lifestyles, all have impacted on methods and tools of communication with stakeholders. Nowadays, enterprises have a vast array of tools and methods of communication with stakeholders. This ranges from direct communication, to traditional media and social media popular in recent years. Of course, none of them is perfect and is not able to fully meet the expectations of all stakeholder groups. One of the oldest and most informative is organizational reporting. However, as economic reality is not static, organizational reporting is also changing.

E. Walińska [12] distinguished five stages of development in the company reporting, starting with a simple financial statement containing only financial data, supplementing it with non-financial information as supplementary information, developing an annual report covering additionally non-financial information, supplementing the annual report with the comments of the management board, ending with the preparation of additional reports, i.e., environmental, social and sustainable development report. All these phases of development were connected with changes in social expectations and the emergence of modern management concepts, such as the concept of corporate social responsibility or the concept of sustainable development, which in their scope also included information to stakeholders about the positive and negative effects of the entire organization's activity [7]. An extensive reporting of an organization, consisting of multiple reports covering only parts of the activity and containing data that are not interconnected, does not fulfil its purpose of communicating with a wide range of stakeholders in an appropriate and effective way. Multiplication of reports and their growing size lead to communication disturbances with stakeholders, while the large dispersion of information makes it difficult to find and analyze it.

The literature on the subject is widely criticized for the traditional form of organizational reporting, in particular financial reporting, which is an obligatory element. E. Szczepankiewicz points out that financial reporting in its current form does not provide information on the business perspectives, i.e., on types of risks and on threats related to the organization's activities, as well as to the investments and projects carried out, environment, social responsibility and environmental activities and intangible factors affecting the value of the company [13].

E. Walińska et al. [6] points out that traditional financial reporting does not provide the key information expected by investors, i.e., increased transparency, the scope of information presented, information enabling the preparation of financial situation forecasts, descriptive

---

[1] The issues of integrated reporting were discussed in the literature of the subject, by among others such as: I.C. Chersan [1], R.G. Eccles, M. Krzus, S. Ribot [2], B. Bek-Gaik [3], D. Garstecki [4, 5], E. Walińska, B. Bek-Gaik, J. Gad, B. Rymkiewicz [6], J. Samelak [7], I. Matuszyk [8], B. Bek-Gaik, B. Rymkiewicz [9], B. Rymkiewicz, I. Matuszyk [10, 11]

information and voluntary disclosures, and does not meet the expectations of other stakeholder groups.

Key weaknesses of the traditional financial statements pointed out by ICAEW [14] are: failure to adequately meet user needs, focus on historical data and on indicators rather than on non-financial indicators of future financial success, focus on reliable asset measurement rather than on intangible assets and knowledge that underpins the success of a modern enterprise, focus on aggregated data and focus on the short-term approach, the existence of a gap between data held by managers and information available to investors.

According to B. Bek-Gaik [3] the drawbacks of traditional financial reporting include: failure to meet the information needs of all stakeholders, limited information on factors influencing the creation of value, difficulties in interpretation, complexity, lack of coherence, limited recognition of some resources, lack of a business model, lack of information on key success factors and strategic plans, inability to determine the market value of the company.

The criticism of the organization's traditional reporting, presented in the literature of the subject, which is based on mandatory financial statements, indicates that changes are necessary. Until now the reporting format has required a number of changes in order to adapt it to the information needs of a wide range of stakeholders. A modern concept of integrated reporting can be the solution to address many, if not all, of the problems and shortcomings of traditional reporting by an organization.

3 The concept of integrated reporting

As mentioned earlier, integrated reporting is a modern solution that allows to eliminate existing weaknesses in the organization's traditional reporting. It is aimed at communication with a wide range of stakeholders [13], which is a response to the most frequent allegations that the information needs of a wide range of stakeholders are not met. This is an undeniable evolution of the approach, since the organizations traditional reporting has been focused mainly on large entities, investors and shareholders, while integrated reporting is addressed to all stakeholders, both internal (employees, shareholders) and external (counterparties, customers, social organizations, local communities, etc.).

The Integrated Report integrates and interconnects financial and non-financial information in a single comprehensive report. This allows to link the most relevant (from the point of view of both the company and a wide range of stakeholders) information about the entity, which until now has been dispersed in many different reports (e.g., management commentary, social report or environmental report) [6].

The main objectives of integrated reporting are: increasing the quality of information available to capital donors (improving the effectiveness of capital allocation), promoting a more coherent and effective approach to corporate reporting (in particular by disclosing information relevant for the organization's ability to create value over time), increasing responsibility for the capital base (not only in terms of financial capital, but also production, intellectual, human, social, relational and natural capital) and supporting integrated thinking (focusing on actions leading to creation value over time) [15]. Focusing on the value creation process is a particularly important feature of integrated reporting. The literature also describes an integrated report as a 'value report' [12].

In order to unify the concept of integrated reporting, the International Integrated Reporting Council in December 2013 developed the The International <IR> Framework [15]. The basic objective of the Framework is to "establish Guiding Principles and Content Elements that govern the overall content of an integrated report, and to explain the fundamental concepts that underpin them". The Framework includes the basic principles of preparing the report, i.e., strategic focus and future orientation, connectivity of information, stakeholder relationships, materiality, conciseness, reliability and completeness and consistency and comparability.

These principles concern two fundamental issues, the quality of the information disclosed (e.g., materiality, conciseness, reliability and completeness) and the content of the report (e.g., strategic focus and future orientation, connectivity of information, stakeholder relationships), which should be defined by taking into account the information needs of a wide range of stakeholders. All of these principles are of key importance for proper communication with stakeholders. In addition to the principles, the Framework also sets out the basic elements that should be adequately covered by an integrated report. These elements include: organizational overview and external environment, governance, business model, risk and opportunities, strategy and resource allocation, performance, outlook, basis of preparation and presentation, general reporting guidance. All these elements should be interconnected, which is undoubtedly an innovative and conducive solution for the analysis of the whole organization. The most important element of the report's content is the business model, which presents in the greatest degree the value-creation process over time.

The Integrated Reporting Framework is a guideline for enterprises on how to properly prepare an integrated report. The postulates presented in the Framework significantly eliminate imperfections of traditional reporting, in particular regarding the improvement of communication with stakeholders.

4 Integrated reporting as a modern tool for communication with stakeholders

It has already been demonstrated that the organization's traditional reporting cannot fully meet the information needs of a wide range of stakeholders. Seeking a new way of communicating with stakeholders has led to the creation of an integrated reporting concept. However, can integrated reporting became a universal and modern tool for communication with stakeholders? We should consider what the main advantages of this form of communication are and if it is not without disadvantages.

Consideration of the benefits of integrated reporting should start with the scope of disclosures. The key issues are non-financial information, in particular information referring to the concept of management aimed at balancing social and environmental economic effects, i.e. the concept of corporate social responsibility and sustainable development. The disclosure of information on social, economic and environmental activities undertaken, the implementation of the concept of corporate social responsibility or sustainable development, management systems and other non-financial information may be an element of building the economic and social value of the enterprise to a much greater extent than traditional material resources [13]. The data on traditional material resources cannot represent the full potential of the organization. Integrated reporting aims to present not only financial, production and natural capital, that is capital that can be easily recorded and described, but also human, intellectual, and social and relational capital, where the true value of a company is hidden. However, the organization without the knowledge and experience of employees, know-how, procedures, patents, etc., is unable to compete effectively on the market.

It should be noted, therefore, that the disclosure of such information is an advantage on the one hand, and a disadvantage on the other. First of all, it is extremely difficult and time-consuming to collect, analyze and process data on intellectual, human and relational capital for the purpose of preparing a report. Some elements are difficult to identify and describe. There is also a problem with their inclusion in the accounts (although the integrated reporting vision requires linking financial and non-financial data). Secondly, such information is usually heavily guarded by the company against market opponents, because it constitutes an important element in building a competitive advantage.

Nowadays in the era of enormous amount of information, the access to information is not the most important element. Much more important than quantity is its quality. As indicated by

E. Szczepankiewicz, an integrated report prepared in an appropriate manner, especially with regard to the quality and reliability of information (reliable, timely, comprehensive, clear and comprehensible) and presenting information on activities for the benefit of society may constitute a basic tool for communication with stakeholders [13].

The key to ensuring the quality of the information published, especially its reliability and veracity, should be their audit, preferably by independent external certified bodies. This is currently problematic, as the audit of integrated reports is not regulated by law. It seems that the introduction of such an obligation could significantly increase confidence in the published data and at the same time increase the interest in integrated reporting.

A major problem with the concept of integrated reporting may arise from its early implementation. Managers have to face the challenge of obtaining information for the purpose of preparing a report. The management and accounting systems not always can provide all the necessary data [13]. The solution to this problem may be the introduction of the XBRL standard (Extensible Business Reporting Language), which allows automatization of reporting processes by linking it to existing financial and management accounting systems of the company[2]. Using the XBRL standard in the process of preparing the integrated report can also significantly decrease its costs by reducing the number of people involved in the process of preparing the report and time. Simultaneously, it enhances the communication possibilities of the report, making it easier to search and analyze data, as well as making them more comparable.

Disclosure of business model data may cause concern to managers, because is sensitive data that it is often a business secret. The decision to disclose this type of data may in extreme cases involve legal liability[3].

From the perspective of communication with stakeholders, the integrated reporting concept has the great advantage of integrating financial and non-financial data into a single document. This increases the effectiveness of research and analysis of data by stakeholders, who until now had to search through numerous corporate publications.

Undoubtedly, the most important feature of the integrated report is the involvement of stakeholders in the process of defining the content of the integrated report. In preparing an integrated report, it is necessary to identify key stakeholders and their information needs. This is important because the company does not create value in isolation, but rather through relations with stakeholders [15]. Involving stakeholders in the process of defining the content of the report changes the role of the organization's reporting from a one-sided communication tool with stakeholders into a two-way tool – the report is an integrated response to legitimate information needs of stakeholders.

In conclusion, it is necessary to point out that the concept of integrated reporting has both advantages and disadvantages. Undoubtedly, it eliminates numerous imperfections of the organization's traditional reporting, however, some of the demands may turn out to be a threat to the company's existence. However, despite of this, it should be recognized that the integrated report is now the most comprehensive and accessible tool for communication with stakeholders.

5 Summary

[2] More about the XBRL standard for integrated reporting: B. Rymkiewicz, I. Matuszyk [10].

[3] For example, an act of unfair competition under Article 11 (1) of the Act of 16 April 1993 on Combating Unfair Competition (Journal of Laws 2003 No. 153 item 1503 with late amendments) is the transfer, disclosure or use of someone else's information constituting a business secret if it threatens or damages the interests of an entrepreneur.

Nowadays, proper communication with stakeholders is becoming the basis for the existence of every organization. Rapid technological development, progress of globalization and social changes have led to the emergence of many modern forms of communication between the company and its stakeholders. Traditional organizational reporting has been slowly adapting to meet the legitimate information needs of a wide range of stakeholders. However, reporting is not a static creation, it constantly adapts to the needs of the environment. At present, the last step in the evolution chain of organization reporting is an integrated report that integrates and links financial and non-financial data into a single comprehensive document. Integrated reporting eliminates many of the shortcomings of traditional reporting, in particular by providing easier access to information, linking data, improving comparability, involving stakeholders in the process of defining the content of a report and many others. All these features are essential for building a proper channel of communication with stakeholders and bilateral communication, because by involving stakeholders in the process of defining the content of a report, an enterprise can respond to their legitimate expectations. Of course, integrated reporting is not without defects. Undoubtedly, disclosing an excessive amount of information, especially on the value creation process, may pose a certain threat to the functioning of the organization. However, the weaknesses identified in the future can be overcome by further evolution of the organization's reporting.

References

1. I.C. Chersan, Study on Practices and Tendencies in Integrated Reporting, Audit Financiar, 13 Issue 129, 91–101 (2015)
2. R.G. Eccles, M. Krzus, S. Ribot, Model of Best Practice in Integrated Reporting 2015, Journal of Applied Corporate Finance 27, no 2 (2015)
3. B. Bek-Gaik, Integrated Reporting – Selected Issues, Zeszyty Naukowe Uniwersytetu Szczecińskiego, Finanse, Rynki Finansowe, Ubezpieczenia 77, 479–491 (2015)
4. D. Garstecki, Integrated reporting in theory and practice of Polish companies [in:] J. Krasodomska, K. Świetla, Współczesne uwarunkowania sprawozdawczości i rewizji finansowej, Uniwersytet Ekonomiczny w Krakowie, 129–143 (2015)
5. D. Garstecki, Financial Data Disclosure in Chosen Companies Integrated Reports, Zeszyty Naukowe Uniwersytetu Szczecińskiego 873, Finanse, Rynki Finansowe, Ubezpieczenia 77, 503–512 (2015)
6. E. Walińska, B. Bek-Gaik, J. Gad, B. Rymkiewicz, Sprawozdawczość przedsiębiorstwa jako narzędzie komunikacji z otoczeniem. Wymiar finansowy i niefinansowy, Wydawnictwo Uniwersytetu Łódzkiego, Łódź (2015)
7. J. Samelak, Conceptual Framework of Integrated Report as a Form of CSR Reporting, Prace Naukowe Uniwersytetu Ekonomicznego we Wrocławiu 314, 155–165 (2013),
8. I. Matuszyk, Reporting Integrated in the Communication of the Enterprise with Surroundings, Zeszyty Naukowe Uniwersytetu Szczecińskiego 855, Finanse, Rynki Finansowe, Ubezpieczenia 74, no 2, 153–161 (2015)
9. B. Bek-Gaik, B. Rymkiewicz, Trends in the Development of Integrated Reporting in the Practice of Polish Listed Companies, Zeszyty Naukowe Uniwersytetu Szczecińskiego, Finanse, Rynki Finansowe, Ubezpieczenia 1 (79), 767–783 (2016)
10. B. Rymkiewicz, I. Matuszyk, The use of XBRL standard in integrated reporting [in:] L. Petrisin, P. Lebkovs'kij, Computer science, information techonologies and management systems: proceedings of the International scientific young scientist conference: 2017, April, 10th to 12th, Ivano-Frankivsk, Ukraine, 194–199 (2017)

11. B. Rymkiewicz, I. Matuszyk, Integrated reporting as modern tool of communication with stakeholders [in:] V.L.Trusko, Problemy nedropol'zovaniâ: meždunarodnyj forum-konkurs molodyh učenyh: 19-21 aprelâ 2017 g.: sbornik naučnyh trubov, čast' I, 61–64 (2017)
12. E. Walińska, Integrated Report – Beginning of the End of Financial Statements?, Zeszyty Teoretyczne Rachunkowości, 82 (138), 151–165 (2015)
13. E. Szczepankiewicz, Evolution of Reporting Enterprises – Problems of Ensuring Comparability in Integrated Report on Sustainability and Socially Responsible Companies, Zeszyty Naukowe Uniwersytetu Szczecińskiego nr 832 Finanse, Rynki Finansowe, Ubezpieczenia 71, 135–148 (2014)
14. ICAEW, New Reporting Models For Business. Information for Better Markets initiative, https://www.icaew.com/-/media/corporate/files/technical/financial-reporting/information-for-better-markets/ifbm/new-reporting-models-for-business-2010-version.ashx (02.08.2017) (2010)
15. IIRC, The International <IR> Framework, https://integratedreporting.org/wp-content/uploads/2013/12/13-12-08-THE-INTERNATIONAL-IR-FRAMEWORK-2-1.pdf (03.08.2017) (2013)

AUTHOR INDEX

EDP Sciences – Web of Conferences
17, Avenue du Hoggar
Parc d'Activité de Courtabouf
BP 112
F-91944 Les Ulis Cedex A
France

ISBN 978-1-5108-6005-6

9 781510 860056

CURRAN ASSOCIATES INC.
proceedings.com